Natural Wastewater Treatment Systems and Sustainability

Related Titles by Nicholas F Gray

Fixed-Film Reactors in Wastewater Treatment
ISBN: 978-1-78634-924-8

Biology of Wastewater Treatment
Second Edition
ISBN: 978-1-86094-328-7
ISBN: 978-1-86094-332-4 (pbk)

Natural Wastewater Treatment Systems and Sustainability

NICHOLAS F GRAY

Trinity College, University of Dublin, Ireland

World Scientific

NEW JERSEY • LONDON • SINGAPORE • BEIJING • SHANGHAI • HONG KONG • TAIPEI • CHENNAI • TOKYO

Published by

World Scientific Publishing Europe Ltd.
57 Shelton Street, Covent Garden, London WC2H 9HE
Head office: 5 Toh Tuck Link, Singapore 596224
USA office: 27 Warren Street, Suite 401-402, Hackensack, NJ 07601

Library of Congress Cataloging-in-Publication Data
Names: Gray, N. F., author.
Title: Natural wastewater treatment systems and sustainability / Nicholas F Gray.
Description: First. | Hoboken : World Scientific, 2022. |
Includes bibliographical references and index.
Identifiers: LCCN 2021034630 | ISBN 9781800610835 (hardcover) |
ISBN 9781800610842 (ebook other) | ISBN 9781800610859 (ebook other)
Subjects: LCSH: Sewage--Purification--Biological treatment. |
Land treatment of wastewater. | Sewage lagoons. | Water--Pollution.
Classification: LCC TD755 G74 2022 | DDC 628.3--dc23
LC record available at https://lccn.loc.gov/2021034630

British Library Cataloguing-in-Publication Data
A catalogue record for this book is available from the British Library.

For any available supplementary material, please visit
https://www.worldscientific.com/worldscibooks/10.1142/Q0320#t=suppl

Desk Editors: Gregory Lee/Amanda Yun

Typeset by Stallion Press
Email: enquiries@stallionpress.com

Printed in Singapore

Dedicated to Lucy, Cat and Becky.

'To you it's just crap, to me it's bread and butter.'

Spike Milligan
Recollections of the latrine orderly

Preface

In 1989 I finally published the textbook *Biology of Wastewater Treatment*. I was and have remained primarily a wastewater engineer with a strong interest in biology. At that time there was no single reference work that encompassed my areas of interest and research. The book, written during the pre-Internet era, was very much based on research literature and my own experience. It ended up being 828 pages long. When Imperial College Press published the last edition in 2004, it had grown to a rather unwieldy 1,440 pages. When approached by the publishers for a new edition, it proved just too difficult a task to achieve within the confines of a single textbook. So, it was decided to break down the original book into a series of key processes and to produce a series of smaller, more focused textbooks. This new text is the revised edition of *Biology of Wastewater Treatment* that deals with natural treatment systems and the challenges the water industry faces in dealing with sustainability and reaching "net zero" by 2050, if not before. The United Kingdom (UK) has set a target of 2030 to achieve this target and is already well on its way to achieving this.

Open access to journal articles has frequently led to early key studies and research being overlooked. In this series of books, I have tried to refer back to the researchers who originally made key observations or discoveries that can be used to track the development of processes and new developments through citation indices.

Rivers, lakes and groundwaters are all under threat, with freshwater ecosystems now facing unprecedented levels of contamination, even after a century of ever-stricter legislation and regulation. In Europe, this is in part due to agricultural practice and, in particular, field size and lack of hedges and riparian vegetation that are important factors in the infiltration of water that provides extra treatment of the runoff, especially of nutrients. But the overriding factor is our increase in population and, especially in urbanization without sufficient planning and investment, to ensure adequate wastewater collection and treatment. This, coupled with the need to reduce carbon footprints associated with wastewater treatment and to improve standards even further to prevent river pollution, is leading to a crisis in wastewater treatment in many countries, especially the UK. We cannot ignore the effects that climate change is also having on wastewater treatment, which is increasing surface water discharge to combined sewers and the frequency of combined sewer overflow, as well reducing dilution capacity in receiving waters and oxygen levels — all causing serious and often long-lasting damage to river flora and fauna.

The use of nature-based solutions is generally regarded as a recent innovation, but many of the techniques have been used by stakeholders for generations. This has been enshrined in the circular economy principles, now well established as the framework for sustainable development, which includes wastewater treatment. Natural treatment systems that use plants and soil microorganisms are very much nature-based solutions and, where appropriate, can offer sustainable and low emissions options for a range of wastewater problems. Combined with on-site treatment and with sustainable urban drainage systems (SuDS), such as swales, infiltration basins, and retention-detention ponds, surface waters are being protected as well as new habitats being created to support and enhance wildlife diversity. In terms of circularity, natural treatment systems have the potential to produce a staggering array of useful and valuable byproducts, including high-value compounds for the pharmaceutical industry.

Conventional energy-demanding treatment systems such as activated sludge are constantly being modified and retrofitted in order to treat ever higher loads or improve effluent standards that often result in even greater carbon emissions. Also, the number of treatment plants reporting failures resulting in raw or partially treated sewage being discharged to surface waters has never been so high. So, conventional treatment is at a crossroads where new sustainable solutions are required if we are to protect the quality of our rivers, estuaries and seas.

One last word about the quote in the frontispiece, which appears in all the editions and reprints of *Biology of Wastewater Treatment*. I was working late one evening hand-drawing the illustrations for the first edition. While listening to the radio, I heard Spike Milligan utter that phrase, which seemed to apply so aptly not only to myself but to all wastewater engineers and scientists. When I wrote to ask him for permission to use it, he was pleased that he belonged to a greater brotherhood and so allowed the phrase to be freely reproduced.

This book is intended for all those working and researching in the area of natural wastewater treatment and sustainability. Although it is not exhaustive nor a design manual, it intends to give a detailed and practical overview of the processes and some guidance to the key research that has underpinned it.

Nick Gray
July 2021

Acknowledgments

I am very grateful to the many people who have kindly provided me information. I would especially like to thank all the publishers, organizations and companies who have given me their permission to reproduce or modify their material in the text.

My special thanks go to Amanda Yun, Senior Editor at World Scientific Publishing, for all her help and patience, and also to Gregory Lee, who has been a such great assistance in getting this book into print.

Contents

Preface vii

Acknowledgments xi

Part I Introduction to Sustainable Natural Wastewater Treatment

1 Sustainability and Water 3
1.1 Defining Sustainability and Sustainable Development 3
1.2 Sustainability and the Water Industry 9
1.2.1 The hydrological cycle 9
1.2.2 The global water crisis 13
1.2.3 Water footprints 15
1.3 Sustainability in Wastewater Treatment 20
References 29

2 The Circular Economy 33
2.1 What is the Circular Economy? 33
2.2 The Circular Economy and the Water Industry 39
2.2.1 Introduction 39
2.2.2 The water pathway 40
2.2.3 The materials pathway 44
2.2.4 The energy pathway 47
References 50

3 Global Warming and Climate Change 53
3.1 Introduction 53
3.2 GHG Emissions and the Water Industry 55
References 61

4 Natural Treatment Systems 63
4.1 The Concept of Natural Treatment 63
4.2 Selection of Processes 67
References 71

Part II Land Treatment

5 The Purification Process 75
5.1 Introduction to Land Treatment 75
5.2 The Process of Treatment 78
5.3 Evapotranspiration 80
5.4 Salinity and Sodicity 84
5.5 Contaminant Problems 89
5.6 Pathogen Removal 96
References 109

6 Subsurface Wastewater Infiltration Systems 117
6.1 Introduction 117
6.2 Design of the SWIS 120
6.3 Biomat Formation and Treatment 126
References 131

7 Slow-rate Irrigation Processes 135
7.1 Introduction to Slow-rate Land Treatment 135
7.2 Process Design 137
7.3 Crop Irrigation 140
7.3.1 Performance 140
7.3.2 Salinity intolerance 142
7.4 Forestry 155
References 162

8 Rapid Infiltration Processes 167
8.1 High-rate Irrigation 167
8.2 Irrigation–percolation 168
8.2.1 Rapid infiltration basins (RIBs) 168
8.2.2 Soil aquifer treatment (SAT) systems 172
8.2.3 Problems with RIB and SAT processes 178
References 180

9 Overland Flow 183
9.1 Introduction 183
9.2 The Treatment Process 185
9.3 Operational Management and Performance 187
References 194

Part III Constructed Wetland Systems

10 The Use of Microalgae and Macrophytes in Wastewater Treatment 199
10.1 Introduction 199
10.2 Microalgae 202
10.3 Submerged Macrophytes 205
10.4 Floating Macrophytes 210
10.4.1 Introduction 210
10.4.2 Water hyacinth 214
10.4.3 Duckweeds and water ferns 224
References 238

11 High-rate Algal Ponds and Photobioreactors 251
11.1 Design and Operational Basis of HRAPs and PBRs 251
11.2 The Role of Microalgae in Wastewater Treatment 257
11.2.1 Chlorophytes (green algae) 258
11.2.2 Cyanobacteria (blue-green algae) 261
11.2.3 Microalgal aggregates 263

11.3 Algal Productivity, Harvesting Methods and Uses of the Algal Crop 265
11.3.1 Algae as a source of protein and other nutrients 265
11.3.2 Algae as a biofuel 270
11.3.3 Harvesting algae 275
11.4 Microalgae and Sustainability 279
References 282

12 Wetlands 289
12.1 Introduction 289
12.2 Natural Wetlands 290
12.3 Removal Processes 292
12.4 Natural Wetlands for Wastewater Treatment 298
12.5 Conclusions 302
References 303

13 Constructed Wetlands Incorporating Emergent Vegetation 307
13.1 Introduction 307
13.2 Free Water Surface Treatment Wetlands 314
13.3 Subsurface Flow Treatment Wetlands 320
13.3.1 Design of horizontal flow reed bed treatment systems 323
13.3.2 Design of vertical flow reed bed treatment systems 331
13.3.3 Design of hybrid reed bed treatment systems 335
13.4 Pathogen Removal 338
References 340

Part IV Waste Stabilization Ponds

14 Introduction to Waste Stabilization Ponds and Lagoons 349
14.1 Development of Waste Stabilization Ponds (WSPs) 349
14.2 Case Studies 354
14.2.1 Medium to large scale WSP treatment system 354
14.2.2 Small scale WSP treatment system 359

14.3 Design 362
References 365

15 Anaerobic Lagoons 369
15.1 The Process 369
15.2 Microbiology 371
15.3 Application 377
15.4 Design of Anaerobic Lagoons 380
15.5 Gas Emissions from Lagoons 383
15.5.1 Methane and carbon dioxide 383
15.5.2 Ammonia 388
15.5.3 Odors and hydrogen sulfide 391
15.5.4 Denitrification 394
15.6 Covered Lagoons for Biogas Collection 396
References 398

16 Oxidation Ponds 405
16.1 Introduction to Oxidation Ponds 405
16.2 Facultative Ponds 405
16.2.1 Basis of operation 405
16.2.2 Ecology 407
16.2.3 Factors affecting performance 417
16.2.4 Design 425
16.2.5 Application 427
16.3 Maturation Ponds 430
16.4 Fishponds 435
16.5 River Purification Lakes 439
16.6 Pathogen Removal in Waste Stabilization Ponds 444
16.7 Contaminant Removal in Waste Stabilization Ponds 454
References 456

17 Partially and Completely Mixed Aerated Ponds 465
17.1 Introduction 465
17.2 Partially Mixed Aerated Ponds 466
17.3 Completely Mixed Aerated Ponds 467
References 473

Part V The Future

18 Sustainable Wastewater Treatment Systems 477
18.1 Assessing Sustainable Options 477
18.2 Emerging Technologies 484
18.3 The Role of Natural Treatment in a Sustainable Water Industry 488
References 492

Index 495

Part I

Introduction to Sustainable Natural Wastewater Treatment

1

Sustainability and Water

1.1 Defining Sustainability and Sustainable Development

The terms "sustainability" and "sustainable development" are now commonly used and while they are often used interchangeably, they do have specific meanings. Sustainability is the endpoint where civilization can thrive within the limits posed by only having one planet, while sustainable development is the process of getting from here and now to a point of sustainability. There are hundreds, if not thousands of definitions of sustainable development, but the most influential and quoted is the one produced by the Brundtland Commission in 1987:

> Sustainable development is a development that meets the needs of the present without compromising the ability of future generations to meet their own needs.

However, the actual definition in the report is subtly different:

> Humanity has the ability to make development sustainable — to ensure that it meets the needs of the present without compromising the ability of future generations to meet their own needs. (World Commission on Environment and Development, 1987)

The Nobel economist Sir John Hicks first conceptualized sustainability in 1946 in terms of income being ***"the amount, whether***

natural or financial capital, one could consume during a period and still be as well off at the end of that period". It was not until 1972 that it was first used in the context of the future of humankind in the book *Blueprint for Survival.* But it would be another 15 years before the concept took on global significance with the publication of the Brundtland Report. This resulted in a global discourse on what sustainability was and how to define it. The Nobel-prize-winning economist Robert Solow (1991) defined sustainability as: ***"an obligation or injunction to conduct ourselves so that we leave to the future the options and the capacity to be as well off as we are, not to satisfy ourselves by impoverishing our successors".*** This definition incorporates the word "obligation", and with it, brings the moral responsibility that we all must use our planet wisely, fairly and unselfishly. The concept of sustainability continues to evolve as our understanding of the complex relationship between economic development and the environment unfolds. So the need to define and pursue sustainability has become increasingly urgent as the environmental crisis deepens (Gray, 2015). Sustainable development has to tackle not only the challenge of global warming-induced climate change, but also existing and emerging problems, including nutrient, metal and microorganic pollution that affects water quality that impacts the environment, ourselves, and of course, other species, as well as increasing problems of water scarcity, food production and preserving the biodiversity of plants, animals and habitat diversity. Yet how can this be achieved, as sustainability lacks precise structures or systems to achieve the desired outcomes, even if we knew exactly what those outcomes should be? In simple terms, economic growth increases the rate of production and consumption of both goods and services. This, in turn, leads to an increase in the use of resources, and an increase in the production of waste, byproducts, and a wide range of pollutants. Therefore, if the mechanisms of economic growth are not controlled or altered, they impact all of us in an increasingly negative manner through the overexploitation of natural resources, the ability of natural systems to assimilate waste, and an increasingly degraded environment (physical, chemical and biological) (Gray, 2015).

Sustainability is perceived to comprise three interdependent dimensions, the so-called economy-ecology-social nexus. All three dimensions have to be addressed simultaneously if sustainable solutions are to be found. These are:

> **Economic Sustainability**, which focuses on the portion of natural resources (both renewable and non-renewable) that provides the physical input into the production process for goods and services (i.e., economically, the maintenance of the man-made capital).
>
> **Environmental Sustainability**, which focuses on the maintenance and protection of environmental services and resources. Often referred to as the life support system but it is much more than this.
>
> **Social Sustainability**, which addresses poverty and human development. The maintenance of the life support systems is the predominant prerequisite for social sustainability that secures social-cultural needs and wellbeing.

The relationship between these three dimensions was illustrated at the 2005 World Summit by three interlocking circles (UN General Assembly, 2005) (Fig. 1.1). The social-economic interactions should be equitable, the economic-environmental relations must be viable and the environmental-social relationship must be bearable. The theory is that sustainability is an equal balance with each sector of equal importance. The economy-ecology-social nexus suggests that there are no limits to growth and that there are always more free resources and capacity to assimilate waste on which to create further growth, which is not the case. In practice, the economic dimension dominates through continuous growth with the environmental dimension being rapidly depleted.

As the environment is limiting and its resources cannot be expanded, then society must flourish within these limits and the economy must then reflect and service the needs of society within those limits. Therefore, to create a sustainable society, the environmental dimension must gain more importance and be reliably protected (Fig. 1.2). The reality of the economy–ecology–social nexus is that the

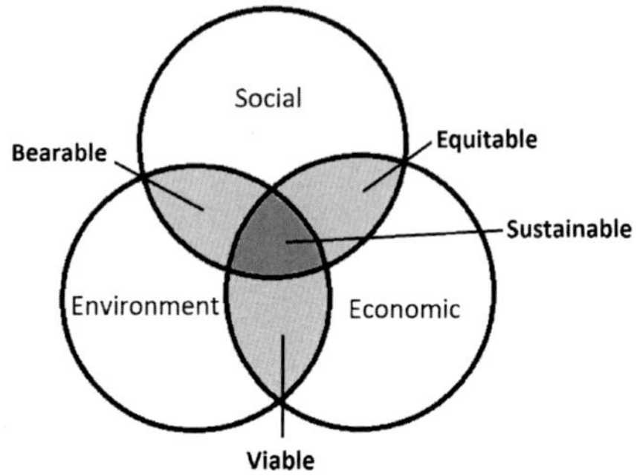

Fig. 1.1. The economy–ecology–social nexus formed the basis of the early environmental sustainability theory.

Fig. 1.2. The economy–ecology–social nexus has become distorted, controlled primarily by economic and social expansion without regard to the biocapacity of the Earth to support it.

economy and social demands far exceed the Earth's biocapacity. Therefore, a better model requires the economy to operate within the limits set by society (e.g. to reflect values such as fairness, justice and liberty). Society flourishes within limits set by the environment, so that the three elements are not equal but serve each other (Fig. 1.3) (Gray, 2015). This is not unique and the concept has been widely adopted in water supply using a new management approach known as demand-side management, the opposite to supply-side management, where the expansion of water demand has to be satisfied within a limited available

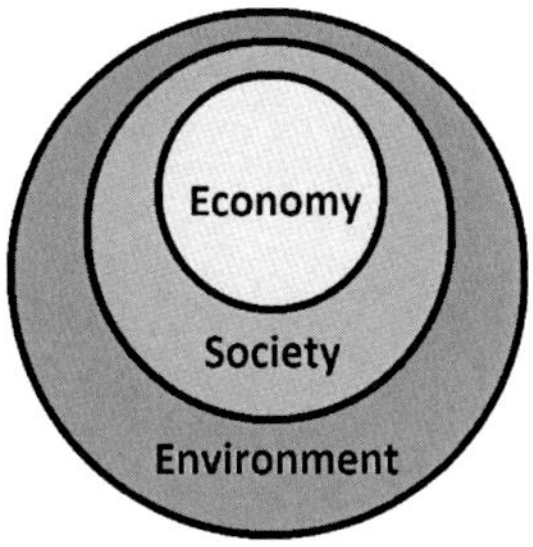

Fig. 1.3. A more sustainable economy-ecology-social nexus design.

volume of water so that any expansion has to be achieved through the conservation of supplies and their better management rather than by further exploitation of water resources (e.g. reservoir construction or increasing abstraction rates) (Gray, 2014).

Sustainability is now seen by all stakeholders as the nucleus on which all future development should be based, thus ensuring equal development opportunities for everyone (i.e., both low and high-income countries) and especially, the future generations to come. Sustainability is core to the *UN 2030 Agenda for Sustainable Development*, which sets 17 Sustainable Development Goals (SDGs). Adopted by all UN Member States in 2015, this is a global blueprint for a sustainable planet where all people can live in peace and prosperity. The SDGs are listed in Table 1.1 and include *Goal 6: Ensure availability and sustainable management of water and sanitation for all* (UN, 2015; 2021). The SDGs form an interconnected framework on which sustainable development can be achieved. Goal 6 has three threads: access to water supply and sanitation, sustainable use and development of water resources, and improved water quality and wastewater management. An EU Sustainable Development Strategy was adopted in 2001, which now includes all the SDGs (EC, 2021).

Sustainable development requires finding suitable solutions using a range of existing and emerging sustainable technologies that may differ according to local circumstances and culture. Such technologies must be both reliable and effective with minimum costs to the

Table 1.1. The UN sustainable development goals.

Goal 1.	End poverty in all its forms everywhere.
Goal 2.	End hunger, achieve food security and improved nutrition and promote sustainable agriculture.
Goal 3.	Ensure healthy lives and promote wellbeing for all at all ages.
Goal 4.	Ensure inclusive and equitable quality education and promote lifelong learning opportunities for all.
Goal 5.	Achieve gender equality and empower all women and girls.
Goal 6.	Ensure availability and sustainable management of water and sanitation for all.
Goal 7.	Ensure access to affordable, reliable, sustainable and modern energy for all.
Goal 8.	Promote sustained, inclusive and sustainable economic growth, full and productive employment and decent work for all.
Goal 9.	Build resilient infrastructure, promote inclusive and sustainable industrialization and foster innovation.
Goal 10.	Reduce inequality within and among countries.
Goal 11.	Make cities and human settlements inclusive, safe, resilient and sustainable.
Goal 12.	Ensure sustainable consumption and production patterns.
Goal 13.	Take urgent action to combat climate change and its impacts.*
Goal 14.	Conserve and sustainably use the oceans, seas and marine resources for sustainable development.
Goal 15.	Protect, restore and promote sustainable use of terrestrial ecosystems, sustainably manage forests, combat desertification, and halt and reverse land degradation and halt biodiversity loss.
Goal 16.	Promote peaceful and inclusive societies for sustainable development, provide access to justice for all and build effective, accountable and inclusive institutions at all levels.
Goal 17.	Strengthen the means of implementation and revitalize the Global Partnership for Sustainable Development.

*Acknowledging that the United Nations Framework Convention on Climate Change is the primary international, intergovernmental forum for negotiating the global response to climate change.

natural, economic and social environments. They can be both low or high-tech in design, so long as they are not detrimental to either the quantity or quality of resources, including biodiversity (van der Vleuten-Balkema, 2003).

1.2 Sustainability and the Water Industry

1.2.1 *The hydrological cycle*

Water is constantly moving within the hydrological cycle driven by solar energy. The Sun causes evaporation from the oceans, which forms clouds and precipitation. Evaporation also occurs on land from surface waters and the soil, with plants contributing significant quantities of water to the atmosphere by evapotranspiration. While about 80% of precipitation falls back into the oceans, the remainder falls onto land. It is this water that replenishes the soil and groundwater, feeds the rivers and lakes, and provides all the water needed by terrestrial and freshwater plants, animals, and, of course, humans (Fig. 1.4). It is important to remember that while the total volume of water in the world remains constant, its chemical quality and availability are constantly changing.

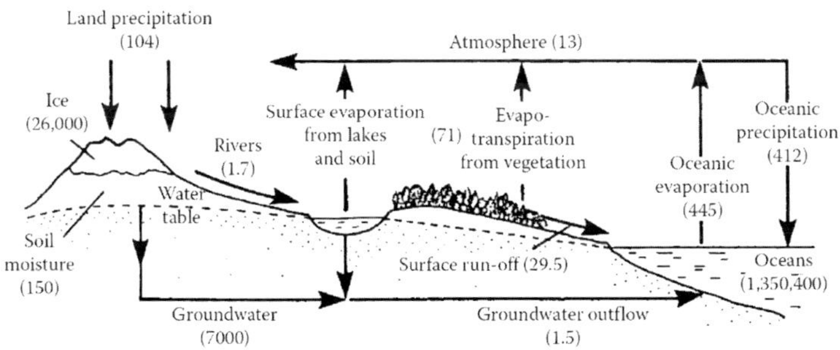

Fig. 1.4. Hydrological cycle showing the volume of water stored and the amount cycled annually. All volumes are expressed as 10^3 km^3.

Table 1.2. Water in the hydrological cycle in terms of area covered, volume and percentage of total water.

Type of water	Area (10^3 km^2)	Volume (10^3 km^3)	Percentage total water (%)
Water vapour (atmospheric)	510,000 (at sea level)	12.9	0.00092
Saline/brackish water			
Oceans	362,033	1,338,000	96.4
Groundwater	—	12,870	0.92724
Lakes (saline)	822	85	0.00612
Freshwater			
Ice caps, permanent snow cover and glaciers	17,000	26,064	1.87783
Groundwater	—	10,530	0.75865
Permafrost and ground ice	21,000	300	0.02161
Lakes	1,240	91	0.00656
Soil moisture	131,000	16.5	0.00119
Wetlands	2,680	11.5	0.00083
Rivers	—	2.12	0.00015
Biological water	131,000	1.12	0.00008
Total water		1,387,984	100
Cyclic water			
Annual evaporation			
From oceans		445	
From land		71	
Total evaporation		516	
Annual precipitation			
Onto oceans		412	
Onto land		104	
Total precipitation		516	

While there is a huge volume of water on the planet, some 97.5% of it is saline with 99.99% in the oceans and the remainder making up the salt lakes (Table 1.2). So, only 2.5% of this volume is freshwater, with 75% locked up as ice caps and glaciers, although this is slowly changing in response to climate change. A further 24% is stored in aquifers. This means that less than 1% of the total freshwater on our planet is found in surface waters (i.e., rivers and lakes) and the soil. So, because 0.01% of the world's water budget is present in surface water with another 0.01% present as soil moisture, there is, in reality, very little freshwater that is readily available for use by an ever-increasing human population, although availability is dependent on many factors such as distribution, volume, seasonality, and quality. Water availability is not even across the planet and is often lowest where population and hence demand is greatest.

The hydrological cycle is continuous, thus making water a renewable resource. In essence, the more it rains, the greater the flow in the rivers and the higher the water table rises as the aquifers fill with water that percolates downwards through the soil into the permeable rocks below. As water supplies ultimately depend on rain, when the amount of rain decreases, then the water supply will also decrease, and in cases of severe drought, may fall to nothing. Therefore, to provide sufficient water for supply all year round, careful management of resources is required.

When precipitation falls into a catchment, one of the following three fates befalls it:

(a) It may remain on the ground as surface moisture and be eventually returned to the atmosphere by evaporation. Alternatively, it may be stored as snow on the surface until the temperature rises sufficiently to melt it. Storage as snow is an important source of drinking water in some regions. For example, throughout Scandinavia, lagoons are constructed to collect the runoff from snow as it melts, and this provides the bulk of the drinking water during summer.

(b) Precipitation flows over the surface into small channels to become surface runoff into streams and lakes. This is the basis of all surface water supplies, with the water eventually evaporating into the atmosphere, percolating into the soil to become groundwater, or continuing as surface flow in rivers back to the sea.

(c) The third route is for precipitation to infiltrate the soil and slowly percolate into the ground to become groundwater that is stored in porous sediments and rocks (Gray, 2017).

Groundwater may remain in these porous layers for periods ranging from just a few days in Karst systems to possibly millions of years in deep confined aquifers (average 1,400 years). Eventually, groundwater is removed by a number of natural and artificial means. These include natural upward capillary movement to the soil surface, plant uptake, groundwater seepage into surface rivers and lakes or directly to the sea, or more likely removal by artificial pumping from wells and boreholes. From a management perspective, the hydrological cycle at this point can be represented as

$$P = E + R + I, \tag{1.1}$$

where P is the precipitation, E is the evapotranspiration, R is the runoff, and I is the infiltration. In France, approximately 55% of precipitation evaporates, 25% becomes surface runoff, and 20% infiltrates into the ground. However, this varies enormously between regions.

The water in the oceans, ice caps and aquifers is all ancient, acting as sinks for both minerals and pollutants. All pollution, whether emitted as a gas, liquid or solid, eventually ends up in the cycle and will ultimately find its way into one of these sinks.

As rain falls through the atmosphere it dissolves gases (e.g. CO_2, NO_x and SO_x), making all rainfall slightly acidic and in some localized areas, highly acidic, so that as it flows over and through the Earth's surface it is constantly dissolving minerals and other materials from the soil and bedrock, forming a chemical record of its passage from the clouds. Therefore, water supplies have a natural variation in quality, which depends largely on their source. Once it enters a surface water body (i.e., stream, river, lake) and to a lesser extent, an aquifer, it becomes vulnerable to contamination by anthropogenic compounds, which generally have a negative effect on water quality (e.g. nutrients, metals and organic contaminants such as pesticides, pharmaceuticals and estrogen-mimicking compounds).

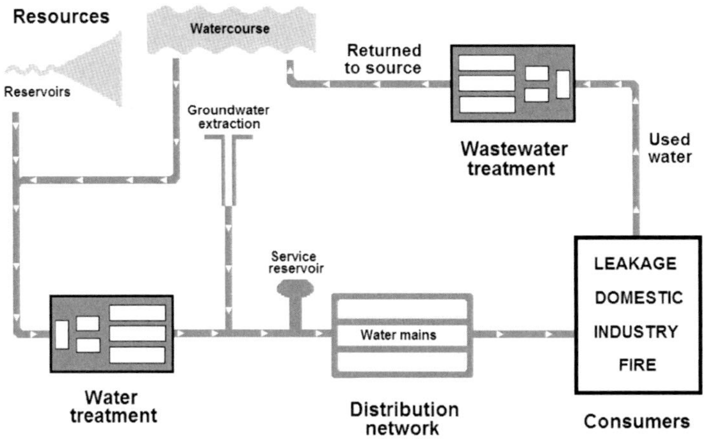

Fig. 1.5. Water services cycle. Water is intercepted within the hydrological cycle for use and is returned to the hydrological cycle after treatment. Ideally, the water quality of the returned water should be close to what it was on abstraction.

In many countries, the cycle is highly managed in order to provide sufficient water for industry, agriculture and domestic use. This requires the management of surface water and groundwater resources, the treatment and supply of water and its subsequent collection, cleansing, and return to the cycle (Fig. 1.5). The hydrological cycle is increasingly affected by climate change through rising surface temperatures increasing the rates of precipitation, evaporation and evapotranspiration. This has resulted in some regions becoming wetter and some sub-tropical areas becoming even dryer, increasing the incidence of droughts. At higher latitudes, changes in snowfall patterns continue to lead to springtime flooding, with the proportion of water stored in the cycle as snow and ice slowly declining, with sea levels rising to accommodate the extra free water.

1.2.2 *The global water crisis*

The world is currently suffering the worst water emergency ever recorded. Worldwide, over two billion people are subject to continuous severe water stress (UN, 2018), and four billion suffer from severe water scarcity for at least one month per year (Mekonnen and Hoekstra, 2016). This is particularly problematic in rural areas (FAO, 2020a), where an estimated 700 million people worldwide will possibly be displaced by intense water scarcity from arid and semi-arid places by 2030. Disease exposure is exacerbated by inadequate water supply and sanitation, with 2.4 billion people lacking access to adequate sanitation. Even in this modern highly connected global society, two million children die each year from preventable waterborne diarrheal diseases. UNICEF (2017) estimates that one in four of the world's children under 18 (i.e., 600 million) will be living in areas of extremely high water stress by 2040. Nearly half the global population are already living in potential water-scarce areas at least one month per year and this could increase to some 4.8 to 5.7 billion by 2050, with about 70% of those affected living in Asia and agricultural irrigation a major factor (FAO, 2020a). Hence, the importance for water to be included as one of the UN SDGs (Table 1.1).

Water stress is an indicator used to measure progress in achieving Goal 6 of the SDGs. It is the ratio of freshwater withdrawn to the total renewable freshwater resources available for supply in a defined area after the volume required to sustain the natural environment is taken into consideration (Damkjaer *et al.*, 2017) (Eq. [1.2]). It is expressed as a percentage, where no stress is equivalent to 0–25%, low stress 25–50%, medium 50–75%, high 75–100%, and critical stress >100%.

$$Water\,Stress(\%) = \frac{TFWW}{TRWR - EFR} \times 100, \tag{1.2}$$

where *TFWW* is the total freshwater withdrawal, *TRWR* is the total renewable freshwater resources, and *EFR* is the environmental flow requirements (FAO, 2020b).

Over abstraction and pollution not only impacts humans but aquatic species and other species dependent upon water resources. The health of our water resources is a critical indicator of the threat from ever-increasing demands of an increasing population and escalating waste disposal, as well as climate change which itself is a major stressor on water ecosystems due to its changing patterns and intensity of rainfall and increased water temperatures. It is not only the surface waters that are under threat — Richey *et al.* (2015) estimate that a third of the world's largest groundwater systems are in danger of being critically overexploited. This is based on measuring the renewable groundwater stress (RGS), which is the ratio of groundwater use to groundwater availability where stress levels are represented as 0–0.1 (low stress level), 0.1–0.2 (moderate stress level), 0.2–0.4 (high stress level), and >0.4 (extreme stress level).

1.2.3 *Water footprints*

Water footprints are used to calculate the total amount of water consumed by an individual, a single business or country, and is the total volume of water used directly and indirectly (i.e., the volume of water used in the production of goods and services used) (Fig. 1.6) (http://www.waterfootprint.org).

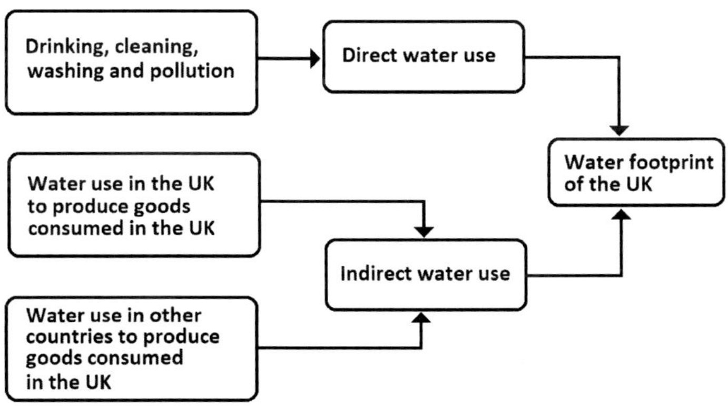

Fig. 1.6. Calculation of the UK water footprint is broken down into direct and indirect water use (Chapagain and Orr, 2008). Reproduced with permission of the WWF-UK, Godalming, Surrey, under a Creative Commons License (https://creativecommons.org/licenses/by-nc/4.0/).

At the business level, direct water usage is the total water used by the company, which can be easily measured using a water meter. Indirect water usage is far more complex to measure and requires the use of a life cycle analysis. Indirect water is expressed as the **virtual water** content of a product (i.e., m^3 of water per ton of a product produced) and is further broken down as **Internal** (water used to manufacture products made in one's country), **External** (water used to manufacture products imported into the country) or **Net Imported**, which is the difference between the external and internal water footprints.

Large volumes of water are used in the manufacture of goods and the provision of services, so water use, like carbon emissions, is linked to consumerism and consumer choice (Gray, 2015). Examples of the amount of water used in the production of unit volumes or weights of common household items are given in Table 1.3. The virtual water footprint can be further scrutinized in terms of the source and the volume of water polluted during production. There are three categories:

BLUE	The volume of ground or surface water used.
GREEN	The volume of rainwater used.
GREY	The volume of freshwater polluted

Water used by the agricultural sector accounts for nearly 92% of the annual global freshwater consumption. As crops need water, water footprints are increasingly used to compare and optimize production methods. The best scenario is where crops are grown solely using rainwater, which would be the case with cereal crops grown in the British Isles. However, this is in contrast to many other crops, including vegetables, where groundwater and surface water may be required for irrigation. Figure 1.7 gives the total water footprint for the production of 500 mL of a soft drink, which varies according to the type and origin of the sugar where the most sustainable sources are those using the most rainwater (green) and causing the least pollution (grey).

The water footprint for the United Kingdom (UK) is explored in Table 1.4 and is equivalent to 4,645 liters per person per day

Table 1.3. The amount of water used in the production of common household items per unit volume or weight.

Per liter	
➢ Water	1 L
➢ Bottled water	4 L
➢ Beer	300 L
➢ Milk	1,000 L
Per cup or glass	
➢ Tea	120 L
➢ Orange juice	850 L
➢ Wine	960 L
➢ Coffee	1,120 L
Meat per kg	
➢ Lamb	6,100 L
➢ Beef	15,000–70,000 L
➢ Eggs	3,300 L
Main foodstuffs per kg	
➢ Bread	1,300 L
➢ Rice	3,400 L
➢ Tea	9,200 L
➢ Roasted coffee	21,000 L
Household items (each)	
➢ Pair jeans	10,850 L
➢ Cotton shirt	4,500 L
➢ Cotton sheet (1kg)	11,000 L
➢ Disposable nappy	810 L
➢ The average car	400,000 L

(L $cap^{-1}d^{-1}$), with agricultural products comprising 73% and household consumption making up just 3% overall. Sixty-two percent of the UK's water footprint comes from outside the country, making it a net importer of virtual water. In comparison, China has a water footprint

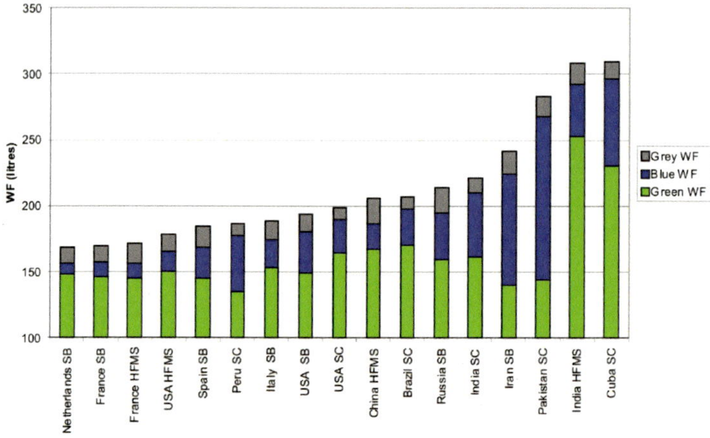

Fig. 1.7. Water volume and sources for the production of a 0.5 L PET-bottle soft drink. Sugar types used are SB (Sugar Beet), SC (Sugar Cane) and HFMS (High Fructose Maize Syrup) (Ercin *et al.*, 2011). Reproduced with permission from Springer, Berlin, Germany.

Table 1.4. The UK water footprint showing the origin of water used by different sectors in gigacubic meters per year (Gm^3yr^{-1}) (Chapagain and Orr, 2008). Reproduced with permission from the WWF-UK under a Creative Commons License (https://creativecommons.org/licenses/by-nc/4.0/).

	Internal	**External**	**Total**	**Total footprint**
		Gm^3 yr^{-1}		**%**
Agricultural products	28.4	46.4	74.8	73
Industrial products	6.9	17.2	24.0	24
Household water use	3.3	—	—	3
Total footprint (Gm^3 yr^{-1})	38.6	63.6	102.1	100
Total footprint (%)	38%	62%	100%	

of 2,932 L cap^{-1}d^{-1} with just 10% from external sources and the United States (US) at 7,781 L cap^{-1}d^{-1} with 20% from external sources. Food is a major import-export commodity and as agriculture uses the largest amount of water, then this virtual water is constantly on the move with Brazil, Mexico and Japan (the major importers), and the US, by far the largest exporter of virtual water (Table 1.5).

Table 1.5. The top six agricultural net importers (A) and exporters of water (B) in Gm^{3}yr^{-1} (Chapagain and Orr, 2008). Reproduced with permission from the WWF-UK under a Creative Commons License (https://creativecommons.org/licenses/by-nc/4.0/).

Country	Export	Import	Net import
A. Net virtual importers of water			
Brazil	91	199	107
Mexico	19	103	84
Japan	4	86	83
China	55	133	78
Italy	38	88	50
UK	15	55	40
Country	**Export**	**Import**	**Net export**
B. Net virtual exporters of water			
US	298	137	161
Australia	71	10	62
Argentina	58	4	54
Canada	70	27	44
Thailand	52	9	43
India	66	24	42

The complexity of the movement of agricultural virtual water is illustrated in Fig. 1.8 for the UK. Water footprints can also be developed for wastewater treatment plants (Morera *et al.*, 2016; Gómez-Llanos *et al.*, 2018; 2020).

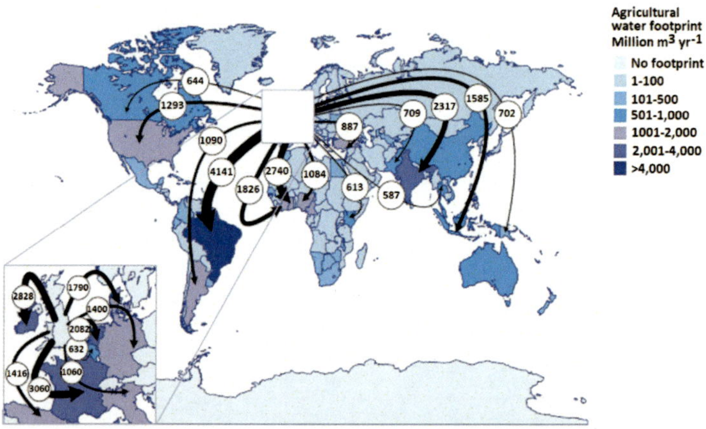

Fig. 1.8. Example of a water movement. The UK External Water Footprint for Agriculture (Chapagain and Orr, 2008). Reproduced with permission from the WWF-UK under a Creative Commons License (https://creativecommons.org/licenses/by-nc/4.0/).

1.3 Sustainability in Wastewater Treatment

Wastewater management goes beyond the treatment plant, encompassing wastewater generation, sewerage and treatment, which should include the recovery of all possible byproducts, including water, nutrients (i.e., nitrogen and phosphorus) and biosolids and their reuse (Bakir, 2001), using the principles of the circular economy (Chap. 2). Successful sustainability in wastewater treatment is dependent on location, the scale of operation, wastewater characteristics, including the amount of contaminants, level of investment, availability of technologies and expertise, and the vulnerability of receiving waters and land. Sustainable wastewater treatment systems are particularly well-suited to small to medium-sized plants, especially those in warmer climates.

Sustainability is normally assessed under three strands: the economic, environmental and social dimensions (Sec. 1.1), although in practice, adopted technologies and operation are based on the optimal use of natural resources such as water and fossil fuels while minimizing the environmental impact of both the treated effluent and biosolids produced (Chang *et al.*, 2015). Balkema *et al.* (2002) have shown that sustainable wastewater treatment objectives can be varied and multiple, for example, minimizing costs, energy use (i.e., greenhouse gas [GHG] emissions), the land area required, loss of nutrients, and waste production, while maximizing products (e.g. reusable water, biogas, biomass, fertilizers, compost) and social acceptance. These objectives can make the selection of technologies under particular cultural and climatic conditions difficult and sometimes conflicting. To overcome the problems of design and operation of wastewater treatment, some form of sustainability assessment is necessary to ensure that the most appropriate technologies and operational management systems are selected at the planning stage and that these goals are realized and possibly improved during operation. This is achieved by sustainability indicators that measure key factors influencing the functionality and sustainability of the wastewater treatment plant.

Economic indicators in the past have been the deciding criteria for the selection of treatment technologies. They are normally measured in terms of investment costs (i.e., land and construction costs) and also operation and maintenance costs. Often, where funding comes from separate sources, cheaper investment costs are selected regardless of operation and maintenance costs. This leads to an inability to sustain the operation of the plant, a scenario commonly seen in low-income countries where infrastructure is supported by overseas aid. A very important factor that is often overlooked is the lifespan of the infrastructure, which must be taken into account in both indicators.

Environmental indicators are assessed in terms of effluent quality, energy and other resources used, and overall impact on the environment and public health. While effluent quality is measured in terms of organic matter (BOD or COD), suspended solids, nutrients (total nitrogen and total phosphorus), metal and other contaminant removal rates, these need to be directly linked to environmental quality objectives (EQOs) set for receiving waters, which may already be limited in terms of assimilation capacity for certain components of the wastewater. Energy consumption is often used as a measure of carbon footprint, which depends on the electrical production mix supplied to the plant; however, due to the potential release of other greenhouse gases, a total greenhouse gas emission value should be calculated for the plant in carbon dioxide equivalents. The recovery and reuse of components in wastewater is a key indicator of sustainability. This includes the reuse of both the water and solids (i.e., biosolids or sewage sludge) content of the wastewater. Biosolids production is also an important indicator, both in terms of volume and contaminant levels, which controls its reuse value and possible environmental impacts — both positive and negative.

Land requirement is dependent on hydraulic loading and the technology selected. In high-income countries, intensive high-energy systems are extremely efficient in terms of both organic and hydraulic loading, making them very compact. In contrast, the types of low-energy systems, particularly in medium to low-income countries, tend to require large areas of land, making direct comparisons difficult, especially in relation to land price. It is debatable whether land area should be considered as an environmental or economic indicator between countries, where land price can be used as a unit of measurement. The reliability of treatment systems is linked to the robustness of the technology and its complexity to operate and maintain. This is an important indicator that can be measured in terms of system failures. As system failures usually result in pollution, then it is included in the environmental dimension; however, some researchers believe it should be considered under the separate category of technical indicators (Cossio *et al.*, 2020).

Social indicators connect wastewater treatment and the physical plant to the community. This will also include the effect of effluent and biosolids disposal. Unlike the other dimensions, social indicators require qualitative assessment in the planning phase of public acceptance, including visual impact and after construction indicators such as odor and noise. Smaller treatment plants and those employing natural treatment systems, especially in low to middle-income countries, tend

Table 1.6. Cassio *et al.* (2020) have listed 40 wastewater treatment sustainability indicators from 43 selected studies. Those in capital letters are considered to be standard or essential indicators. The percentage and number of studies in which the indicator was used and whether it was for the planning and/or operational phase are shown in the columns on the right. Reproduced with permission from Elsevier Publishing.

Dimensions and indicators	**%**	**No. of studies**	**Planning**	**Operation**
Environmental dimension				
ENERGY USE	47	20	14	6
GLOBAL WARMING POTENTIAL	47	20	12	8
REMOVAL OF BOD, TSS, TN, TP AND FC	44	19	9	10
LAND AREA	44	19	16	3
QUALITY OF EFFLUENT AND SLUDGE	37	16	9	7
POTENTIAL RECYCLING	35	15	10	5
Sludge production	30	13	7	6
Use of natural resources (water and raw materials)	30	13	7	6
Eutrophication potential	19	8	5	3
Economic dimension				
INVESTMENT COSTS	63	27	18	9
OPERATING AND MAINTENANCE COSTS	42	18	12	6
Tariff	16	7	1	6
Cost effectiveness	9	4	2	2
Affordability	9	4	2	2
Ensure economic sustainability of the utility	7	3	1	2
Technical dimension				
RELIABILITY	47	20	14	6
COMPLEXITY OF CONSTRUCTION AND O&M	37	16	12	4

(*Continued*)

Table 1.6. (*Continued*)

Dimensions and indicators	%	No. of studies	Planning	Operation
Durability	19	8	5	3
Robustness	16	7	5	2
Adaptability	14	6	4	2
Upgrading, extending, or modifying the scope of future development	9	4	3	1
Quantity of treated wastewater as a percentage of the total volume of wastewater	5	2	—	2
Actual people equivalent (PE) as a percentage of design PE	2	1	—	1
Load generated by pollutants entering the WWTP (per inhabitant connected; per catchment area; per population density)	2	1	—	1
Treatment level	2	1	—	1
Social dimension				
PUBLIC ACCEPTANCE	37	16	10	6
AESTHETICS	37	16	9	7
Public health risk	26	11	4	7
Coverage	23	10	1	9
Participation	21	9	5	4
Occupational health and safety	21	9	4	5
Local employment	16	7	5	2
Staff requirements	16	7	5	2
Satisfaction (employees)	14	6	1	5
Stimulation of sustainable behavior	12	5	3	2
Awareness	12	5	2	3
Expertise	7	3	1	2
Institutional dimension				
Interactions	14	6	1	5
Institutional capacity	14	6	—	6
Information	5	2	—	2

to adopt a broader approach, which includes local participation with an emphasis on public health and reuse.

From the assessment of 43 separate studies, Cossio *et al.* (2020) identified 12 commonly used indicators from a possible 40 used to assess the sustainability of wastewater treatment using five sustainability dimensions: Environmental, Social, Economic, Technical, and Institutional (Table 1.6). While these 12 indicators are valid for all plants, they do not take into account local factors such as affordability, the need for public engagement, local environmental quality needs, etc. This is illustrated by Cossio and his colleagues, who devised a specific set of indicators for small-scale wastewater treatment plants in Bolivia, adding locally relevant indicators to the standard sustainability indicators (Table 1.7).

Table 1.7. Suggested sustainability indicators selected by Cossio *et al.* (2020) for small-scale wastewater treatment plants in Bolivia, classed under the sustainability dimensions: Environmental, Social, Economic, Technical, and Institutional. Selected local indicators are in italics while standard indicators are shown in normal text, all with possible units of measurements. Many indicators, especially social and institutional, are assessed by specifically designed scoring systems. Adapted from Cossio *et al.* (2020) and reproduced with permission from Elsevier Publishing.

Dimension/ Indicator	Units	Description of sustainability indicator
Environmental		
Energy use	kW pe^{-1}	Total energy use (all fuels).
Global warming	kg CO_2 eq. pe^{-1} $1yr^{-1}$ or kg CO_2 eq. m^{-3}	Total of all greenhouse gas emissions.
Removal of BOD, TSS, TN, TP and FC	%	Efficiency of the WWTP according to monitoring parameters.
Land area	m^2pe^{-1}	Physical footprint of plant and infrastructure.
Quality of effluent and sludge	mg L^{-1} kg $pe^{-1}yr^{-1}$	Effluent quality according to monitoring parameters (e.g. organic matter and pathogens).

(*Continued*)

Table 1.7. (*Continued*)

Dimension/ Indicator	Units	Description of sustainability indicator
Potential recycling	g pe^{-1} yr^{-1} or %	Acceptable quality of effluent and sludge for reuse.
Eutrophication potential	kg N or P pe^{-1} yr^{-1}	Environmental impact on the recipient water body due to release of nutrients.
Social		
Public acceptance	Scoring system	Acceptance of the WWTP before implementation following consent by the population.
Aesthetics	Scoring system	Acceptance of the nuisance level of odors from the WWTP, influenced by its location and distance to the nearest households.
Public health risk	Number of outbreaks per unit population	Number of waterborne disease related cases in population
Participation	Scoring system	Level of user participation before the project is implemented (i.e., communication), validation of demand for the project, water board members coordinating with local leaders, users carrying out operational tasks if necessary.
Staff requirements	Number per flow rate to plant	Staff available to operate the WWTP (technical expert, operator, administrator, social communicator).
Employee satisfaction	Scoring system	Staff at the WWTP with a proper salary.
Awareness	Scoring system	Level of responsibility of inhabitants and a sense of the need for a WWTP to reduce contamination.

Table 1.7. (*Continued*)

Dimension/ Indicator	Units	Description of sustainability indicator
Expertise	Number of experts and skilled staff	Technical and social expertise exists, and there is a transfer of knowledge to new members of staff or water board members (if necessary).
Economic		
Investment costs	\$ $pe^{-1}yr^{-1}$ or \$ m^{-3}	Investment into infrastructural maintenance and development.
Operation and maintenance costs	\$ m^{-3}	Operation and maintenance costs per inhabitant-year.
Tariff	\$ per month	Appropriate tariff calculated to cover the technical requirements (e.g. technical expert, operator, materials) and which are socially accepted by the users.
Cost effectiveness	Scoring system	Percentage of income from tariffs allocated to a WWTP, operating costs, and monthly balance after expenditure.
Affordability	\$ m^{-3} or % of average income per m^3	Service cost affordable to users linked to the willingness to pay and reflected in the percentage of the users' debt.
Technical		
Reliability	Scoring system	Assessment of reliability in performance and operation, number of unit process failures, etc.
Complexity of construction and O&M	Scoring system	Appropriate technology designed in line with the local context and available resources (e.g. technical and economical to operate, maintain, replace or upgrade).

(*Continued*)

Table 1.7. (*Continued*)

Dimension/ Indicator	Units	Description of sustainability indicator
Sewerage functionality	Number of failures per month	Integrity of sewerage network including leaks, pipe collapse and pumping failures.
Institutional		
Interactions	Number of events per year	Collaboration between the water association and the municipality, local universities and health institutions.
Institutional capacity (water association)	Scoring system	Water association strengthened and with the capacity to fulfil its functions to manage the WWTP (e.g. institutional organigram, monitoring capacity, regulations and functions, O&M plan, awareness of environmental limits, fulfilment of agreements, capacity to transfer knowledge).
Institutional capacity (higher level)	Scoring system	Experts available at municipal, departmental and national level to carry out their respective tasks in the management of the WWTPs (e.g. norms, laws, regulations).
Information	Scoring system	Information/training to users during the pre-investment and construction of the project. General information distributed to users connected on how to use the sewage network. Technical information for operators to manage the wastewater treatment plant.

Sustainability indices can be used singly or in groups by sustainability dimension, or individually weighted to give a single cumulative score, so that treatment plants can be compared and ranked. An example is given in Chap. 18.

References

Bakir, H.A. (2001) Sustainable wastewater management for small communities in the Middle East and North Africa. *Journal of Environmental Management*, **61**, 319–328.

Balkema, A.J., Preisig, H.A., Otterpohl, R. and Lambert, F.J.D. (2002) Indicators for the sustainability assessment of wastewater treatment systems. *Urban Water*, **4**, 153–161.

Chang, C.-C., DiGiovanni, K., Zhang, G., Yang, X. and You, S.-H. (2015) Sustainability. *Water Environment Research*, **87**, 1208–1255.

Chapagain, A.K. and Orr, S. (2008) *UK Water Footprint: The Impact of the UK's Food and Fibre Consumption On Global Water Resources, Volume 1*. WWF-UK, Godalming, UK. https://waterfootprint.org/media/downloads/Orr_and_Chapagain_2008_UK_waterfootprint-vol1.pdf

Cossio, C., Norrman, J., McConville, J., Mercado, A. and Rauch, S. (2020) Indicators for sustainability assessment of small-scale wastewater treatment plants in low and lower-middle class income countries. *Environmental and Sustainability Indicators*, **6**, 100028.

Damkjaer, S. and Taylor, R. (2017) The measurement of water scarcity: Defining a meaningful indicator. *Ambio*, **46**, 513–531. https://doi.org/10.1007/s13280-017-0912-z

EC. (2021) The 2030 agenda for sustainable development and the SDGs. European Commission, Brussels. https://ec.europa.eu/environment/sustainable-development/SDGs/index_en.htm

Ercin, A.E., Martinez-Aldaya, M. and Hoekstra, A.Y. (2011) Corporate water footprint accounting and impact assessment: The case of the water footprint of a sugar-containing carbonated beverage. *Water Resources Management*, **25**, 721–741.

FAO. (2020a) *The State of Food and Agriculture: Overcoming Water Challenges in Agriculture*. Food and Agricultural Organization of the United Nations, Rome. www.fao.org/documents/card/en/c/cb1447en/

FAO. (2020b) Indicator 6.4.2. Food and Agriculture Organization of the United Nations, Rome. https://unstats.un.org/sdgs/metadata/files/Metadata-06-04-02.pdf

Gómez-Llanos, E., Durán-Barroso, P. and Matías-Sánchez, A. (2018) Management effectiveness assessment in wastewater treatment plants through a new water footprint indicator. *Journal of Cleaner Production*, **198**, 463–471. https://doi.org/10.1016/j.jclepro.2018.07.062

Gómez-Llanos, E., Matías-Sánchez, A. and Durán-Barroso, P. (2020) Wastewater treatment plant assessment by quantifying the carbon and water footprint. *Water*, **12**(11), 3204. https://doi.org/10.3390/w12113204

Gray, N.F. (2014) *Drinking Water Quality: Problems and Solutions.* Cambridge University Press, Cambridge.

Gray, N.F. (2015) *Facing Up to Global Warming: What is Going On and How You Can Make a Difference?* Springer, Cham, Switzerland.

Gray, N.F. (2017) *Water Science and Technology: An Introduction.* CRC Press, Boca Raton, FL.

Mekonnen, M.M. and Hoekstra, A.Y. (2016) Four billion people facing severe water scarcity. *Science Advances*, **2**(2), 1500323. https://advances.sciencemag.org/content/2/2/e1500323

Morera, S., Corominas, L.I., Poch, M., Aldaya, M.M. and Comas, J. (2016) Water footprint assessment in wastewater treatment plants. *Journal of Cleaner Production*, **112**, 4741–4748. https://doi.org/10.1016/j.jclepro.2015.05.102

Richey, A.S., Thomas, B.F., Lo, M-H., Reager, J.T., Famiglietti, J.S., Voss, K., Swenson, S. and Modell, M. (2015) Quantifying renewable groundwater stress with GRACE. *Water Resources Research*, **51**(7), 5217–5238. https://agupubs.onlinelibrary.wiley.com/doi/full/10.1002/2015WR017349

Solow, R. (1991) Sustainability: An economist's perspective. The 18th Edward J. Seward Johnson Lecture to the Marine Policy Center Woods Hole Oceanographic Institution, Woods Hole Massachusetts. http://cda.morris.umn.edu/~kildegac/Courses/Enviro/3008/Readings/Solow.pdf

UN. (2015) Transforming our world: The 2030 agenda for sustainable development. A/RES/70/1. United Nations, New York. https://sustainabledevelopment.un.org/content/documents/21252030 Agenda for Sustainable Development web.pdf

UN. (2018) SDG 6 Synthesis Report 2018 on water and sanitation. United Nations, New York. https://www.unwater.org/publication_categories/sdg-6-synthesis-report-2018-on-water-and-sanitation/

UN. (2021) Sustainable Development Goals. United Nations, New York. https://www.unenvironment.org/explore-topics/sustainable-development-goals/about-sustainable-development-goals

UN General Assembly. (2005) 2005 World Summit Outcome, Resolution A/60/1, adopted by the General Assembly on 15 September 2005. United Nations, New York.

van der Vleuten-Balkema, A.J. (2003) Sustainable wastewater treatment, developing a methodology and selecting promising systems. Proefschrift, Technische Univesiteit Eindhoven, Eindhoven.

World Commission on Environment and Development. (1987) *Our Common Future: The Report of the World Commission on Environment and Development.* Oxford University Press, Oxford. http://www.un-documents.net/wced-ocf.htm

2
The Circular Economy

2.1 What is the Circular Economy?

The circular economy is an enabling concept that starts to provide a real framework to sustainable development as discussed in Sec. 1.1. It is finding ways to change from a linear flow economy dominated by high consumption rates and waste production, where resources are generally only used once (resulting in substantial and increasingly irreversible environmental degradation), to a circular economy where production and waste streams are closed (ensuring a circularity of resources), and waste outputs are significantly reduced or preferably eliminated, thereby allowing ecosystems to recover.

The Ellen MacArther Foundation (2021) has been pivotal in the practical development of the concept, defining it as:

> Looking beyond the current take-make-dispose extractive industrial model, a circular economy aims to redefine growth, focusing on positive society-wide benefits. It entails gradually decoupling economic activity from the consumption of finite resources and designing waste out of the system. Underpinned by a transition to renewable energy sources, the circular model builds economic, natural and social capital. It is based on three principles: design out waste and pollution, keep products and materials in use, and regenerate natural systems.

The core objectives of all sustainable industries, including the water industry of which wastewater treatment is a key component,

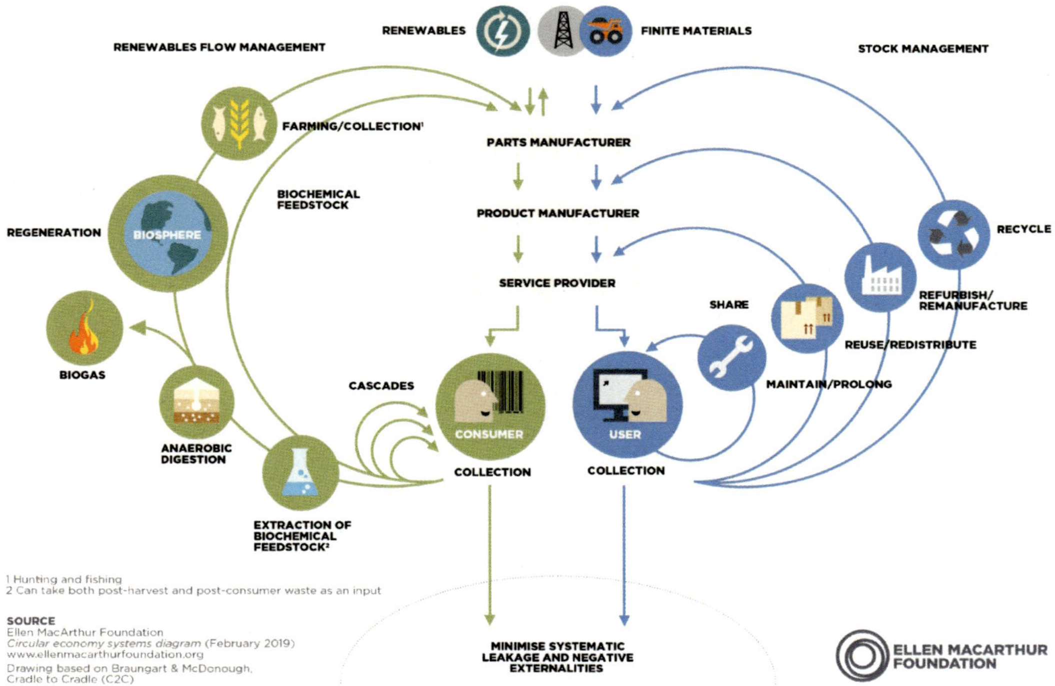

Fig. 2.1. The Circular economy butterfly diagram. Reproduced with permission from the Ellen MacArthur Foundation (https://www.ellenmacarthurfoundation.org).

are: (i) incrementally reducing waste and pollution through innovative design, (ii) continuously recovering and recycling products and materials, eliminating the need for use of finite and scarce resources, and (iii) improving the interface between an activity and the natural world, and preserving and regenerating natural ecosystems and processes. This is represented by the universally recognized butterfly representation of the circular economy by the Ellen MacArthur Foundation (Fig. 2.1).

The circular economy has evolved out of the best available technologies and practices, though they have also set new standards that will require new and challenging technological development. The core concepts of the circular economy have been listed by Gladek *et al.* (2015) and summarized by Jazbec and Turner (2018) as:

- Materials are incorporated into the economy in such a way that they can be cycled at continuous high value and are never dissipated into the environment in unrecoverable form.
- All energy is based on renewable sources.
- Biodiversity is structurally supported and enhanced through all human activities in a circular economy.
- Human society and culture are preserved through human activities.
- The health and wellbeing of humans and other species should be structurally supported through the activities of the economy.
- Human activities should generate value in measures beyond financial aspects. Material and energy are not currently available in infinite measure, so their use should be an intentional and meaningful contribution.
- Optimize for geographically short material cycles.
- Optimize the time scale of material cycles for material demand.
- Match the quality of resource availability to the type of demand.
- Preserve complexity and diversity in social, ecological and physical flows.
- Balance overall material input and output of ecologically relevant flows.
- Focus on key impact reduction as a priority (e.g. health impacts in dense zones).

They also identified the key barriers to achieving a circular economy, and stress the importance of policy in their removal. These include: **Laws and regulations** — policy often leads to unforeseen consequences in changing market conditions (e.g. legal definition of waste); **Culture** — as the circular economy requires co-operation between sectors and chains, existing conservative cultures and vested interests in sectors can be obstacles to forming efficient and successful cooperation; **Market** — including existing incentives, resource flows and quality, external pricing and financing of circular initiatives; and finally **Technology** — this includes the scaling up of pilot-scale technologies and the complexity of bringing together independent technologies, often for the first time.

How the circular economy could be applied to the human water cycle has been explored by the engineering company Arup together with the Ellen MacArthur Foundation (Tahir *et al.*, 2019), who have summarized below how the key circular economy principles can be applied to the water industry.

Principle 1: Design out waste externalities:

- Optimize the amount of energy, minerals and chemicals used in the operation of water systems in concert with other systems.
- Optimize the consumptive use of water within a sub-basin in relation to adjacent sub-basins (e.g. use in agriculture or evaporative cooling).
- Use measures or solutions that deliver the same outcome without using water.

Principle 2: Keep resources in use

- Optimize resource yields (water use and reuse, energy, minerals, and chemicals) within water systems.
- Optimize energy or resource extraction from the water system and maximize their reuse.
- Optimize the value generated in the interfaces of the water system with other systems.

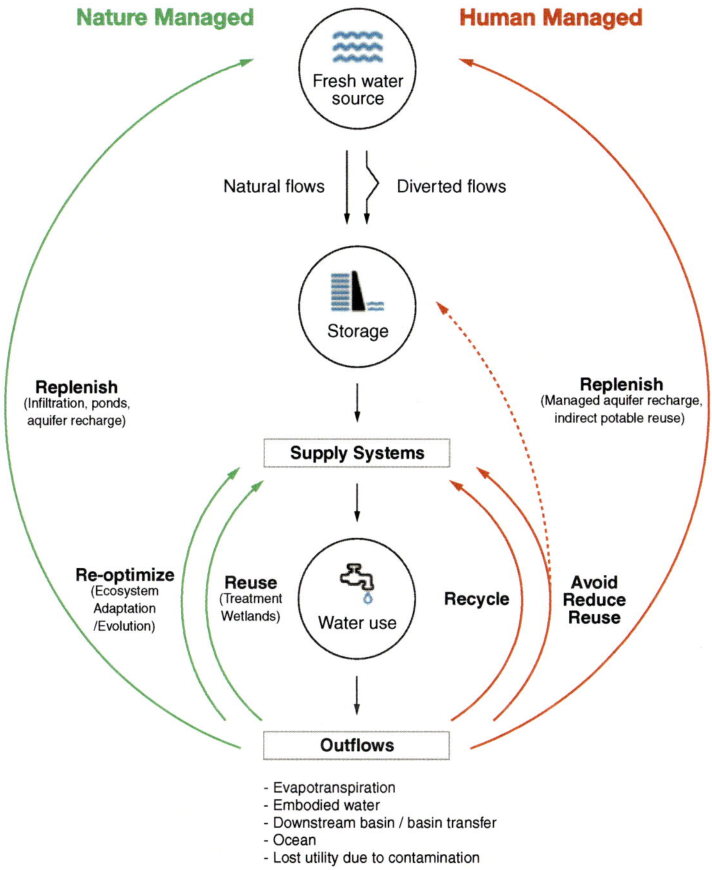

Fig. 2.2. Water specific circular economy systems diagram (Tahir *et al.*, 2019). Reproduced with permission from ARUP and the Ellen MacArthur Foundation.

Principle 3: Regenerate natural capital

- Maximize environmental flows by reducing consumptive and non-consumptive uses of water.
- Preserve and enhance the natural capital (e.g. river restoration, pollution prevention, quality of effluent, etc.)

- Ensure minimal disruption to natural water systems from human interactions and use.

They have also produced a water-specific circular economy systems diagram for water (Fig. 2.2). The diagram represents the water cycle within a single catchment or watershed and shows how the human water cycle (Fig. 1.5) can become better aligned to the natural water (hydrological) cycle (Fig. 1.4). The nature-managed half of the diagram is broken down into three key processes — re-optimize, reuse and replenish — to optimize water quality and volume within the hydrological cycle. **Re-optimize** refers to the need to sustain adequate flows in rivers and volumes within aquifers to protect, and wherever possible, to enhance these resources and ecosystems, including their biodiversity. Within the hydrological cycle, water changes its chemical composition as it passes over and through soils and permeable rocks, as well as when it interacts with both flora and fauna (i.e., **reuse**). As it passes through soil or flows downstream in rivers, it undergoes the process of self-purification by microorganisms and plants. For the nature side to work, water must eventually be allowed to **replenish** the cycle by evapotranspiration, infiltration or by surface runoff into streams and rivers in as clean a condition as possible.

It is how the human-managed side is operated that really controls the level of sustainability achieved by water services in relation to the environment. Globally, the major issues affecting the natural water system are over-abstraction, water loss from water resources, especially via irrigation, and pollution. Not only does this result in serious damage to rivers, lakes and groundwaters, but the current approach is also both non-sustainable and financially expensive. The key mechanisms are based on demand-side management techniques such as **avoiding** water use through better manufacturing design and **reducing** usage through improved use of efficiency, conservation and management. There is enormous potential as described earlier for the **reuse** of water, both within the industry and the home, creating closed loops where water is used multiple times during production before being treated on-site so that it can be reused again, which also includes water **recycling** within operations. Finally, water can be treated back to its former quality and used to replenish the water catchment.

2.2 The Circular Economy and the Water Industry

2.2.1 *Introduction*

To achieve the transition from finite to renewable resources requires a detailed analysis of exiting pathways, which have been largely linear in design (i.e., resource – production – consumption – waste). This is to help identify circular pathways with an emphasis on the reuse of resources, thereby helping to decouple the economy from growth and development, putting the environment at the core of the economic, societal and environmental nexus (Fig. 1.3). In terms of the water industry, this means seeing wastewater as a resource rather than waste and at the same time tackling the key global issues of energy use, water scarcity, and nutrient release to receiving waters. The need to tackle these issues is not new and the industry has seen immense changes over the past 30 years, in particular, using a variety of management and technological solutions largely in response to more stringent environmental legislation. Examples include water demand management (e.g. employing conservation measures such as education, leak reduction, metering, and efficient water-use appliances, such as low-flush toilets, and in some areas, water efficiency labeling) (Gray, 2017), controlling contaminants entering the waste stream, biological nutrient removal, better operational management and control (e.g. especially in the areas of aeration demand and prevention of NO_x and CH_4 formation and loss), advanced anaerobic digestion (AAD) using pretreatment steps such as thermal and enzymatic hydrolysis to increase the reliability of biogas production, and many more. So, to a great extent, many of the tools required to achieve circularity in the waste industry already exist; what it lacks is coherent policy and a strong regulatory environment in which to develop long-term, viable and secure markets, which is difficult within a critical and possibly vulnerable service industry. The International Water

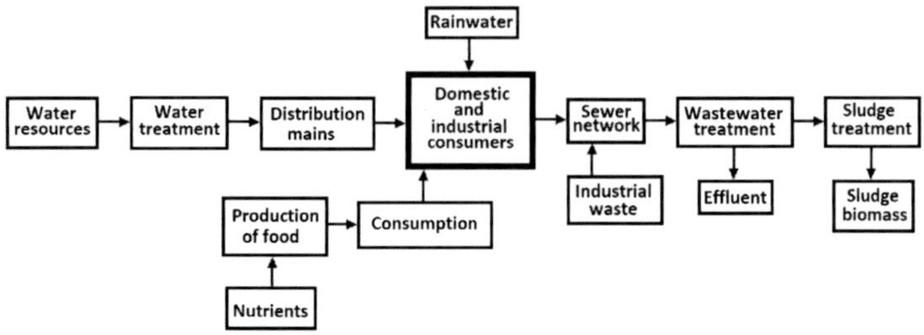

Fig. 2.3. Management pathways in the water services cycle.

Table 2.1. Typical breakdown of water use in a UK household, although these figures vary widely for different families and individuals.

Use	%	Use	%
Toilet	35	Wash basin	8
Kitchen sink	15	Outside use	6
Bath	15	Shower	5
Washing machine	12	Dishwasher	4

Association (IWA, 2016) has broken down the circular economy (as it relates to the water sector) into the three integrated pathways — water, material and energy — considered below, raising the concept of wastewater treatment plants from simple detoxification systems into fully integrated resource recovery plants. Examples of studies of the application of circular economy pathways in wastewater treatment are given by Sugam *et al.* (2016) and Jazbec and Turner (2018).

2.2.2 *The water pathway*

There are enormous possibilities to close the linear pathway where water is simply abstracted, treated, used by consumers or lost via leaks, discharged to the sewer, or treated at a wastewater treatment plant and then released back to the catchment or watershed in a modified state (Fig. 2.3). Within the system, all the abstracted water is treated to drinking water quality even though the majority of this will be used for non-consumptive purposes. The per capita consumption of water in the United Kingdom (UK) and Ireland is between 150 and 180 liters per day (L d^{-1}), with less than 5% used for either drinking or food preparation, with the largest percentage used for flushing the toilet or increasingly personal hygiene (Table 2.1). Outside the house, hosepipes can use considerable quantities of water with a garden sprinkler, for example, using on average 1,000 liters per hour (L h^{-1}).

The circular economy requires the water pathway to become a closed-loop system starting with reducing consumption. This is achieved by using water demand management principles such as leakage control and metering, both of which can reduce the overall demand for natural water resources (Gray, 2018). Water treatment uses both energy and a wide array of chemicals, and this can be reduced by investing in the natural infrastructure upstream of abstraction. The cleaner the water abstracted, the fewer treatment steps are required, thus reducing both capital investment and operational costs. It also makes sense to invest in the natural infrastructure rather than water treatment, as it also improves and protects aquatic ecosystems and possibly increases biodiversity (Fig. 2.4).

Traditionally, rainwater harvesting has been widely used for watering gardens, which is seasonal. However, it can also be used to supplement certain household functions such as toilet flushing. With different levels of on-site treatment, it can be used for laundry, showers and even consumption. Flooding in urban areas can be exacerbated due to runoff from household and commercial roofs, which can be mitigated by harvesting and storage that also reduces nutrient and chemical losses from soil entering surface waters or the sewer network. At the community level, sustainable urban drainage systems (SuDSs), using swales and detention/retention ponds, can prevent water from paved areas from direct entry into local streams and rivers which can cause pollution, or into sewers activating combined sewer overflows into surface waters. SuDSs can also provide treatment of water before natural infiltration to groundwater, temporary storage

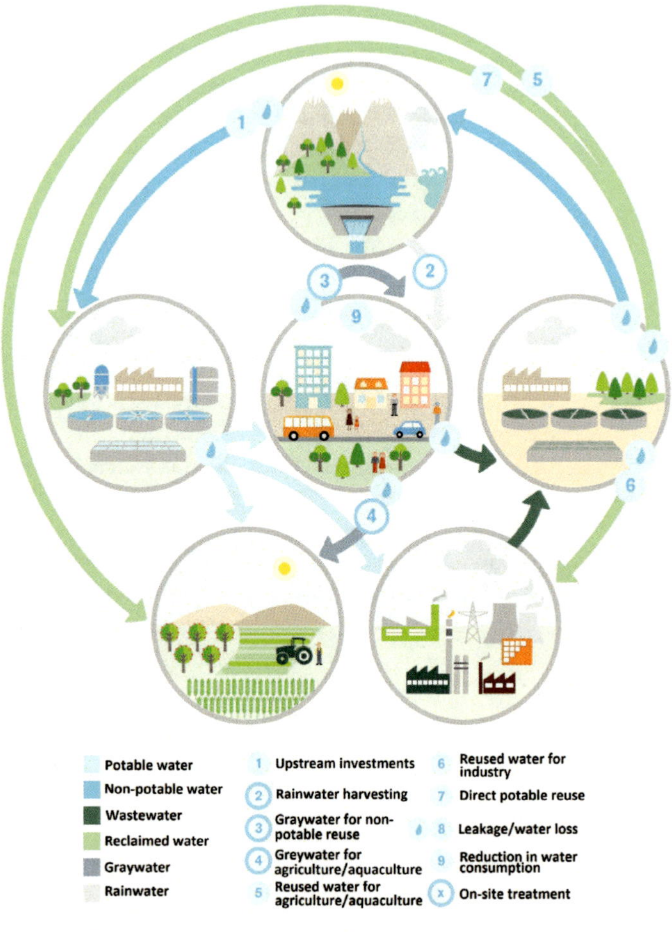

Fig. 2.4. The IWA water pathway model (IWA, 2016). Reproduced courtesy of the International Water Association (www.iwa-network.org), London.

Table 2.2. Potential for water reuse of wastewater at household or local community level.

Type of water	Content	Potential use
Black	Urine and feces	None
Brown	Feces only	None unless dry composted over several years
Yellow	Urine only	Can be used as fertilizer in garden
Grey	Washing water	Flushing toilets
White	Runoff/rainfall from roofs	Unfiltered but debris-free: Flushing toilets Filtered: Laundry, hot water

before release, and significantly reduce the risk of flooding (Woods Ballard *et al.*, 2015). From supermarkets to sports stadia, water is now collected from roofs and car parking areas, stored and treated, if necessary, and reused for toilet flushing and grass irrigation. This reduces water demand and protects the natural infrastructure. In fact, all gray water can potentially be reused (Table 2.2). Gray water is largely pathogen-free, although it contains contaminants that will need to be reduced at the source or removed by treatment — a major limiting factor is detergents and pharmaceutical and personal care products (PPCPs). Treated water can be used for irrigation and as a cheap source of nutrients, although this works best at the local level for agriculture, horticulture and even aquaculture. The industry already reuses water to a high degree with different processes using the water as it becomes increasingly contaminated when it eventually needs to be retreated. Large industrial water users are often limited by water availability, so using on-site treatment could allow water to be reused continuously.

Large rivers are frequently used for abstraction with the treated wastewater returned close to where it was originally abstracted (i.e., indirect potable reuse [IPR]). This is a common occurrence on the Rivers Severn and Thames in the UK, with water reused multiple times. Certain contaminants that cannot readily be removed by wastewater or subsequent water treatment (e.g. sodium chloride, estrogen, and estrogen-mimicking compounds) can increase with reuse, affecting taste and safety. However, where water is scarce, treated wastewater can be used as a supply stream for water treatment plants (i.e., direct potable reuse [DPR]), which can be both cost-effective and reduce associated greenhouse gas (GHG) emissions compared to alternatives such as desalination, which is energy-intensive. Examples of DPR are found in areas of high water scarcity and heavily populated areas of the United States (US), Canada and Singapore (Gerling, 2016). DPR is projected to become increasingly important in Western US where there are high levels of water scarcity (Scruggs and Thomson, 2017). While it is already feasible to create circularity in the water pathway, there are problems with potential chemical and microbial contaminants, which will require new enabling legislation, regulations and standards, as well as investment.

2.2.3 *The materials pathway*

At the core of the circular economy is the continuous recycling of materials. Ideally, finite resources such as metals should never be released into the environment because it renders them unrecoverable. In the water services cycle (Fig. 2.3), treatment processes generate significant amounts of biosolids (i.e., water treatment solids, preliminary screenings, primary and secondary sewage sludge, tertiary treatment sludge), the majority of which is organic but highly contaminated with both potentially toxic elements and also a wide range of human, animal and plant pathogens. Embedded in these solids and the effluent is a wide range of valuable raw materials such as fats, proteins, carbohydrates, nutrients (nitrogen and phosphorus), vitamins, metals, and other compounds. Some are recovered in their pure state or transformed using additional biological processes into more manageable forms such as fungi, algae or other forms of biomass. Sidestream processes are often incorporated into wastewater treatment plant design, allowing other organic wastes to be incorporated or inorganic compounds to be produced, such as calcium phosphate (CaP) or struvite ($NH_4MgPO_4 \cdot 6H_2O$). The process of anaerobic digestion is of key importance, especially in the recovery of carbohydrates in the form of biogas (CH_4) and phosphorous as struvite ($NH_4MgPO_4 \cdot 6H_2O$), and the elimination of pathogens by using pre-digestion sludge

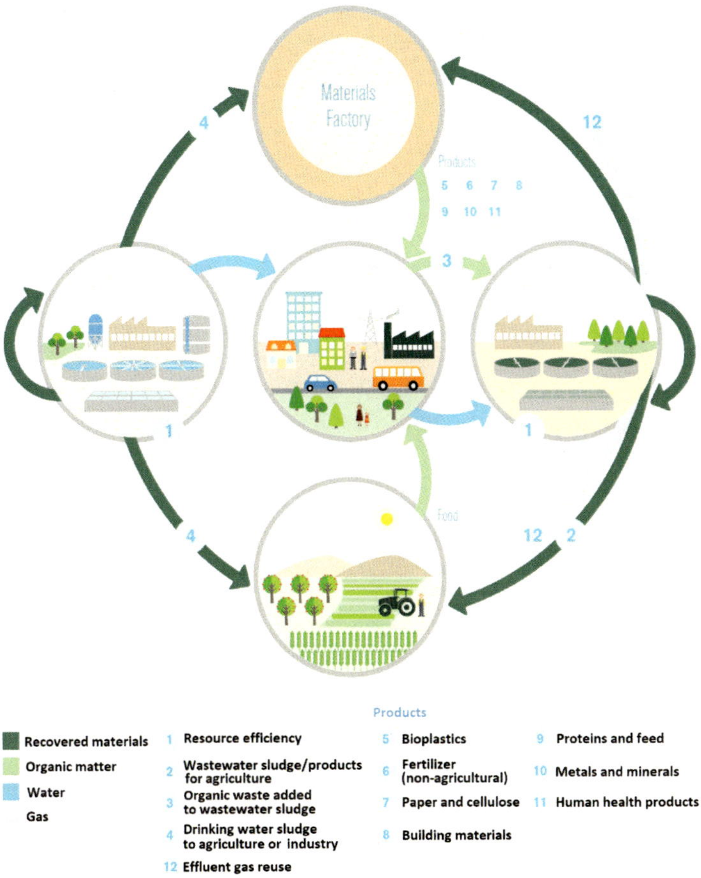

Fig. 2.5. The IWA material pathway model (IWA, 2016). Reproduced courtesy of the International Water Association (www.iwa-network.org), London.

hydrolysis or post-digestion thermal drying. The major output is the production of stabilized solids that can be used both as a fertilizer and soil conditioner in agriculture or further modified, often by adding other waste products, to produce high-quality compost for horticulture. The key material outputs are being investigated by a major EU project named Smart-Plant (2021), which lists the key outputs and major emphasis placed on a sustainable operation. As with water conservation, the efficient use of resources is vital to ensure that as little raw materials and energy are used during water and wastewater treatment, thereby minimizing material losses. Contaminants can also be reduced by eliminating or minimizing them at the source of production, and by encouraging minimal discharge of contaminants into the waste stream directly from households, businesses and industry through education, replacement with more sustainable products, and better management. The material pathway is summarized in Fig. 2.5.

A good example of circularity in the materials pathway that has important possible environmental implications is the production of biodegradable polymers from wastewater. Polyhydroxyalkanoates (PHAs) have very similar properties to certain petroleum-based

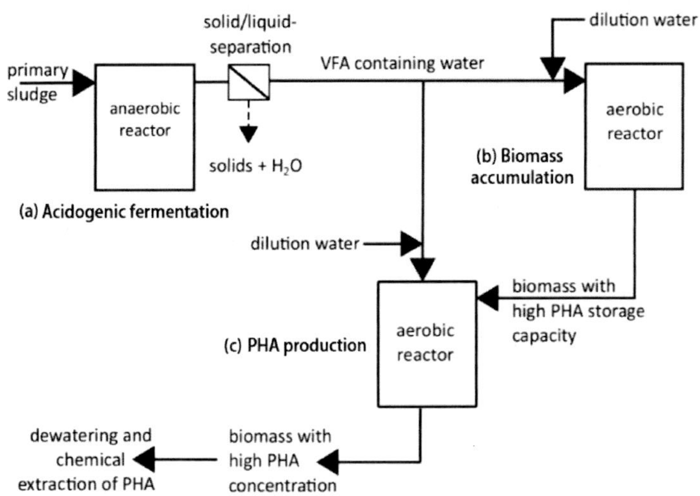

Fig. 2.6. The production of polyhydroxyalkanoates (PHAs) from wastewater for the production of biodegradable plastics (Valentino *et al.*, 2017). Reproduced with permission from Elsevier Publishing.

products used for making plastics. PHA-rich biomass can be made using mixed cultures of bacteria in open reactors, rather like activated sludge, using digester leachate which is rich in volatile fatty acids (VFAs) as a carbon source, with the polymer extracted from the enriched biomass (Morgan-Sagastume *et al.,* 2014). The process is made up of three stages: (a) digestion of sewage sludge to generate a VFA-rich output by acidogenic fermentation, (b) an aerobic reactor to generate robust mixed microbial community with high PHA-storing potential, and (c) a second aerobic reactor specifically for PHA production (Fig. 2.6). The process also has the advantage of reducing sludge volumes for disposal (Valentino *et al.,* 2017). Pittmann and Steinmetz (2017) found that PHA accumulation within cells could be as high as 28.4% of dry weight, which gives details of loading rates and optimum operating conditions. The recovery of cellulose directly from toilet paper at the preliminary treatment stage at wastewater treatment plants is also being trialed. The high cellulose content and natural bacterial polymers in domestic wastewater can be used to produce bioplastics and a range of biocomposite products on-site as a sidestream production facility.

The problems facing the material pathway in wastewater treatment are quality control, finding markets, scale, and reliability of production and economical price. A key challenge is the variable nature of wastewater, especially those containing industrial wastes and different wastewater treatment processes used resulting in varying levels of contamination and quality of products. Many manufacturing and food processing wastewaters are ideal for high-quality product outputs, such as single-cell protein, cellulose, metals, and the production of a wide range of high-value pharmaceutical products (Sec. 11.3).

2.2.4 *The energy pathway*

The water industry uses enormous amounts of energy, and a major factor in achieving sustainability and "net zero" emissions is reducing this dependence by improving operational management to reduce fossil fuel consumption, seeking low-energy alternatives, increasing the use of renewable energy and using processes as well as

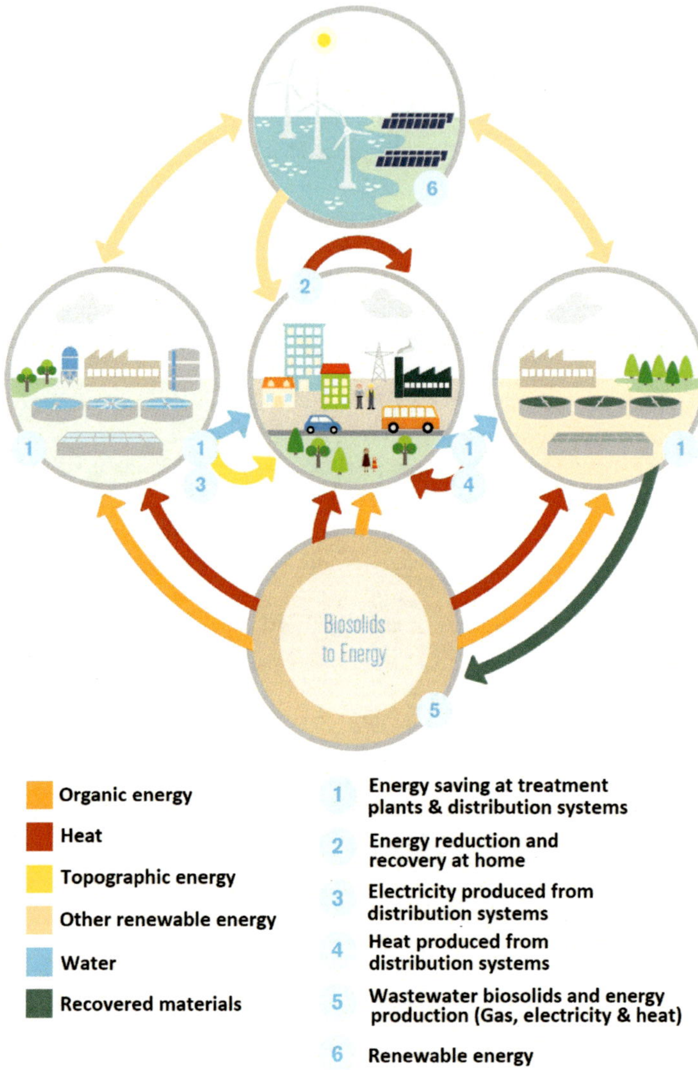

Fig. 2.7. The IWA energy pathway model (IWA, 2016). Reproduced courtesy of the International Water Association (www.iwa-network.org), London.

infrastructure to generate energy on-site. The major use of energy in the home after space heating is heating water for washing, food preparation and drinks. Water companies can help consumers use water more efficiently by reducing demand, thereby saving energy during production, transport and disposal. This includes giving advice and support on the use and selection of low-water-use appliances (e.g. laundry and dishwashers), with water efficiency labeling core to this, and by promoting water-related conservation measures in the home (e.g. water-efficient showers and faucets, only heating the required volume of water, better cooking practices, etc.) (Gray, 2017).

Within the energy pathway, there are many opportunities to save as well as generate energy (Fig. 2.7). Where inclines are to be found in water distribution networks, then micro-turbines can be installed to generate electricity, with schemes in France already generating 4.5×10^6 kWh yr^{-1}. These do not work in sewerage pipes due to the presence of solids but there are opportunities for heat recovery, especially in the home or business premises, for example, from laundry and other warm wastewater streams. The closer to the point of discharge, the greater the potential for heat recovery. However, a balance has to be struck in removing heat from wastewater as sewage can often be ten degrees warmer than the ambient temperature in winter, which increases the rate of biological oxidation and reduces energy usage.

The high carbon content of biosolids makes it an ideal fuel after further treatment; these include biogas production, gasification, pyrolysis or simple biomass combustion after dewatering and drying. The remaining solids and ash are rich in nutrients and trace elements and can be used as a fertilizer or further processed to recover specific materials such as metals.

Water and wastewater treatment plants are often located on large sites, as are water resources. In recent years, increasing use has been made of these sites for renewable energy production using wind turbines or solar panels, and to a lesser extent, geothermal energy. Increasingly, wastewater treatment plants are becoming green power stations, combining biogas production and renewable energy generation to become self-sufficient in energy use and even net exporters of electricity or biogas becoming a "net-zero emitter" of GHGs (Chap. 3).

References

Ellen MacArthur Foundation. (2021) *The Circular Economy.* https://www.ellenmacarthurfoundation.org/

Gerling, A. (2016) Potable reuse 101: An innovative and sustainable water supply solution. American Water Works Association, Denver, CO. https://www.awwa.org/Portals/0/AWWA/ETS/Resources/PotableReuse101.pdf?ver=2018-12-12-182505-710

Gladek, E., van Odijk, S., Theuws, P. and Herder, A. (2015) Circular buiksloterham: Vision and ambition: Transitioning Amsterdam to a circular city. Metabolic, Studioninedots & DELVA Landscape Architects, Amsterdam. https://issuu.com/delvalandscape/docs/circularbuiksloterham_eng_full_repo

Gray, N.F. (2017) *Water Science and Technology: An Introduction, 4th ed.* CRC Press, Boca Raton, FL.

Gray, N.F. (2018) *Drinking Water Quality: Problems and Solutions.* Cambridge University Press, Cambridge. https://doi.org/10.1017/CBO9780511805387

IWA. (2016) Water utility pathways in a circular economy. International Water Association, London. https://iwa-network.org/wp-content/uploads/2016/07/IWA_Circular_Economy_screen-1.pdf

Jazbec, M. and Turner, A. (2018) Creating a circular economy precinct, report prepared by the Institute for Sustainable Futures. University of Technology Sydney, for Sydney Water, Sydney, NSW. https://www.sydneywater.com.au/web/groups/publicwebcontent/documents/document/zgrf/mja2/~edisp/dd_206045.pdf

Morgan-Sagastume, F., Valentino, F., Hjort, M., Cirne, D., Karabegovic, L., Gerardin, F., Johansson, P., Karlsson, A., Magnusson, P., Alexandersson, T., Bengtsson, S., Majone, M. and Werker, A. (2014) Polyhydroxyalkanoate (PHA) production from sludge and municipal wastewater treatment. *Water Science and Technology*, **69**(1), 177–184.

Pittmann, T. and Steinmetz, H. (2017) Polyhydroxyalkanoate production on waste water treatment plants: Process scheme, operating conditions and potential analysis for German and European municipal waste water treatment plants. *Bioengineering,* **4**(2), 54.

Scruggs, C.E. and Thomson, B.M. (2017) Opportunities and challenges for direct potable water reuse in arid inland communities. *Journal of Water Resource Planning and Management*, **143**(10), 04017064.

Smart-Plant. (2021) Scale-up of low-carbon footprint material recovery techniques for upgrading existing wastewater treatment plants. EU Smart-Plant Project. https://www.smart-plant.eu/

Sugam, R.K., Jain, A. and Neog, K. (2016) Circular economy pathways for municipal wastewater management in India: A practitioner's guide. 2030 Water Resources Group, Washington DC. https://www.2030wrg.org/wp-content/uploads/2017/01/Circular-Economy-Pathways-India.pdf

Tahir, S., Shouler, M. and Steichen, T. (2019) Water and circular economy: A white paper. Ellen MacArthur Foundation, Cowes, UK.

Valentino, F., Morgan-Sagastume, F., Campanari, S., Villano, M., Werker, A. and Majone, M. (2017) Carbon recovery from wastewater through bioconversion into biodegradable polymers. *New Biotechnology*, **37A**, 9–23.

Woods Ballard, B., Wilson, S., Udale-Clarke, H., Illman, S., Scott, T., Ashley, R. and Kellagher, R. (2015) The SuDS manual (C753). CIRIA, London.

3

Global Warming and Climate Change

3.1 Introduction

The Kyoto Protocol lists seven key greenhouse gases (GHGs) that directly contribute to global warming-induced climate change. These are carbon dioxide (CO_2), methane (CH_4), nitrous oxide (N_2O), all of which are released during water and in particular wastewater treatment, and gases used in refrigeration (i.e., hydrofluorocarbons [HFCs]) or for the production of semiconductors, LCD displays, and certain solar panels (i.e., perfluorocarbons [PFCs], sulfur hexafluoride [SF_6] and nitrogen trifluoride [NF_3]). To standardize emissions over any reporting period, all GHGs are expressed in terms of Carbon Dioxide Equivalent (CO_2e), which takes into account the global warming potentials of the different gases based on the fourth assessment report of the Intergovernmental Panel on Climate Change (IPCC) over a 100-year-period (Gray, 2015).

The term Global Warming Potential (GWP) is used to total up the contribution of all the individual GHGs where CO_2 has an assigned GWP of 1 and is used as the baseline unit (i.e., the reference gas) to which all other GHGs are compared (Table 3.1). So the GWP is used to express the impact of all GHGs using the same units (CO_2e). All GWP values depend on the time span over which the potential is calculated. Therefore, short-lived GHGs initially have large effects that become less significant over time relative to CO_2. Methane, for example, has a GWP of approximately 25 over 100 years but 62 over 20 years. So in order to collectively calculate the effects

Table 3.1. Comparison of the Global Warming Potential (GWP) of some greenhouse gases. Those used in standard accounting are highlighted in bold. The table also compares their atmospheric lifetime and GWP over 20, 100 and 500-year time frames.

Greenhouse gas	Chemical formula	Atmospheric lifetime (years)	Global Warming Potential (GWP) for given time horizon		
			20-yr	100-yr	500-yr
Carbon dioxide	CO_2	30–100	1	1	1
Methane	CH_4	12	72	**25**	7.6
Nitrous oxide	N_2O	114	289	**298**	153
CFC-12	CCl_2F_2	100	11,000	**10,900**	5,200
HCFC-22	$CHClF_2$	12	5,160	**1,810**	549
Tetrafluoromethane	CF_4	50,000	5,210	**7,390**	11,200
Hexafluoroethane	C_2F_6	10,000	8,630	**12,200**	18,200
Sulfur hexafluoride	SF_6	3,200	16,300	**22,800**	32,600
Nitrogen trifluoride	NF_3	740	12,300	**17,200**	20,700

of all GHGs, a similar time frame must be used, which is why the Kyoto Protocol uses the GWP time frame of 100 years, which is the lifespan of CO_2 itself. If a climate policy is enacted to limit long-term temperature increases, the effects of short-lived emissions may be overestimated if the time horizon chosen is too short. On the other hand, a time horizon of 100 years versus one of 20 years does, to a certain extent, underestimate the importance of short-lived emissions. This is shown in Table 3.1, which also highlights the importance of preventing the release of some of these high GWP GHGs, many of which are vital for industrial processes. However, although they have high GWPs, they are released in very small amounts compared to CO_2 (Gray, 2015). Using these values, the United Kingdom (UK) Government has produced two standard values for carbon emissions from water and wastewater treatment, including the distribution of water via mains and the collection of wastewater via sewerage, respectively. Values for 2019 were calculated at 0.344 and 0.708 kg CO_2e m^3 for water and wastewater treatment, respectively (BEIS, 2020).

Globally, wastewater treatment is responsible for 3% of all carbon emissions.

3.2 GHG Emissions and the Water Industry

GHG emissions are calculated using the Greenhouse Gas Protocol, which is a standard accounting and reporting methodology developed by the World Resources Institute and the World Business Council for Sustainable Development (WRI/WBSCD, 2004).

Under this protocol, emissions are allocated into three categories, known as Scope 1, 2 and 3 standards, for accounting purposes, from which a total volume of CO_2e emitted can be calculated. The **Scope 1** standard covers all direct emissions arising from the activities of an organization from sources that it either owns or controls. This includes on-site fuel combustion such as gas boilers, the use of company vehicles, and fugitive emissions such as leaks from air-conditioning systems. The **Scope 2** standard covers the indirect emissions created during the generation of the electricity purchased and used by the organizations. The **Scope 3** standard covers all other indirect emissions from activities of organizations, from sources not under their direct control. These include emissions from waste, water, business travel, and procurement (Carbon Trust, 2021) (Fig. 3.1). The methods for calculating Scope 3 emissions are fully explained in the Greenhouse Gas Protocol technical guidance manual (WRI/WBSCD, 2013).

Emissions from the water and wastewater treatment are associated with the release of CO_2, CH_4 and N_2O (Table 3.2). Pumping, aeration and sludge drying are large energy users, linked primarily to the use of grid electricity, making the water industry the second-largest user of electricity in the UK. These emissions are classified under Scope 3; however, these emissions are a consequence of water demand and usage by individuals, households, businesses, and industries, so ultimately, total emissions are controlled by consumption rates, although wastewater emissions are also related to water quality standards imposed on effluents. How water is used is not included in the water industry's carbon footprint, but emissions are affected overall.

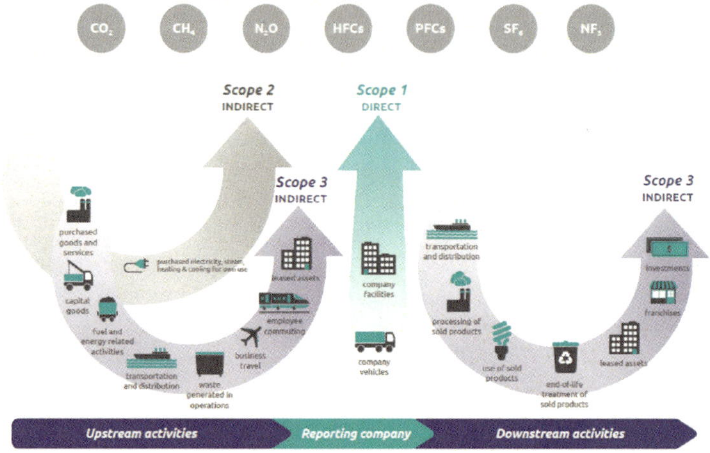

Fig. 3.1. Scope 3 emissions are made up of 15 different reporting categories from which greenhouse gas emissions are calculated (WRI/WBSCD, 2013). Reproduced with permission from the World Resources Institute and World Business Council for Sustainable Development.

In the UK, current water-related carbon emissions per household are equivalent to 2.6 kg CO_2e per day. Of this, 80 to 90% is associated with use in the home with the remainder associated with supply and treatment. WaterWise (2021) has estimated that a 6% reduction in use could reduce overall emissions by 1.3 Mt CO_2e per year. However, the heating of water is a critical issue, with more efficient food and drink preparation, water-efficient water appliances, and low-energy showers all reducing energy use.

Globally, water and wastewater treatment utilities are a diverse mixture of private, state and semi-state entities that vary in size from serving a single community to a whole country. The two services are usually combined, allowing greater control of water resources and environmental quality at the catchment level. They also share a common drive toward ever-greater efficiency and the need to conserve water resources. As a sector, they have been very proactive in tackling

Table 3.2. Methane and nitrous oxide emission routes associated with wastewater treatment and discharges (Doorn and Irving, 2006). Reproduced with permission from the Intergovernmental Panel on Climate Change.

Types of treatment and disposal				CH_4 and N_2O emission potentials
Collected	Untreated		River discharge	Stagnant, oxygen-deficient rivers and lakes may allow for anaerobic decomposition to produce CH_4. Rivers, lakes and estuaries are likely sources of N_2O.
Collected	Untreated		Sewers (closed and under ground)	Not a source of CH_4/N_2O.
Collected	Untreated		Sewers (open)	Stagnant, overloaded open collection sewers or ditches/canals are likely significant sources of CH_4.
Collected	Treated	Aerobic treatment	Centralized aerobic wastewater treatment plants	May produce limited CH_4, from anaerobic pockets. Poorly designed or managed aerobic treatment systems produce CH_4. Advanced plants with nutrient removal (nitrification and denitrification) are small but distinct sources of N_2O.
Collected	Treated	Aerobic treatment	Sludge anaerobic treatment in centralized aerobic wastewater treatment plant	Sludge may be a significant source of CH_4 if emitted CH_4 is not recovered and flared.
Collected	Treated	Aerobic treatment	Aerobic shallow ponds	Unlikely source of CH_4/N_2O. Poorly designed or managed aerobic systems produce CH_4.
Collected	Treated	Anaerobic treatment	Anaerobic lagoons	Likely source of CH_4. Not a source of N_2O.
Collected	Treated	Anaerobic treatment	Anaerobic reactors	May be a significant source of CH_4 if emitted CH_4 is not recovered and flared
Uncollected	Septic tanks			Frequent solids removal reduces CH_4 production.
Uncollected	Open pits/Latrines			Pits/Latrines are likely to produce CH_4 when temperature and retention time are favorable.
Uncollected	River discharge			See above.

climate change, reducing pollution, and preserving surface and groundwater physicochemical and biological quality. In the UK, the Government has made a legal commitment to be "net zero" by 2050. To that effect, water companies made a voluntary agreement to accelerate the rate of GHG reduction to achieve "net-zero emissions" by 2030, reflecting the urgency in tackling global warming.

The concept of net zero is an accounting system to balance the emissions produced against those removed so that the organization or individual is technically not producing excess GHGs. This can be done by eliminating emissions through efficiency in fuel use and the development of low carbon-releasing processes, or more commonly by offsetting, usually through carbon sequestration by soils, peatlands and forestry, although the permanency of these varies (Gray, 2015). Carbon capture and storage have been successfully applied to large industrial units such as electricity generation and cement manufacture but not to wastewater treatment due to the small scale of GHG emissions as point sources. A net zero approach is necessary as there are always residual emissions, which is why on-site offsetting options such as renewable power generation are so important for the water industry.

Emissions from the UK water industry fell from 4.4 Mt CO_2e in 2011 to 2.4 Mt CO_2e by 2019, a reduction of 45%, largely as a result of improvements in energy efficiency and the use of renewable energy and biogas. Part of this reduction is due to the ongoing decarbonization of the electricity sector. Current emissions are summarized in Fig. 3.2. Going beyond energy efficiency to achieve net zero requires extensive investment in new technologies as well as innovative solutions. Collectively, UK water companies have developed a structured approach, the Net Zero 2030 Route Map (Water UK, 2020), which has three pathways: (1) managing and reducing the demand for water and wastewater services, (2) identifying the most energy-intensive areas and developing innovative technological solutions, and (3) using nature-based solutions to increase carbon sequestration, especially on company land. The key mechanisms to do this are:

- Reduction in water use and the production of wastewater
- Rapid deployment of renewable energy deployment
- Greater energy efficiency
- Green procurement ensuring low carbon partnerships and activities

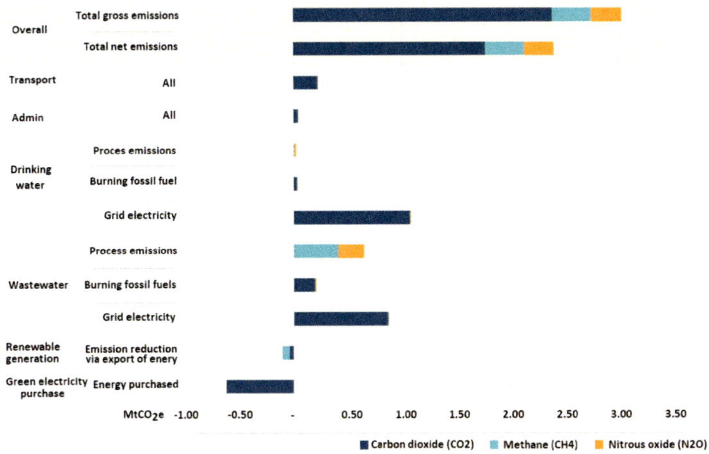

Fig. 3.2. Breakdown of the sources of greenhouse gases from the UK water industry from 2018 to 2019 (Water UK, 2020). Reproduced with permission from Water UK.

- Exporting energy through the development of renewable energy generation
- Replacement of petrol and diesel vehicles with low-energy alternatives
- Tackling process emissions, especially the loss of CH_4 and N_2O from wastewater treatment (Fig. 3.2)
- Using nature-based solutions
- Habitat restoration and tree planting on company land
- Developing offsets for emissions difficult to abate.

One of the most challenging areas will be in dealing with process emissions from wastewater treatment. For example, wastewater treatment is responsible for 60% of the net emissions from a large water company in the UK emitting 248 kt CO_2e per annum, with treatment processes responsible for 17.5% and sludge treatment for 42.5% overall. A major challenge is the release of CH_4 and N_2O from wastewater processes, which have a large impact due to their high global warming potential (Table 3.1).

Aboobakar (2014), based at the Cranfield Water Science Institute, has studied long-term continuous monitoring of GHG emissions and operational variables in nitrifying and biological nutrient removal-activated sludge plants. The study revealed that CH_4 was not only produced where anaerobic conditions were found, including non-aerated zones in activated sludge, but were also emitted from aerobic zones at an emission rate of 0.07% of influent chemical oxygen demand (COD). Nitrous oxide is formed due to incomplete denitrification or during nitrification through hydroxylamine (NH_2OH) oxidation and nitrifier denitrification pathways (Fig. 3.3). Nitrous oxide emissions were a function of the relative nitrogen fractions present with emission rates of 0.05 to 0.72% based on influent total nitrogen (TN) and NH_4^+, but much higher when based on removal rates at 0.13 to 3.9% of TN and NH_4^+ removed.

Process emissions are difficult to control and will require offsets in the medium term until better management and technologies

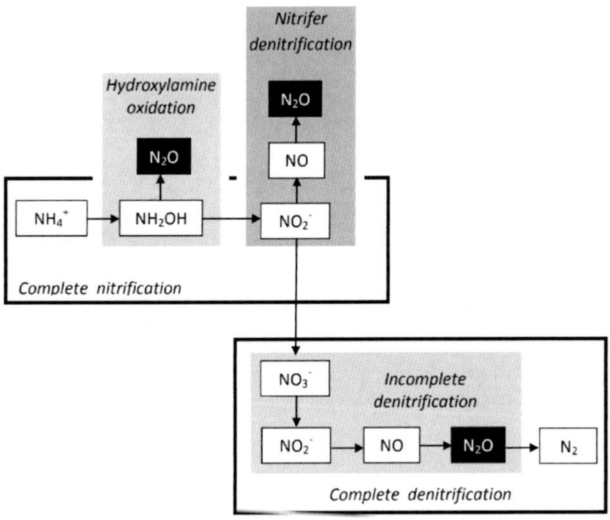

Fig. 3.3. Nitrous oxide, a powerful greenhouse gas, is produced as a result of incomplete nitrification and denitrification (Aboobakar, 2014). Reproduced with permission from Cranfield University.

become available. Offsets do not necessarily have to be found outside the company, For example, water companies are collectively major landowners, and they are keen to optimize nature-based solutions by improving the management of over 20,000 ha of peatlands and grassland to optimize carbon sequestration. They are also planning to plant 11 million trees before 2030. The land around reservoirs and at individual treatment plants, including infrastructure roof space, all provide opportunities for on-site renewable power generation using wind, solar and hydro-generation options. This not only reduces electricity used from the grid but also allows energy to be exported off-site. In the long term, carbon sequestration and electricity generation will become increasingly important factors in achieving net-zero emissions. Biogas generated from sewage sludge also has an important role to play in offsetting but has twice the carbon benefit when used as a replacement for fossil fuels (i.e., as biodiesel) than when used to generate electricity.

References

Aboobakar, A. (2014). Quantification of greenhouse gas emissions in biological wastewater treatment. PhD thesis. Cranfield Water Science Institute, University of Cranfield.

BEIS. (2020) Greenhouse gas reporting: Conversion factors 2019. Version 1.3. Department for Business, Energy and Industrial Strategy and Department for Environmental, Food and Rural Affairs, London. https://www.gov.uk/government/publications/greenhouse-gas-reporting-conversion-factors-2019

Carbon Trust. (2021) Resources. The Carbon Trust, London. https://www.carbontrust.com/resources

Doorn, M.R.J. and Irving, W. (2006) Wastewater treatment and discharge. In *The IPPC Guidelines for National Greenhouse Gas Inventories, Volume 5: Waste*. Intergovernmental Panel on Climate Change, Geneva, Switzerland.

Gray, N.F. (2015) *Facing Up to Global Warming: What is Going On and How You Can Make a Difference*. Springer, Cham, Switzerland.

WaterWise. (2021) Net zero and the role of water efficiency. WaterWise, London. https://www.waterwise.org.uk/knowledge-base/net-zero-and-the-role-of-water-efficiency-waterwise-2021/

Water UK. (2020) Net zero 2030 route map. Water UK, London. https://www.water.org.uk/routemap2030/wp-content/uploads/2020/11/Water-UK-Net-Zero-2030-Routemap.pdf

WRI/WBSCD. (2004) *The Greenhouse Gas Protocol: A Corporate Accounting and Reporting Standard (revised edition)*. The World Resources Institute and World Business Council for Sustainable Development, Washington DC, USA. http://www.wri.org/publication/greenhouse-gas-protocol-corporate-accounting-and-reporting-standard-revised-edition

WRI/WBSCD. (2013) Technical guidance for calculating Scope 3 emissions (Version 1.0). Supplement to the corporate value chain (Scope 3). Accounting and reporting standard. Greenhouse Gas Protocol Initiative. The World Resources Institute and World Business Council for Sustainable Development, Washington DC, USA. https://ghgprotocol.org/sites/default/files/standards/Scope3_Calculation_Guidance_0.pdf

4

Natural Treatment Systems

4.1 The Concept of Natural Treatment

Natural treatment systems are an important component of ecological engineering and sustainable sanitation solutions. Ecological engineering is defined as "the design of sustainable ecosystems that integrate human society with its natural environment for the benefit of both" (Mitsch and Jørgensen, 2003). Conventional wastewater treatment requires large amounts of resources both in construction and their operation. While conventional systems have made a significant difference to both public and environmental health, the ever increasing population and the intensification of agriculture have continued to severely impact surface waters so that all surface freshwaters around the world are seriously under threat. This has placed even greater pressure on wastewater treatment, which are normally point discharges, to significantly improve quality, not only in terms of nutrients but other pollutants such as metals, trace organics, and personal care and pharmaceutical products (PCPPs). Using conventional treatment solutions will result in large increases in energy and other resources to achieve these higher quality standards in many cases at unaffordable costs. However, natural treatment systems combined with nature-based solutions offer a sustainable and affordable alternative, allowing all nations, regardless of wealth, a chance to combat climate change and environmental degradation, and at the same time improve the wellbeing of the community.

Natural treatment systems differ from conventional processes, in that they require comparatively little or no energy either in their construction or operation, making them highly sustainable with a small carbon footprint. They are constructed using the minimum of synthetic materials, particularly concrete, and rely on renewable energy. Oxygen is supplied either by natural diffusion, photosynthesis, or a combination of the two. This results in silent, almost odorless processes that are very robust and simple to construct and operate.

Natural treatment systems include land treatment (e.g. surface irrigation, subsurface irrigation, infiltration-percolation and overland flow), microalgal and macrophyte-based systems including floating macrophytes (e.g. duckweed, hyacinth ponds), those employing wetlands species (e.g. constructed wetlands), and waste stabilization ponds, which are classed as either anaerobic lagoons or oxidation ponds (e.g. facultative, maturation and fish ponds, and also purification lakes). These systems offset the need for energy and expensive electromechanical equipment by utilizing greater areas of land, making them more important in warmer climates where land is comparatively cheap. The advantages and disadvantages are summarized in Table 4.1. To treat the wastewater from a population equivalent of 20,000, producing 3,600 m^3d^{-1} would require an annual energy usage of 1,000,000 kWh yr^{-1} if using activated sludge, 120,000 kWh yr^{-1} if using a rotating biological contactor, and possibly as little as 0 kWh yr^{-1} if using a waste stabilization pond system (Table 4.2). Low-income countries are particularly suited to natural treatment systems, with processes applicable whatever the climate, although there is greater choice and higher efficiencies where it is warmer and there is plenty of sunshine. Key factors are low cost, simple construction and operation, and simple mechanical technology ensuring low-cost and simple maintenance.

An important factor in achieving "net zero" is offsetting those difficult or unavoidable emissions, especially from wastewater treatment processes such as biological nutrient removal (BNR) activated sludge. The use of nature-based solutions offers an opportunity to simultaneously tackle climate change and biodiversity loss. Apart from wastewater, pollution from agriculture, often coupled with

Table 4.1. Comparison of natural treatment systems with conventional wastewater treatment systems (e.g. activated sludge).

	Natural treatment system	Activated sludge
Applicability	Rural, small populations	Urban, large populations
Sustainability factors		
Energy	Natural or renewable	Electricity
Construction materials	Soil bunds, butyl liners	Concrete, steel, plastic
Aeration	Diffusion, photosynthesis	Mechanical aeration or diffusers
Chemical usage	None or very low	Moderate to high
Chemical types	Lime	Lime, metal salts, polyacrylamides
Carbon footprint	Low	Very high
Wildlife value	High	Low
Operational factors		
Sludge production	Low, on-site disposal	High, off-site disposal
Odor	Low	Moderate to high
Noise	None or very low	Moderate to high
Aerosol formation	None	High
Controllability	Low to moderate	High
Design and costs		
Design layout	Dispersed	Compact
Land area	High	Low
Capital cost	Low to moderate	High
Operational cost	Low	High

over-abstraction of surface and ground waters, is a major driver of poor natural water quality where low-cost small-scale treatment solutions are urgently required, especially to deal with nutrient loss to surface waters. As the HM Treasury (2020) net zero review emphasizes, "the conservation and restoration of natural habitats such as peatlands, forests and coastal habitats can reduce greenhouse gas emissions, sequester carbon, build resilience in ecosystems (which protects biodiversity and provides benefits for society) and help

Table 4.2. Comparison of the total energy consumption for typical natural and conventional wastewater treatment systems treating 1 million gallons per day (i.e., 3,785 m^3d^{-1}), achieving US effluent quality standards for biochemical oxygen demand (BOD), total suspended solids (TSS), phosphorus (P) and nitrogen (N). Adapted from Middlebrooks *et al.* (1981).

Wastewater treatment processes	Energy usage (10^3 kWh yr^{-1})
Natural treatment systems	
Land treatment:	
Rapid infiltration (facultative pond)	150
Slow rate, ridge, and furrow (facultative pond)	181
Overland flow (facultative pond)	226
Waste stabilization ponds:	
Facultative pond + intermittent sand filter	241
Facultative pond + microscreens	281
Aerated pond + intermittent sand filter	506
Traditional treatment systems	
Filtration:	
Trickling filter + anaerobic digestion	783
RBC + anaerobic digestion	794
Trickling filter + gravity filtration	805
Trickling filter + N removal + filter	838
Activated sludge:	
Activated sludge + anaerobic digestion	889
Activated sludge + anaerobic digestion + filter	911
Activated sludge + nitrification + filter	1,051
Activated sludge + sludge incineration	1,440
Advanced treatment systems:	
Activated sludge + advanced water treatment	3,809
Physical chemical advanced secondary	4,464

people adapt to the impacts of unavoidable climate change". Natural treatment systems offer a unique opportunity to seamlessly co-exist with such nature-based solutions by supporting and increasing the resilience that ecosystems are currently lacking due to so many external pressures.

4.2 Selection of Processes

Land treatment systems offer some of the most low-energy, sustainable options for wastewater treatment, especially when coupled with water and nutrient reuse. Reuse of wastewaters using subsurface irrigation or slow-rate irrigation offer huge opportunities to sustain forestry and agriculture in water-stressed and semi-arid areas, and to reuse valuable water and nutrients in growing timber and other crops while also treating the effluents, making them important tools in water conservation management (Chap. 7). Forested systems have longer wastewater application seasons than agricultural crops and high hydraulic loadings but relatively poor nitrogen removal, although this is dependent not only on tree species but the age of the trees and, importantly, the soil conditions. Vegetative filter systems such as short rotation willow coppice (SRWC) can treat wastewaters to the same degree of purity as conventional biological secondary wastewater treatment, with even better total nitrogen (TN) and total phosphorus (TP) removal, and at very low costs. Conventional short rotation coppice plantations are less profitable than those irrigated with wastewater, making SRWC not only a sustainable treatment option but a viable biomass for an energy system. Coupling biomass production for energy and phytoremediation is a perfect example of the concept of the circular economy (Sec. 7.4). Land treatment also provides sustainable treatment options in sensitive landscapes, supporting and enhancing biodiversity by the creation of new habitats. As a tertiary step, rapid infiltration systems can be invaluable in recharging aquifers and surface waters, protecting them from over-abstraction, although this requires high-quality effluents to prevent contamination (Sec. 8.2.1). Waste stabilization ponds can be used as pretreatment so long as algae or other solids are removed by sand filtration or microfiltration. Alternatively, soil-aquifer treatment (SAT) systems can be used (Sec. 8.2.2.), which also provides an opportunity to employ waste materials (e.g. granulated tires) in treatment, enhancing nitrogen removal in particular. Overland flow or grass filtration offers secondary treatment capability, with grass harvested either for fodder in the United States (US) or as silage for biogas production in Europe. They have also been used to treat polluted streamwater.

The use of plants, including microalgae, macrophytes and other types, offers a wide range of secondary and tertiary treatment options, including contaminant removal (Sec. 10.1). Such systems require low capital investment and are very cheap to operate, providing a low-energy, highly sustainable, and carbon-neutral wastewater treatment that produces valuable plant or secondary (animal) biomass for energy, protein or other high-value products (Sec. 11.3). Plant biomass can sequester carbon dioxide (CO_2), with duckweed able to remove six times more CO_2 than released by respiration (Sec. 10.4.3). Microalgae are grown either in open ponds, high-rate algal ponds (HRAPs) or transparent walled closed-systems known as photobioreactors (PBRs). HRAPs and PBRs are not designed for optimum purification efficiency but maximum algal production. The latter is used primarily for algae-based biofuel and pharmaceutical production, employing many different systems from simple tubular systems to more complex, bubble columns, flat-panel or air-lift designs. They are best integrated with existing factory waste streams, primarily in the food processing sector, using waste heat and flue CO_2 to optimize algal growth and reduce both carbon emissions and nutrient loads from the factory (Sec. 11.1). For every kilogram of dry weight (DW) of microalgae biomass produced, 1.83 kg CO_2 is sequestered.

While natural wetlands have been used for wastewater treatment in the past, it is now accepted that these fragile and important habitats need to be protected. The increasing disappearance rate of smaller wetlands, in particular, has resulted in extensive legislative protection, which means that wetlands should only be considered for use for wastewater treatment, including tertiary treatment, as a last result (Sec. 12.4). The treatment processes found in natural wetlands have been effectively mimicked in the development of artificial or constructed wetlands, which are highly efficient low-energy systems that can be used for secondary and/or tertiary treatment of sewage, treating stormwater and other wastewaters, including acid mine drainage, landfill leachate, agricultural, and industrial wastewaters. They can be used in all climates and provide valuable habitats for wildlife. Subsurface flow (SSF) systems, more commonly known as reed beds, using emergent vegetation are primarily used in temperate zones due to climate conditions, while open free water surface (FWS) systems are mainly used for tertiary treatment often coupled with fish ponds (Secs. 13.2 and 13.3).

Waste stabilization ponds have become a major treatment system, globally being a sustainable alternative to conventional biological treatment processes. The advantage of such a system is its ease of operation and that no skilled labor nor mechanical parts or power are required. They now represent the major treatment process in countries where climatic conditions are ideal, including Australia, the southern US, Africa, and Asia. In Europe, waste stabilization ponds, being robust, low maintenance, and low energy systems, are increasingly used as a sewage treatment option for small rural communities. Capital costs are about 30% and 50% that of a conventional system and reed bed system, respectively. Waste stabilization ponds are classified according to the biological activity (i.e., whether the breakdown of organic matter is primarily anaerobic or aerobic, and the importance of algae in the process). Anaerobic lagoons are used for the pretreatment of high-strength organic wastewaters (Chap. 15), while facultative ponds are a traditional secondary aerobic treatment system in which oxygen for bacterial oxidation is supplied by algae (Sec. 16.2). Maturation ponds are used for the tertiary treatment of effluents from secondary treatment, including facultative ponds, that eliminate pathogens and remove residual biochemical oxygen demand (BOD), nutrients and suspended solids (Sec. 16.3).

Facultative ponds are especially popular in the rural areas of the US, serving 90% of the communities that have a population of fewer than 10,000 people, with nearly half of all wastewater treatment systems in Canada of this type. Facultative ponds in southern Europe are used for small communities (<2,000 PE [population equivalent]), whereas larger-sized plants are used for tourist areas where there is a large increase in population during the summer. Maturation ponds are widely used throughout the world as a tertiary treatment process. All waste stabilization ponds have low construction and operational costs and use little energy, resulting in low carbon emissions. However, unlike anaerobic lagoons which are now covered to recover methane, facultative ponds are open structures so that any gases produced are released directly to the atmosphere. Thus, careful operation is required to minimize the release of any potential greenhouse gases.

The selection of the most sustainable options for wastewater treatment depends on numerous factors, and not just the smallest carbon footprint or cost (Table 1.6). The volume and nature of wastewater, location, condition of the receiving water, potential for recovery, and reuse of wastewater components are all important issues to be considered. There are a number of ways that a selection can be made with the most popular, based on environmental impact and economic feasibility assessments incorporating life cycle analysis (Meneses *et al.*, 2015), or by using multi-criteria decision-making analysis using models (Ouyang *et al.*, 2015; Ren and Liang, 2017). Life cycle assessment is a key mechanism in the quantification of all impacts and stages associated with wastewater treatment, by using the "cradle to grave" concept of a range of standard techniques outlined in the ISO 14000 series (ISO, 2006a; b; Friedrich *et al.*, 2007; Corominas *et al.*, 2013).

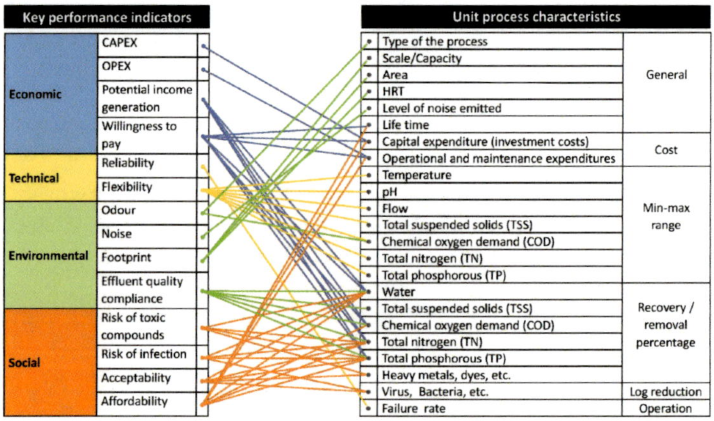

Fig. 4.1. Example of how sustainability indices are used to compare wastewater treatment processes. Capital and operational expenditure are abbreviated as CAPEX and OPEX, respectively (van Schaik *et al.,* 2021). Reproduced with permission under the Creative Commons License CC-BY.

Zang *et al.* (2015) list the main impact indicators from wastewater treatment as eutrophication potential, global warming potential, toxicity-related impact, acidification potential, photochemical oxidation potential, ozone layer depletion, energy use, and other emerging indicators (e.g. water use, land use). While the carbon and water footprint analysis has also been used (Chen *et al.*, 2020), the use of sustainability indicators has proven to be a particularly effective and straightforward method to compare individual wastewater treatment plants and technologies (Sec. 1.3) and is increasingly used in the selection of the most appropriate technologies (Fig. 4.1). The process of selecting the most sustainable treatment systems is considered further in Sec. 18.1.

References

Chen, K-H., Wang H-C., Han, J-L., Liu, W.Z., Cheng, H.Y., Liang, B. and Wang, A-J. (2020) The application of footprints for assessing the sustainability of wastewater treatment plants: A review. *Journal of Cleaner Production*, **277**, 124053.

Corominas, L., Foley, J., Guest, J.S., Hospido, A., Larsen, H.F., Morera, S. and Shaw, A. (2013) Life cycle assessment applied to wastewater treatment: State of the art. *Water Research*, **47**, 5480–5492.

Friedrich, E., Pillay, S. and Buckley, C. (2007) The use of LCA in the water industry and the case for an environmental performance indicator. *Water South Africa*, **33**(4), 443–452.

HM Treasury. (2020) Net zero review: Interim report. HM Treasury, London.

ISO. (2006a) Environmental management: Life cycle assessment principles and framework. International Standard 14040. International Standards Organisation, Geneva.

ISO. (2006b) Environmental management: Life cycle assessment requirements and guidelines. International Standard 14044. International Standards Organisation, Geneva.

Meneses, M., Concepción, H., Vrecko, D. and Vilanova, R. (2015) Life cycle assessment as an environmental evaluation tool for control strategies in WT plants. *Journal of Cleaner Production*, **107**, 653–661.

Middlebrooks, E.J., Middlebrooks, C.H. and Reed, S.C. (1981) Energy requirement for small wastewater treatment systems. *Journal of the Water Pollution Control Federation*, **53**(7), 1172–1198.

Mitsch, W.J. and Jørgensen, S.E. (2003) Ecological engineering: A field whose time has come. *Ecological Engineering*, **20**(5), 363–377.

Ouyang, X., Guo, F., Shan, D., Yu, H. and Wang, J. (2015) Development of the integrated fuzzy analytical hierarchy process with multidimensional scaling in selection of natural WT alternatives. *Ecological Engineering*, **74**, 438–447.

Ren, J. and Liang. H. (2017) Multi-criteria group decision-making based sustainability measurement of wastewater treatment processes. *Environmental Impact Assessment Review*, **65**, 91–99.

Van Schaik, M.O., Sucu, S., Cappon, H.J., Chen, E-S., Martinson, D.B., Ouelhadj, D. and Rijnaarts, H.H.M. (2021) Mathematically formulated performance indicators for design and evaluation of treatment trains for resource recovery from urban wastewater. *Journal of Environmental Management*, **282**, 111916. https://doi.org/10.1016/j.jenvman.2020.111916

Zang, Y., Li, Y., Wang, C., Zhang, W. and Xiong, W. (2015) Towards more accurate life cycle assessment of biological wastewater treatment plants: A review. *Journal of Cleaner Production*, **107**, 676–692.

Part II
Land Treatment

5
The Purification Process

5.1 Introduction to Land Treatment

An estimated 18 million hectares of land are irrigated with wastewater at present, especially in areas most affected by global warming, which are increasingly resorting to water reuse (Verbyla *et al.*, 2016). Land treatment of wastewater dates back to the 16th century and became widely used in Europe during the 19th century. At the time of the British Rivers Pollution Prevention Act of 1876, new methods of sewage treatment were being investigated — with treatment the only alternative method of sewage disposal, apart from discharging it directly to a watercourse. During this time land treatment underwent significant improvement. Instead of simply flooding the land, the soil was plowed deeply, and the sewage allowed to flow into the furrows (Fig. 5.1). The waste matter decayed and made the soil very fertile, allowing vegetables of excellent quality to be grown on top of the furrows, hence the term "sewage farms". Sewage farming became a lucrative business, with the rights to a town's sewage often costing farmers considerable amounts of money. However, because of the amount of land required, the occurrence of unpleasant smells and the problem of flies in warmer months, sewage farms were gradually superseded by other biological treatment methods. Various improvements in land treatment were introduced in an attempt to reduce the area of land required. The most successful was the "Latham soil-filter system", which comprised three layers of soil separated by layers of

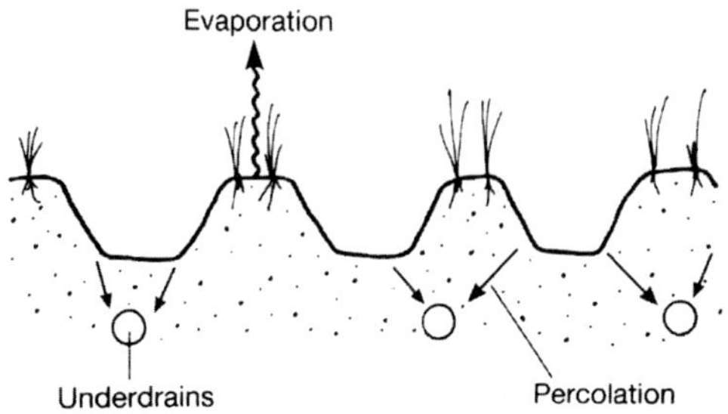

Fig. 5.1. Land treatment by ridge and furrow irrigation.

coke breeze overlying land drains. This method reduced the area of land required by up to a factor of 10 when compared with traditional sewage farm methods. However, even with these new developments, the loading rate to land was restricted to a maximum of 0.05 $m^3m^{-2}d^{-1}$, which is roughly equivalent to 2,500 people per hectare. In 1900, the City of Birmingham (in the United Kingdom [UK]) used 1,133 ha of land to treat its sewage, an area that stretched for 4 km along the banks of the River Tame. The area available for treatment had to be increased at a rate of 2.6 ha per month to keep pace with the growing population (Bruce and Hawkes, 1983). However, sewage farms were slow to disappear and several major European cities still relied partly on such systems up to the 1990s. The large area of flooded land also proved an excellent feeding ground for many birds, especially waders, and studies have suggested that the demise of sewage farms may have been closely linked to the reduction in the population of certain birds (Fuller and Glue, 1980; 1981). Although not widely practiced now on a large scale in Europe for domestic wastewaters, spray irrigation of animal slurry and some food processing wastewaters remain common. Spray irrigation of dairy wastewaters is widely practiced in New Zealand, whereas in Germany, at the Braunschweig Land Treatment Plant, the treated effluent from a population of about 270,000 people is reused on 300 ha of farmland by sprinkler irrigation (Boll and Kayser, 1986).

Wastewater-based irrigation of agricultural land is practiced throughout the world where water is in short supply, especially in arid tropical and semi-tropical areas. It is particularly common in the

Southern Mediterranean, the south-west of the United States (US), South America, Australia, and India. Land treatment using on-site subsurface infiltration or percolation areas is still widely used in conjunction with septic tank systems. There are four distinct types of land application of wastewater: on-site subsurface infiltration (Chap. 6), slow-rate land application (Chap. 7), rapid infiltration (Chap. 8) and overland flow (Chap. 9); however, there is some overlap between these categories (USEPA, 2006). The important factors in the design of land treatment systems are land availability, permeability and the depth of soil, type of bedrock and nature of groundwater, climate including precipitation, evapotranspiration and temperature, plant cover, topography, wastewater type, and required final effluent quality (Johnson, 1973). Vegetation plays an important role in land treatment through evapotranspiration, nutrient uptake, preventing erosion, and as a support medium for microbial activity. In slow rate filtration, the vegetation can also be a commercial crop such as animal fodder, food or biomass.

All four types of systems are widely used in the US, with irrigation systems most widely used for secondary treatment and rapid infiltration used primarily for groundwater recharge. In Europe, land treatment is less widely used and is generally restricted for secondary or tertiary treatment of wastewater from small communities. Bankside infiltration is practiced in some European countries as a method of tertiary treatment for secondary-treated effluents that are discharged to surface waters subsequently used for water treatment abstraction. However, while the production of wastewater is continuous, the demand for irrigation is cyclic, resulting in periods where wastewater must be stored for later use. Land treatment is now also widely used in sustainable urban drainage schemes (SuDS) to control surface and road runoff by using vegetative strips, swales and retention-detention ponds (Woods Ballard *et al.,* 2015).

There are a number of potential concerns relating to wastewater irrigation (Bond, 1998; Rowe and Abdel-Magid, 1995); these include:

- **Suspended solids**, which can add to the humic layer of soils, but can also clog distribution systems and the capillary pores in the soil layer, thus reducing infiltration.

- **Trace elements and organic compounds,** including heavy metals common in domestic wastewater (e.g. Cu, Zn, Ni, Cd), which can be accumulated in the soil and become toxic to plants or bioaccumulated in crops or fodder. They may also be leached to groundwater.
- **Nutrient leaching to groundwater**, including nitrate, phosphorus and sodium, which can adversely affect the water quality for abstraction for public supply.
- **Salinity and the effect of increasing soil sodicity.**
- **Potential public health risks.**

5.2 The Process of Treatment

Wastewater is purified as it percolates over the plant surface, and through the root zone and soil, involving a wide range of physical, chemical and biological processes (Fig. 5.2). These include filtration, soil adsorption, chemical precipitation, microbial breakdown,

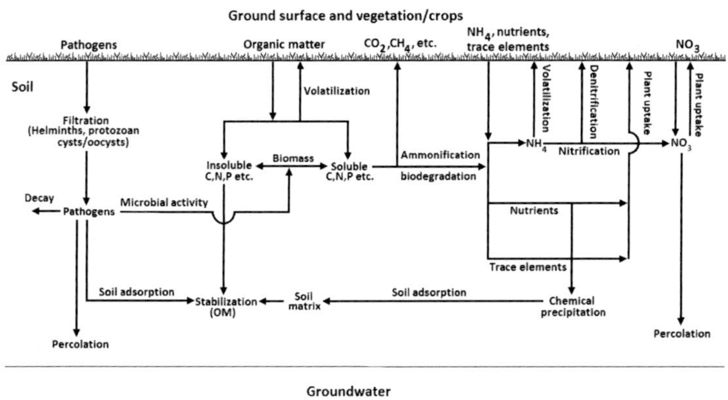

Fig. 5.2. Wastewater irrigation uses the self-purification capacity of soils and plants to ensure that groundwater or recovered water is protected from contaminants and pathogens (redrawn from [Paranychianakis *et al.*, 2006]). Reproduced with permission from the Taylor & Francis Group.

predation by soil mesofauna, natural die-off, nitrification-denitrification, volatilization, and, importantly, plant uptake (Paranychianakis *et al.*, 2006). The key role of land treatment is the oxidation of organic matter, removal of nutrients, and the inactivation and/or destruction of pathogens. Evapotranspiration linked to hydraulic loading ultimately controls the amount of water passing through the soil system, with plants playing a vital role in water balance, infiltration rates, nutrient utilization, metal and trace organic uptake, and microbial community structure.

Suspended material, including microorganisms, is physically filtered out of the solution as the water percolates through the upper soil layer. Organic material trapped in the soil is rapidly utilized by the high density of heterotrophs present. For example, a single plate count using just one gram of soil yielded 107 bacteria, 106 actinomycetes, and 105 fungi, which, in turn, is supporting a large and diverse micro and macrofauna (Miller, 1974). The humus, silt and clay particles that comprise the soil provide a very large surface area for ion exchange with mineral ions, especially cations, which become strongly bonded onto soil particles. In practice, this results in a high removal efficiency of metals, phosphorus and ammonia from the water. Any nitrogen present is normally fully utilized by plant uptake and subsequent harvest, nitrification–denitrification or volatilization (Figs. 7.5 and 8.5). Where there is a low carbon:nitrogen (C:N) ratio, plant roots can provide a supplementary carbon source for denitrification. The relative importance of physical, chemical and biological treatments vary between irrigation systems, which are examined in each chapter. For example, biofilm development in overland flow treatment on the vegetation and soil surface is important in the oxidation of both organic and nitrogen present in the applied wastewater.

Overloading land treatment sites can result in the land becoming clogged with organic matter. This is unsightly and produces strong and unpleasant odors, and is often associated with fly nuisance, but as the soil is anaerobic, the pH gradually falls until minerals are released and leaching occurs.

5.3 Evapotranspiration

Evapotranspiration is a key variable in land treatment controlling the amount of irrigation water that passes into the soil and eventually enters the groundwater or is collected by subsurface drainage for reuse. It also determines the required irrigation rate to maintain healthy crops. Evapotranspiration is the total water loss in milliliters per unit time (e.g. mm yr^{-1}) by surface evaporation from both soil and plant surfaces and via plant transpiration. The more soil surface that is exposed, the greater the evaporation of water. So, when plants are first sown and start to grow, evapotranspiration is almost entirely composed of surface evaporation; however, as the plants develop, they create a canopy of leaves that shades the soil surface, slowing down water loss through evaporation (Fig. 5.3). Maximum evaporation rates occur immediately after irrigation or precipitation when the water is both on plant and soil surfaces and then slowly declines until water is added again. The type of irrigation used also affects the rate of evaporation. Drip irrigation reduces soil evaporation, as only a small area is wetted at a time, while subsurface irrigation eliminates evaporation almost entirely. Where only the plant canopy and soil surface is wetted, for example, during a short shower of rain, nearly all the water is lost by evaporation without any measurable soil

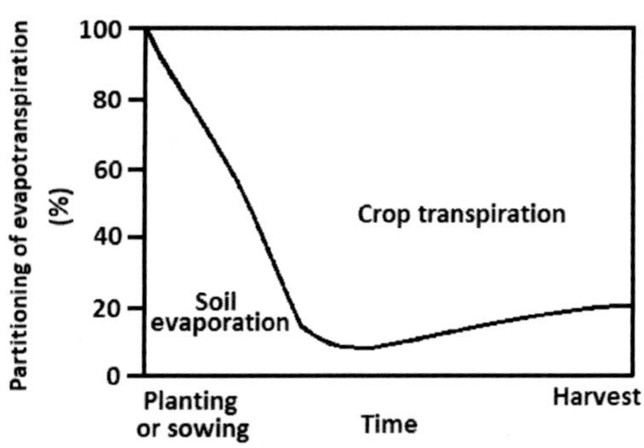

Fig. 5.3. Variation in the relative contribution of crop and soil evaporation to total evapotranspiration.

infiltration. For this reason, surface sprinklers should not be used for frequent small applications of wastewater as this will almost all be lost by evaporation. To reduce evaporation losses, sprinklers using higher loadings should be used for long periods in the summer, or at night when temperatures are lower.

Very little of the water taken up by plant roots is used for metabolism (<1%) with the remainder lost to the atmosphere via the stomata on plant leaves (i.e., transpiration). The rate of transpiration varies between plants and their state of growth and is controlled by air temperature and humidity, soil moisture content, and wind strength. As soils dry out, the stomata conserve water by slowing the rate of transpiration. In contrast, transpiration increases as the air become hotter and drier, or windier.

Knowing the rate of evapotranspiration is pivotal to calculating irrigation rates and carrying out water budgets. Crop evaporation (ET_c) is calculated as:

$$ET_c = ET_0 \times K_c, \tag{5.1}$$

where ET_0 is the defined or reference crop evapotranspiration rate and K_c is the specific crop coefficient. When the irrigation rate exceeds ET_c for a specific vegetation or crop cover, the excess is the amount of water percolating deep into the soil.

The ET_0 is the total evaporation and transpiration for a specific crop (usually grass) grown under ideal conditions. It is normally calculated directly from weather data using the FAO Penman–Monteith method (Allen *et al.*, 1998). Today, there is a range of models that can give real-time values of the ET_0, taking into account changes in weather conditions. This value is used to calculate the actual evapotranspiration of the crop being grown by multiplying it with the K_c, which varies according to the crop type, stage of growth (leaf or ground cover), and climate, and is expressed as a number between 0.1 and 1.2. The stage of growth is normally broken down into four stages: (1) initial growth phase (10% ground cover), (2) crop development (60–80% ground cover), (3) mid-season stage (>80% ground cover), and (4) full maturity until harvest (>80% ground cover), which, in turn, significantly alters the K_c for the crop (Fig. 5.4).

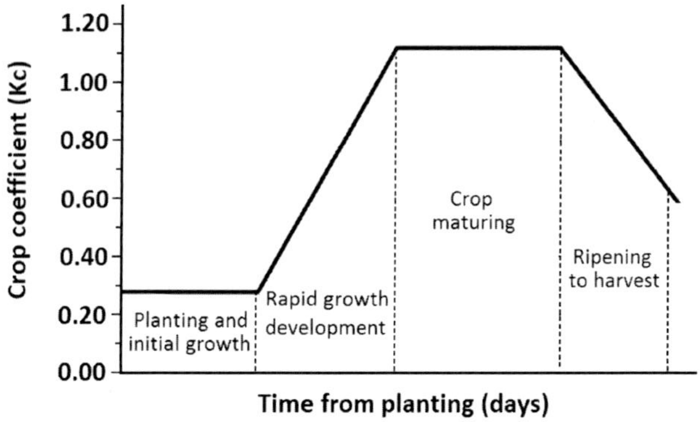

Fig. 5.4. Variation in crop coefficient K_c during the life cycle of the crop.

Table 5.1. Comparison of crop coefficient K_c for humid and dry conditions for perennial forage crops (Doorenbos and Pruitt, 1977). Reproduced with permission from the Food and Agricultural Organization of the United Nations.

	Condition	
Crop	**K_c Humid, light to moderate wind**	**K_c Dry, light to moderate wind**
Alfalfa		
Minimum	0.50	0.40
Mean	0.85	0.95
Peak	1.05	1.15
Grass for hay		
Minimum	0.60	0.55
Mean	0.80	0.90
Peak	1.05	1.10
Clover, grass legumes		
Minimum	0.55	0.55
Mean	1.00	1.05
Peak	1.05	1.15
Pasture		
Minimum	0.55	0.50
Mean	0.95	1.00
Peak	1.05	1.10

K_c (minimum) represents conditions just after cutting.
K_c (mean) represents value between cuttings.
K_c (peak) represents conditions before harvesting under dry soil conditions. Under wet conditions increase values by 30%.

For example, the difference between K_c for the first and third stages (i.e., ~10 and 80% ground cover), respectively, is 0.30 and 1.15 for barley, 0.30 and 1.15 for cotton, 0.30 and 1.20 for maize, and 0.40 and 1.15 for soybeans. Doorenbos and Pruitt (1977) give other examples for perennial and seasonal crops in Tables 5.1 and 5.2, respectively. Actual daily required irrigation rates could be calculated using Eq. (5.1) (see Table 5.3).

Table 5.2. Crop coefficient K_c for optimum growth conditions for seasonal crops (for mid to late season) (Doorenbos and Pruitt, 1977). Reproduced with permission from the Food and Agricultural Organization of the United Nations.

Crop	Crop stage	K_c Humid[a]	K_c Dry[b]
Alfalfa[c]	1–4	0.85	0.95
Barley	3	1.05	1.15
	4	0.25	0.20
Clover	1–4	1.00	1.05
Corn	3	1.05	1.15
	4	0.55	0.60
Cotton	3	1.05	1.20
	4	0.65	0.65
Grain	3	1.05	1.15
	4	0.30	0.25
Grapes	3	0.80	0.90
	4	0.65	0.70
Oats	3	1.05	1.15
	4	0.25	0.20
Pasture grass	1–4	0.95	1.00
Rice	3	1.1	1.25
Sorghum	3	1.00	1.10
	4	0.50	0.55
Soybeans	3	1.00	1.10
	4	0.45	0.45
Sugar beets	3	1.05	1.15
	4	0.90	1.00
Wheat	3	1.05	1.15
	4	0.25	0.20

[a]Humidity 70%, light wind 0–16 miles per hour.
[b]Humidity 20%, light wind 0–16 miles per hour.
[c]Peak factors are 1.05 for humid conditions and 1.15 for dry conditions.

Table 5.3. Example of the variation in the daily crop water usage (ET_c) for a maize crop during the mid-season growth stage over the period of a week.

Day	ET_o	K_c	ET_c (mm d^{-1})
Monday	4.8	1.20	5.8
Tuesday	4.2	1.20	5.0
Wednesday	4.7	1.20	5.6
Thursday	3.9	1.20	4.7
Friday	4.9	1.20	5.9
Saturday	5.2	1.20	6.2
Sunday	5.5	1.20	6.6

In practice, calculating crop evapotranspiration accurately is a specialist task and full details are given in the FAO manual (Allen *et al.*, 1998). Calculated ET_c values are also influenced by other factors such as soil compaction, cultivation and irrigation practices, soil salinity, plant diseases, and pests. Crop ET_c values can be reduced by direct irrigation systems (e.g. subsurface or drip irrigation), reducing wind velocities across irrigation plots by planting windbreaks or using mulches to reduce evaporation from the soil when crops and trees are first planted.

5.4 Salinity and Sodicity

Irrigation with wastewater has the potential to add large quantities of salts to treatment plots, with wastewaters having between 200 to 3,000 mg L^{-1} of total dissolved solids (TDS). If salts are not leached out of soils, perhaps due to high evaporation and transpiration rates, then they accumulate in the soil, causing a reduction in crop yields and even plant toxicity. This poses a significant problem for the reuse of wastewater for land treatment and, in particular, crop production. For example, an application of 1,000 mm of treated wastewater with a TDS concertation of 600 mg L^{-1} would be equivalent to adding 6 tons of salt per hectare of treated land each year. This has been a problem at the Western Treatment Plant at Werribee that serves part of Melbourne in Australia (Sec. 7.2). The salts come not only from domestic wastewater but also from industrial wastewater and infiltration from groundwater sources. While the average TDS in the domestic wastewater was 319 mg L^{-1}, the total municipal concentration sent for treatment was 913 mg L^{-1}, which is equivalent to 444 tons d^{-1} (Ibrahim *et al.*, 2007; Muyen *et al.*, 2011). The TDS of settled wastewater is very similar to that of secondary-treated wastewater, even though the organic matter has been oxidized. The total dissolved solids from irrigation water and fertilizers accumulate in the soil, making it more saline, which can be measured using conductivity. Salinization is a major cause of soil degradation, resulting in desertification.

Wastewater contains a high concentration of sodium (Na^+) ions in the form of NaCl, which is toxic to plants. A discharge of excess Na^+ ions to the land disperses the soil particles and inhibits plant growth (Ellis, 1974) and as it accumulates, percolation is severely impaired and the tilth of the soil is reduced. In contrast, the divalent cations calcium (Ca^{2+}) and magnesium (Mg^{2+}) will rectify the effects of excess Na^+. The ratio between Na^+ and both Ca^{2+} and Mg^{2+} is known as the sodium absorption ratio (SAR), which is used to characterize the suitability of a soil for plant growth:

$$SAR = \frac{Na^+}{\sqrt{\frac{\left(Ca^{2+} + Mg^{2+}\right)}{2}}}, \quad (5.2)$$

where the cation concentrations are given as millequivalents L^{-1} (i.e., concentration in mg L^{-1} multiplied by 0.044, 0.050 and 0.082 for Na^+, Ca^{2+} and Mg^{2+}, respectively).

Sodicity occurs when the content of exchangeable Na^+ ions in the soil increases above a certain threshold (Sec. 7.3.2). These ions are found as crystallized $NaHCO_3$ or Na_2CO_3 salts in both the solid and liquid phases of the soil and as they increase the soil becomes less productive until no plants can thrive. Salts are naturally derived from the weathering of parent rocks within the soils by water, which slowly solubilized them. Salts can also be due to oceanic aerosols, or irrigation with mineralized groundwater. A sodic soil is classically characterized as having an exchangeable sodium percentage (ESP) of ≥15%, although this value depends on local conditions, with values as low as 5% being classified as being seriously damaged. Once formed, these sodium-saturated soils are difficult to correct by irrigation alone. Even if the soluble Na^+ is removed, the cation exchange sites in the soil are saturated with the ion. Normally, the soil receives enough rainfall to leach out any excess Na^+ into groundwater where they eventually end up in the sea, but in arid and semi-arid areas, where treated wastewater is widely used to irrigate recreational areas as well as farmland for crop production, there is always a high risk that sodicity will occur.

Sodium (Na^+) is attached loosely to clay particles along with other cations including Ca^{2+}, Mg^{2+}, potassium, and hydrogen. When >5% of the attached cations are Na^+, then the soil is classed as sodic and problems start with the soil structure. The attached Na ions cause the clay particles to separate. This weakens the soil structure, causing the particles to collapse together, thus reducing soil interstices. This, in turn, reduces porosity and water penetration, and availability. This can also result in surface crust formation to plant roots.

Cations attached to clay particles will swap with other cations, so that sodicity is reversible by cation exchange, using gypsum ($CaSO_4.2H_2O$) or, less commonly, lime (CaO). The calcium replaces the sodium and allows the clay particles to reaggregate, releasing the sodium back into solution, which can then be leached out well below the root zone. It is usual practice to then add organic matter to the soil to help stabilize the structure. This will improve infiltration and percolation as well as crop health and production. The amount of commercial gypsum (2 mm in diameter and approximately 70% purity) required is dependent on the ESP of the soil. Sodic soils with an ESP between 5 and 10 require application rates of 2 to 5 tons per hectare (t ha^{-1}) and 5 t ha^{-1} for soil with ESPs >10 (Shahid *et al.*, 2018a). The cation exchange process that occurs can be described as:

$$\underset{\text{Sodic soil}}{2Na^+ - \text{Clay}} + \underset{\text{Gypsum}}{CaSO_4 \cdot 2H_2O}\ \underset{\text{Normal soil}}{Ca^{2+} - \text{Clay}} + \underset{\text{Soluble salt}}{Na_2SO_4^-} + 2H_2O. \tag{5.3}$$

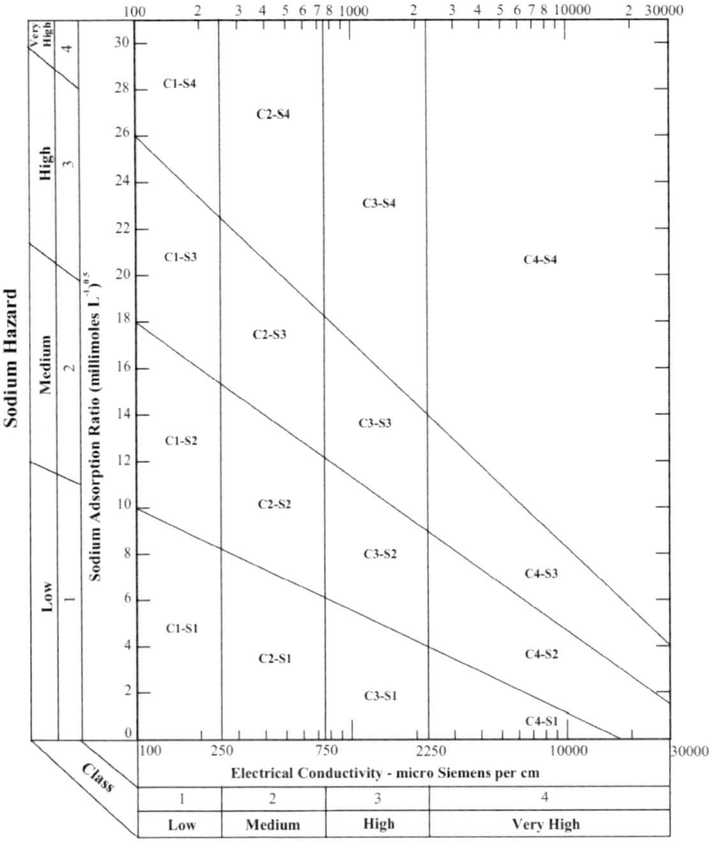

Fig. 5.5. The suitability of wastewaters and effluents for irrigation can be classified using the sodium adsorption ratio (Eq. [5.2]) and the conductivity (millisiemens per meter at 25°C) (Zaman *et al.*, 2018). Waters in areas C1-S1, C1-S2, C2-S2, and C2-S1 are suitable for almost all irrigation purposes; waters in areas C1-S3 and C3-S1 should be avoided if possible and only used in conjunction with expert advice; all other waters are not suitable for irrigation under any circumstances. However, in practice, many wastewaters used for irrigation fall into the areas C3-S1, C3-S2, C4-S1, and C4-S2, resulting in serious sodicity. Reproduced with permission from the International Atomic Energy Agency (https://creativecommons.org/licenses/by/3.0/igo/).

Wastewaters have a SAR of between 4.5 and 7.9, which increases the salinity and ESP of soils. Irrigation of soil with wastewater leads to complex chemical changes, such as precipitation and oxidation-reduction. Immobilization of cations is enhanced at a soil water pH >7 but acidic conditions lead to some mineral solubilization and leaching. Electrical conductivity (EC) is used as a measure of total dissolved solids, which indicates the salinity hazard to crops, while the SAR, which is a measure of the sodium or alkali hazard to crops, is used to assess the suitability of wastewaters and effluents for irrigation (Fig. 5.5) (USDA, 1954; Zaman *et al.*, 2018).

Wastewater irrigation is increasingly being adopted to combat the problems of increasing water scarcity, especially in arid and semi-arid areas throughout the world. These areas experience low rainfall, high evaporation rates, and hence low leaching resulting in soils having naturally higher salt concentrations. Over one billion hectares of cultivated land is currently classed as having either saline or sodic soils, with as much as 25 to 30% of all irrigated land being salt-affected. There are approximately 620 million hectares of sodic soils globally, which are mainly located in Australia (340 million hectares), the

Table 5.4. Area (in millions of hectares) of arid or semi-arid land affected by salt, creating either saline or sodic soils by continent (Zaman *et al.*, 2018). Reproduced with permission from the International Atomic Energy Agency (https://creativecommons.org/licenses/by/3.0/igo/).

	Salt-affected area (mha)		
Continent	**Saline soils**	**Sodic soils**	**Total**
Africa	122.9	86.7	209.6
Australasia	17.6	340.0	357.6
Mexico/Central America	2.0	—	2.0
North America	6.2	9.6	15.8
North and Central Asia	91.5	120.2	211.7
South America	69.5	59.8	129.3
South Asia	82.3	1.8	84.1
Southeast Asia	20.0	—	20.0
Total	**412.0**	**618.1**	**1,030.1**

USSR (120 million hectares), and Argentina (54 million hectares) (Shahid *et al.*, 2018b) (Table 5.4).

The problem of soil salinization and sodicity in irrigated soils is reviewed in detail by Zaman *et al.* (2018), with a useful example of the long-term effects of wastewater irrigation in the development of sodic soils (Muyen *et al.*, 2011). The effect of salinity on wastewater irrigation systems, including crop tolerance, crop sensitivity, and the effect on yields (Tables 7.1 and 7.2) are discussed further in Sec. 7.3.2. Salinity also changes the normal soil microfauna, including the important mycorrhizal fungi due to hyperosmotic and oxidative stresses (Becerracastro *et al.,* 2015).

5.5 Contaminant Problems

Sodium, chloride and boron ions can all cause significant plant injury. Sodium toxicity results in leaf burn or scorch mainly along the edges of leaves. Sodium toxicity can be related to plant tolerance based on their ESP level: sensitive (<15%), semi-tolerant (15–40%), and tolerant (>40%).

While boron is a necessary plant nutrient with at least 0.02 ppm required, concentrations above 0.3 ppm are considered problematic. Like sodium, boron also accumulates in the soil. Signs of toxicity appear in older leaves first as yellowing or spotting, especially at their edges and tips. This leads to the leaf drying up, which progresses along its veins to its center, eventually affecting the whole plant. Crop thresholds for boron in irrigation water are shown in Table 5.5 (Zaman *et al.,* 2018).

Unlike sodium or boron, chloride (Cl^-) ions remain in the solution and so are easily leached out. It accumulates in leaves and causes necrosis at the leaf tip, which spreads down the edges until the leaves drop off the plant. As the concentration of chloride ions increase, so does the impact. A concentration of <7 ppm is considered safe for the majority of crops and vegetative cover plants; 70 to 140 ppm results in sensitive plants showing slight to moderate damage. At concentrations of 141 to 350 ppm, moderately tolerant plants will show signs of slight to significant damage as the concentration increases, but

Table 5.5. Crop tolerance to boron[a] (Zaman *et al.*, 2018). Reproduced with permission from the International Atomic Energy Agency (https://creativecommons.org/licenses/by/3.0/igo/).

Very sensitive <0.5 ppm	**Sensitive 0.5–0.75 ppm**	**Less sensitive 0.75–1.0 ppm**	**Moderately sensitive 1.0–2.0 ppm**	**Moderately tolerant 2.0–4.0 ppm**	**Tolerant 4.0–6.0 ppm**	**Very tolerant >6.0 ppm**
Lemon	Avocado	Garlic	Pepper, red	Lettuce	Tomato	Cotton
Blackberry	Grapefruit	Sweet potato	Pea	Cabbage	Parsley	Asparagus
	Orange	Sunflower	Carrot	Celery	Beet, red	
	Apricot	Bean	Radish	Turnip		
	Peach	Sesame	Potato	Oats		
	Cherry	Strawberry	Cucumber	Corn		
	Plum	Bean, kidney		Clover		
	Grape	Peanut		Squash		
	Walnut			Muskmelon		
	Onion					

[a]Maximum concentrations tolerated in soil-water or saturation extract without yield or vegetative growth reduction. Boron tolerance varies depending upon climate, soil conditions and crop varieties. Maximum concentrations in the irrigation water are approximately equal to these values or slightly less.

concentrations that are >350 ppm result in severe damage and even the death of plants (Ayers and Westcot, 1985).

Irrigation exposes both the soil and plants to a wide range of heavy metals commonly found in wastewater (i.e., Fe, Zn, Pb, Cu, Ni, Cr, and Cr). These accumulate in the soil as well as bioaccumulate in crops, both grains and vegetables, to re-enter the food chain either by direct ingestion of plants, fruit and grains or via livestock fed on contaminated feed (Singh *et al.*, 2010). Metals are phytotoxic, reduce crop yields, and alter the normal soil microbial community structure. The World Health Organization (WHO, 2006) has calculated the maximum permissible pollutant concentrations of metals and organic compounds in the soils above, which could result in the transfer of these contaminants via irrigated crops into the human food chain (Table 5.6).

Table 5.6. Maximum tolerable wastewater irrigation soil concentrations to protect human health (WHO, 2006). Reproduced with permission from the World Health Organization under the WHO CC BY-NC-SA 3.0 IGO license.

Chemical	Soil concentration (mg kg^{-1})
Element	
Antimony	36
Arsenic	8
Barium[a]	302
Beryllium[a]	0.2
Boron[a]	1.7
Cadmium	4
Fluorine	635
Lead	84
Mercury	7
Molybdenum[a]	0.6
Nickel	107
Selenium	6

(*Continued*)

Table 5.6. (*Continued*)

Chemical	**Soil concentration (mg kg^{-1})**
Silver	3
Thallium[a]	0.3
Vanadium[a]	47
Organic compound	
Aldrin	0.48
Benzene	0.14
Chlordane	3
Chlorobenzene	211
Chloroform	0.47
2,4-D	0.25
DDT	1.54
Dichlorobenzene	15
Dieldrin	0.17
Dioxins	0.00012
Heptachlor	0.18
Hexachlorobenzene	1.40
Lindane	12
Methoxychlor	4.27
PCBs	0.89
PAHs (as benzo[*a*]pyrene)	16
Pentachlorophenol	14
Phthalate	13 733
Pyrene	41
Styrene	0.68
2,4,5-T	3.82
Tetrachloroethane	1.25
Tetrachloroethylene	0.54
Toluene	12
Toxaphene	0.0013
Trichloroethane	0.68

[a]The computed numerical limits for these elements are within the ranges that are typical for soils.

Cadmium poses the most serious threat to food chain contamination from crops grown on land irrigated with wastewaters, but is a particular problem in some areas of China, where the land has become so contaminated with heavy metals, including cadmium, that growing crops is no longer feasible (Wang *et al.*, 2001; Liu *et al.*, 2005; Li *et al.*, 2009). In a series of plot experiments irrigated at different salinities, Wang *et al.* (2019) found that salt stress resulted in cadmium becoming more available. The addition of 0.5% sodium salts to irrigation water, comprising NaCl, Na_2SO_4, $NaHCO_3$, and Na_2CO_3 at a ratio of 1:9:9:1, applied over 12 months, resulted in increases in the cadmium concentration in winter wheat grain of 44%, with the soil's Cd availability increasing by 68%. There was also a significant reduction in crop yield.

Bacteria from the phyla Proteobacteria and Actinobacteria are commonly associated with metal-contaminated soils, with the former considered to be the most stress-tolerant bacteria in soils heavily contaminated with heavy metals (Li *et al.*, 2016). Wang *et al.* (2019) observed that saline stress affected the soil bacterial flora in cadmium-contaminated soil, with Bacillaceae, Staphylococcaceae and Pseudomonadaceae becoming more dominant. Sphingomonadaceae (Proteobacteria) were particularly sensitive to Cd, while the Nitriliruptoraceae (Actinobacteria) were the most sensitive to high salt stress, with the authors suggesting their potential use as biomarkers.

The fate of trace organic contaminants in land treatment systems is complex and difficult to predict or control. Irrigation not only adds trace organic compounds but also affects the mobility of those already present or used for crop management such as herbicides, fungicides and insecticides. Within the treatment plots, these organic contaminants will undergo varying degrees of transformation and changes in transport. Key mechanisms include adsorption, degradation, transport, and volatilization (Fig. 5.6).

The pH of the wastewater can affect the ion-exchange capacity of both organic and inorganic components of the soil, especially silicate clays, which are pH-dependent. The wastewater is rich in organic matter, especially dissolved organic matter (DOM), which significantly affects the sorption, movement and degradation of

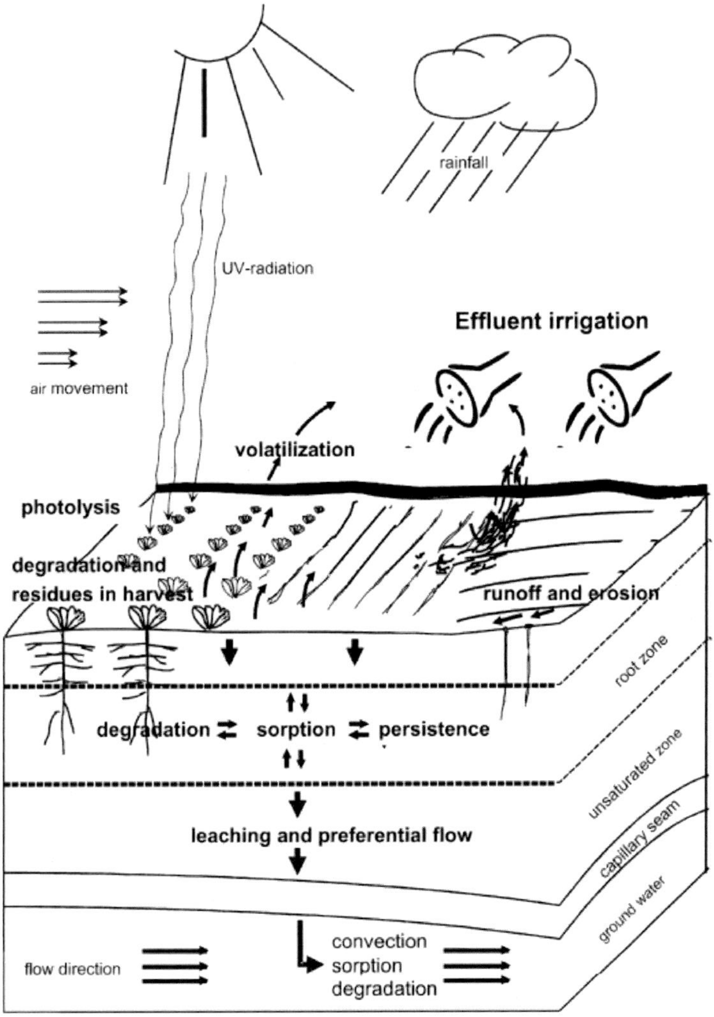

Fig. 5.6. The fate of organic contaminants in wastewater irrigation land treatment systems (Müller *et al.*, 2007). Reproduced with permission from Elsevier Publishing.

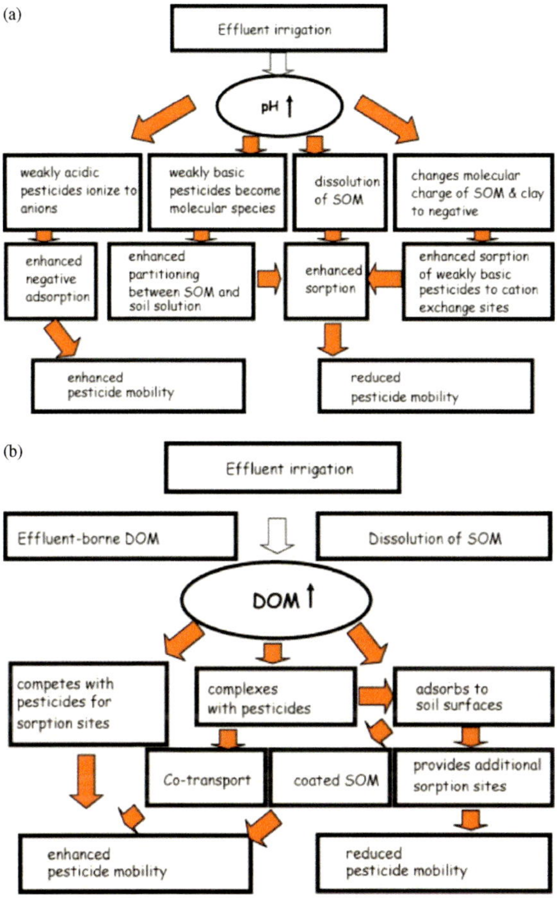

Fig. 5.7. Sorption and degradation of organic contaminants in the soil as affected by: (a) increase in pH, (b) increase in dissolved organic matter (DOM), and (c) transformation and degradation of organic contaminants due to an increase in nutrient supply, pH and DOM (Müller *et al.,* 2007). Reproduced with permission from Elsevier Publishing.

(c)

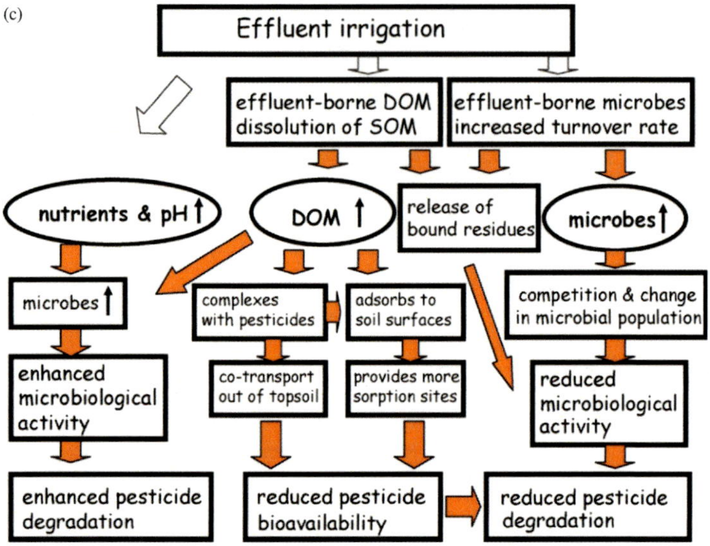

Fig. 5.7. (*Continued*).

organic contaminants present. While providing extra adsorption sites to capture organic contaminants, DOM also competes for the adsorption sites on mineral soil particles, preventing the removal of contaminants and allowing them to pass through the soil layer into groundwater. However, organic contaminants can also form soluble complexes with DOM, which can be adsorbed but with increased mobility. Finally, wastewater promotes microbial activity through the addition of organic matter and other nutrients to the soil, ensuring optimum moisture levels, as well as through the addition of bacteria and fungi, which increases microbial degradation of the contaminants (Fig. 5.7).

5.6 Pathogen Removal

Unlike the chemical contaminants in domestic and municipal wastewaters used for irrigation, the hazards presented by the presence of pathogens are normally short-term but often severe. Wastewater contains a wide range of pathogens including viruses, bacteria, protozoa, nematodes, and cestodes. Infection depends less on their numbers but rather on the probability of their transfer to a host. Contamination of plants and soil by fecal microorganisms and pathogens during irrigation readily occurs, allowing pathogens to complete life cycles by re-infecting humans. Wastewater irrigation also introduces plant pathogens into the soil, and spreads viruses and bacteria that will either infect grazing animals or find their way into the surface or groundwater.

Pathogens can survive for long periods in stored wastewater, on plant surfaces, and in the soil (Table 5.7). Survival depends on many factors, including whether the pathogens can produce resistant cysts,

Table 5.7. The approximate survival time for a range of pathogens associated with agricultural wastewater irrigation (WHO, 2006). Reproduced with permission from the World Health Organization under the WHO CC BY-NC-SA 3.0 IGO license.

	Survival times (days)		
Organism	**Fresh water and sewage**	**Crops**	**Soil**
Viruses			
Enteroviruses[a]	<120, usually <50	<60, usually <15	<100, usually <20
Bacteria			
Thermotolerant coliforms	<60, usually <30	<30, usually <15	<70, usually <20
Salmonella spp.	<60, usually <30	<30, usually <15	<70, usually <20
Shigella spp.	<30, usually <10	<10, usually <5	ND
V. cholerae	ND	<5, usually <2	<20, usually <10
Protozoa			
E. histolytica cysts	<30, usually <15	<10, usually <2	<20, usually <10
Cryptosporidium oocysts	<180, usually <70	<3, usually <2	<150, usually <75
Helminths			
Ascaris eggs	Years	<60, usually <30	Years
Tapeworm eggs	Many months	<60, usually <30	Many months

ND: no data

[a]Poliovirus, echovirus and coxsackievirus.

oocysts (protozoa) or eggs (helminths), with viruses generally surviving longer than bacteria. Other factors include:

- **Humidity:** Higher die-off in drier conditions.
- **Temperature:** High temperatures increase the rate of pathogen die-off, while freezing can also adversely affect some pathogens.
- **UV radiation:** This is highly effective in deactivating pathogens, especially bacteria and viruses.
- **Plant type:** This is critical where plants are eaten raw. This is due to surfaces that retain moisture are protected from UV radiation or are difficult to clean due to surface hair, being sticky or having crevices (Stine *et al.,* 2005). This includes all salad crops like onions, radish, lettuce, zucchini, and cucumber.
- **Soil characteristics:** High organic content can prolong pathogen survival.
- **pH:** Neutral to slightly acidic soils prolong virus survival while neutral to slightly alkaline soils favor bacteria survival.
- **Soil microfauna:** Pathogens enter an alien environment when irrigated onto plants and soil as they have to compete with the natural microbial community. Effects range from direct predation to naturally released metabolites that enhance the die-off of pathogens.

For nearly a century, public health has been protected by using a multiple-barrier technique involving the use of both wastewater and water treatment (Fig. 5.8). When wastewater is used for irrigation, the degree of treatment it receives before it is applied to crops and/or soil plays an important part in the risks associated to those who come into contact with the irrigation water either directly or indirectly (Sano *et al.,* 2016). The more treatment it receives before irrigation, the lower the risks of infection. However, even after advanced treatment, wastewaters still contain bacteria and especially enteric viruses (Gibson, 2014).

The greatest proportion of irrigation water is applied to pasture or grain crops that need to be processed and cooked before being

Source	Fecal coliforms (FC)
Human fecal coliform discharges	1 950 000 000 FC/person/day
	Fecal coliforms/100 mL
Municipal raw sewage	8 260 000 FC cells/100 mL

Sewage treatment reductions

	Cumulative reduction (percent)	FC surviving
Primary	50	4 130 000
Secondary	80	1 652 000
Tertiary	98	165 200
Disinfection	99.99	800

Fig. 5.8. The use of barriers in the control of pathogens (adapted from Geldrich *et al.* [1972]). Reproduced with permission from the American Water Works Association.

eaten. Cotton also responds favorably to irrigation with sewage (Dean and Lund, 1981). The treatment requirements for sewage used for crop irrigation vary significantly and depend on the quality of the raw sewage and type of crop to be irrigated. In California, for example, primary effluent may be used for fodder, fiber, and seed crops, and secondary treatment is the minimum treatment required before sewage can be used on food crops or pasture used for dairy herds (Pettygrove and Asano, 1984). The use of sewage for irrigation of vegetables and fruit, however, is discouraged (Sheikh *et al.*, 1984). But in desert areas, where water is the limiting factor for plant growth, sewage may have to be used for these types of crops, with the nutrients in sewage being an extra economic incentive. Fresh vegetables and fruit in arid areas are a rich source of income to farmers and sewage will inevitably be used. However, the dangers are very real, for example, a cholera outbreak in Jerusalem was traced to lettuce grown by a farmer who had illegally diverted raw sewage to irrigate his crop (Shuval and Fattal, 1980; Fattal *et al.*, 1986). Poliovirus was isolated from lettuce sprayed with sewage up to 4 weeks after application, although survival appeared to be weather-dependent (Larkin *et al.*, 1976). The virus was shown to persist for between 93 and 123 days in the winter when sprayed on the soil used to grow vegetables, but this was reduced to 8 to 11 days in the spring (Tierney *et al.,* 1977). The degree of contamination depends on the quality of the wastewater used. The United States Environmental Protection Agency (USEPA) calculates that the acceptable risk to consumers of contracting a waterborne disease from drinking water complying with current US microbial standards is 10^{-4}, i.e., an individual has a 1 in 10,000 chance of contracting the disease in question. Shuval *et al.* (1997) have examined the health risks of consuming raw salad crops irrigated with wastewater effluents of varying microbial quality. Where WHO guidelines were met, i.e., <1,000 fecal coliforms (FC) per 100 mL^{-1}, the annual risk of contracting hepatitis A was 10^{-6} to 10^{-7} and rotavirus infection was 10^{-5} to 10^{-6}. In contrast, where irrigation water had 10^{7} FC 100 mL^{-1}, the risk of contracting hepatitis A increased to 10^{-2}. Thus, according to the authors, the consumption of salad crops irrigated with wastewaters treated to <1,000 FC 100 mL^{-1} is 1 to 3 orders of magnitude safer than drinking water conforming to the US Safe Water Standards.

In crop irrigation, the excreted pathogens of most concern are human intestinal nematodes and fecal bacteria. Irrigation with untreated wastewater is extremely hazardous to both field workers and consumers; the former are particularly susceptible to helminth infections and the latter to bacterial infections such as cholera and typhoid fever. To reduce the risk of infection, the WHO has set strict guidelines, which were updated in 2006, for the required reduction of pathogens wastewater should receive when used for crop irrigation. These guidelines are based on an integrated assessment procedure known as the Stockholm Framework. This has resulted in health-based guidelines that have been adopted internationally. The approach combines risk assessment and risk management, as pathogens can remain viable for long periods before and after irrigation, depending primarily on the degree of wastewater treatment and irrigation method. The guidelines identify three risk groups: consumers, farm-workers and their immediate families, and communities and properties that are adjacent or nearby irrigation areas (Table 5.8).

Table 5.8. Primary health risks from bacterial, viral, protozoal, and helminth infections by the risk group (WHO, 2006). Reproduced with permission from the World Health Organization under the WHO CC BY-NC-SA 3.0 IGO license.

	Health risks		
Group exposed	**Helminth infections**	**Bacterial/virus infections**	**Protozoal infections**
Consumers	Significant risk of *Ascaris* infection for both adults and children reported with untreated wastewater	Cholera, typhoid and shigellosis outbreaks reported from the use of untreated wastewater; seropositive responses for *Helicobacter pylori* (untreated); increase in non-specific diarrhoea when water quality exceeds 10^4 thermotolerant coliforms/100 ml	Evidence of parasitic protozoa found on wastewater-irrigated vegetable surfaces, but no direct evidence of disease transmission
Farm workers and their families	Significant risk of *Ascaris* infection for both adults and children in contact with untreated wastewater; risk remains, especially for children, when wastewater treated to <1 nematode egg per liter; increased risk of hookworm infection in workers	Increased risk of diarrhoeal disease in young children with wastewater contact if water quality exceeds 10^4 thermotolerant coliforms/100 ml; elevated risk of *Salmonella* infection in children exposed to untreated wastewater; elevated seroresponse to norovirus in adults exposed to partially treated wastewater	Risk of *Giardia intestinalis* infection was found insignificant for contact with both untreated and treated wastewater; increased risk of amoebiasis observed with contact with untreated wastewater
Nearby communities	*Ascaris* transmission not studied for sprinkler irrigation, but same as above for flood or furrow irrigation with heavy contact	Sprinkler irrigation with poor water quality (10^6–10^8 TC/100 ml) and high aerosol exposure associated with increased rates of infection; use of partially treated water (10^4–10^5 thermotolerant coliforms/100 ml or less) in sprinkler irrigation not found to be associated with increased viral infection rates	No data on transmission of protozoan infections during sprinkler irrigation with wastewater

TC, total coliforms

Crops are categorized as being suitable for either restricted or unrestricted irrigation techniques. Restricted irrigation applies to crops not grown for direct human consumption or food crops that are not eaten raw. Unrestricted irrigation is required for vegetables and salad crops that are eaten raw. The health-based targets for irrigation for unrestricted and restricted irrigation are given in Table 5.9 with guideline values for fecal coliforms of <1,000 FC 100 mL^{-1} and a 6 to 7 log_{10} reduction in viruses when used on crops that are eaten raw, such as lettuce or celery (WHO, 1989; 2006). Viruses are often used to model the potential risk of pathogen transfer from irrigation. The WHO has used the rotavirus to calculate the risk in the guidelines, as this virus is most likely to offer a higher risk compared to other pathogen groups. Ito *et al.* (2017) give an example of how risk assessment is calculated for the use of wastewater for irrigation using norovirus as model organisms.

The WHO quality guidelines of <1 egg L^{-1} in wastewater used for both restricted and unrestricted irrigation has been criticized as being too high for restricted irrigation systems. Ayres *et al.* (1992) have proposed that nematode guidelines for restricted irrigation be relaxed to 10 eggs L^{-1}. Waste stabilization ponds are typically used in regions where effluents are reused. For that reason, pond design is normally based on required levels of helminth egg removal. Mara and Pearson (1998) give design examples for waste stabilization ponds where effluents are to be used for restricted and unrestricted irrigation. Where effluents are discharged to fishponds, no trematode eggs are allowed due to the high sexual multiplication of the *Schistosoma* spp. in water snails, the intermediate host.

There are a number of health protection measures that must be considered to ensure that the risk of pathogen infection is minimized. These include: (i) adequate and effective wastewater treatment, (ii) careful selection and possible restriction of the crops grown, (iii) the irrigation application technique should minimize the risk of pathogen transfer such as the use of drip irrigation instead of spraying reducing contamination of plant surfaces as well as the formation of aerosols, (iv) employing periods when no wastewater is applied to ensure sufficient die-off of pathogens, (v) good hygiene practice in fields, during harvesting, transport and at food markets, (vi) health

Table 5.9. Health-based targets for wastewater irrigation in the production of agricultural crops (WHO, 2006). Reproduced with permission from the World Health Organization under the WHO CC BY-NC-SA 3.0 IGO license.

Category	Reuse conditions	Exposed group	Irrigation technique	Intestinal nematodes[b] (arithmetic mean number of eggs per liter[c])	Fecel coliforms (geometric mean number per 100 ml[d])	Wastewater treatment expected to achieve required microbiological quality
A	Unrestricted irrigation A1 For vegetable and salad crops eaten uncooked, sports fields, public parks[e]	Workers, consumers, public	Any	0.1[f]	10^3	Well-designed series of waste stabilization ponds (WSPs), sequential batch-fed wastewater storage and treatment reservoirs (WSTRs) or equivalent treatment (e.g. conventional secondary treatment supplemented by either polishing ponds or filtration and disinfection)
B	Restricted irrigation Cereal crops, industrial crops, fodder crops, pasture and trees[g]	B1 Workers (but no children <15 years), nearby communities	Spray or sprinkler	1	10^5	Retention in WSP series including one maturation pond or in sequential WSTRs or equivalent treatment (e.g. conventional secondary treatment supplemented by either polishing ponds or filtration)
		B2 as B1	Flood/furrow	1	10^3	As for Category A

(*Continued*)

Table 5.9. (*Continued*)

Category	Reuse conditions	Exposed group	Irrigation technique	Intestinal nematodes[b] (arithmetic mean number of eggs per liter[c])	Fecel coliforms (geometric mean number per 100 ml[d])	Wastewater treatment expected to achieve required microbiological quality
		B3 Workers including children <15 years, nearby communities	Any	0.1	10^3	As for Category A
C	Localized irrigation of crops in category B if exposure of workers and the public does not occur	None	Trickle, drip or bubbler	Not applicable	Not applicable	Pretreatment as required by the irrigation technology, but not less than primary sedimentation

[a] In specific cases, local epidemiological, sociocultural and environmental factors should be taken into account and the guidelines modified accordingly.
[b] *Ascaris* and *Trichuris* species and hookworms; the guideline limit is also intended to protect against risks from parasitic protozoa.
[c] During the irrigation season (if the wastewater is treated in WSPs or WSTRs, which have been designed to achieve these egg numbers, then routine effluent quality monitoring is not required).
[d] During the irrigation season (fecel coliform counts should preferably be done weekly, but at least monthly).
[e] A more stringent guideline limit (200 fecel coliforms/100 ml) is appropriate for public lawns, such as hotel lawns, with which the public may come into direct contact.
[f] This guideline limit can be increased to 1 egg/l if (i) conditions are hot and dry and surface irrigation is not used or (ii) if wastewater treatment is supplemented with anthelmintic chemotherapy campaigns in areas of wastewater reuse.
[g] In the case of fruit trees, irrigation should stop two weeks before fruit is picked, and no fruit should be picked off the ground. Spray/sprinkler irrigation should not be used.

awareness and hygiene education for all stakeholders, (vii) washing and cooking all produce as well as disinfection of food preparation areas, and (viii) immunization. For those actively engaged in wastewater irrigation, training should include health awareness and disease transmission prevention. Personal protective equipment, access to clean areas, water and changing areas, personal immunization, and the control of disease vectors and intermediate hosts are all essential requirements (Fig. 5.9). All these can be achieved by carrying out a

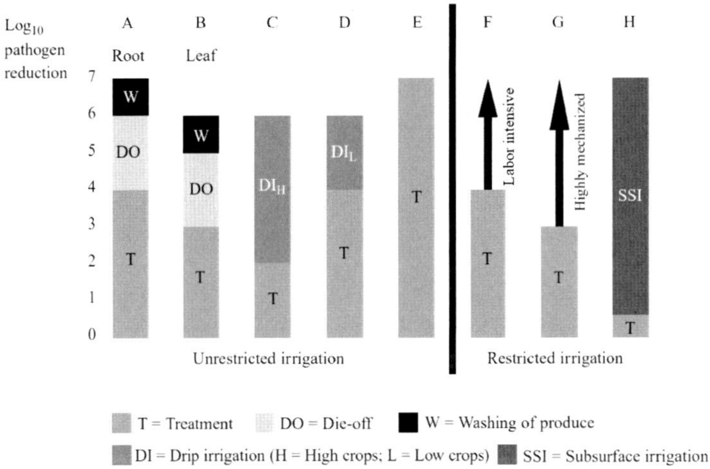

Fig. 5.9. Log_{10} pathogen reduction achieved using different combinations of health protection measures for unrestricted and restricted irrigation as recommended by the WHO (2006). The degree of wastewater treatment required depends on what other health protection measures are available. For example: (a) root crops that are eaten uncooked (e.g. onions), (b) non-root salad or vegetable crops eaten uncooked (e.g. lettuce), (c) high-growing crops (e.g. tomatoes), (d) low-growing non-root crops, (e) advanced wastewater treatment for any crop, (f) labor-intensive restricted irrigation, (g) for highly mechanized agriculture, and (h) single household or small hotel using subsurface irrigation after septic tank treatment. Verification levels of wastewater treatment performance (*Escherichia coli* per 100 mL^{-1}) required for each option are $\leq 10^3$, $\leq 10^4$, $\leq 10^5$, $\leq 10^3$, $\leq 10^1$, and $\leq 10^0$ for the unrestricted irrigation options (a) to (e), respectively, and $\leq 10^4$, $\leq 10^5$ and $\leq 10^6$, for the restricted irrigation options (f) to (h), respectively. Reproduced with permission from the World Health Organization under the WHO CC BY-NC-SA 3.0 IGO license.

proper risk assessment at each location, which should also include the whole life cycle of the crops produced (WHO, 2006).

Similar quality guidelines have been implemented in many countries such as Australia, the European Union, and the US (Environment Protection Agency Victoria, 2003; Marecos do Monte, 2007; The State of Queensland, 2008; Angelakis *et al.*, 2008; Brissaud, 2008; Ortega and Iglesias, 2009; Agrafioti and Diamadopoulos, 2010; USEPA, 2012; Kalavrouziotis *et al.*, 2015; Paranychianaki *et al.*, 2015). A number of different guidelines have been compared by Jeong *et al.* (2006) (Table 5.10), which are constantly being reviewed and updated (Gerba *et al.*, 2017).

Chlorinated secondary wastewater effluents are widely used in southern US for the irrigation of golf courses, school grounds, parks, and cemeteries, although UV disinfection is also used (Bourrouet *et al.*, 2001; Salgot *et al.*, 2012; Rahman *et al.*, 2016). More stringent guidelines are required for such areas of grass, as the public is more likely to come into direct contact with pathogens in these open spaces (DeBoer and Linstedt, 1985). Krishnazwami (1971) had earlier proposed that sewage effluents be treated and disinfected before use and contain <100 coliforms or <20 FC 100 mL^{-1}.

Plant pathogens are spread by the reuse of infected water or sewage, and plant-pathogenic bacteria and fungi, and nematodes can remain viable for many years in the soil (Steadman *et al.*, 1975). Plant pathogens arise from vegetable waste and soil that are discharged with wastewater, which includes viruses, fungi, bacteria, and nematodes. The use of sewage for irrigation, and especially the use of sewage sludge on agricultural land, helps the dissemination of plant pathogens back to agricultural soils. The most important plant pathogens are the cyst-forming nematodes belonging to the genus *Heterodera*. Seven species are regularly isolated from sewage sludge (Table 5.11), the most important being the potato cyst eelworms, *H. rostochiensis* and *H. pallida,* which are endemic in Europe. In the soil, the nematode is present as cysts that contain the eggs of the parasite. Even in the absence of the host plant, these cysts can remain viable for many years and infection is particularly difficult to eradicate. The parasite is easily dispersed via contaminated soil, and in wet weather, when

Table 5.10. Comparison of irrigation water quality standards and guidelines for wastewater irrigation in agriculture (Jeong *et al.*, 2016). Reproduced with permission from the Multidisciplinary Publishing Institute under the Creative Commons CC BY 4.0 license.

Parameters	South Korea[1]		WHO[2]		US EPA		Crprus[3]		France		Greece		Israel[4]	Italy	Portugal		Spain	
Coliform (/100 mL)	Food crops	ND TC	Unrestricted	*E. coli* (cfu) 1000	Food crops	ND FC (median)	Cooked vegetables	FC (MPN) 100	Unrestricted	*E. coli* (cfu) 250	Unrestricted	*E. coli* (cfu) 5 (80%) 50 (95%)	FC (cfu) 10	*E. coli* (cfu) 100 (max) 10 (80%)	Vegetables consumed raw[a]	FC (cfu) 100	Uncooked vegetables	*E. coli* (cfu) 100
	Processed food crops	TC (MPN) 200	Restricted	*E. coli* (cfu) 10,000	Processed food crops	FC (cfu) 200 (median)	Crops for human consumption	FC (MPN) 1000	All crops except those consumed raw	*E. coli* (cfu) 10,000	Restricted	*E. coli* (cfu) 200 (median)			Cooked vegetables	FC (cfu) 1000	Crops for human consumption	*E. coli* (*cfu*) 1000
Turbidity (NTU)	Food crops	2	—(b)		Food crops	2 (average)	—		—		Unrestricted	2 (median)	—	—	—		Uncooked vegetables	10
	Processed food crops	5			Processed food crops	—					Restricted	—		—			Crops for human consumption	—
Suspended solids (mg/L)	—		—		Food crops	—	Cooked vegetables	15	Unrestricted	15	Unrestricted	10(80%)	TSS 10	TSS 10	TSS	60	Uncooked vegetables	20
					Processed food crops	TSS 30	Crops for human consumption	45	All crops except those consumed raw	Varies [c]	Restricted	35					Crops for human consumption	35
BOD (mg/L)	8		—		Food crops	10	Cooked vegetables	15	—		Unrestricted	10(80%)	10	20	—		—	—
					Processed food crops	30	Crops for human consumption	30			Restricted	25					—	—

(*Continued*)

Table 5.10. (*Continued*).

Parameters	South Korea[1]	WHO[2]	US EPA	Crprus[3]	France	Greece	Israel[4]	Italy	Portugal	Spain
COD (mg/L)	—	—	—	—	Unrestricted: 60 All crops except those consumed raw: Varies	—	100	100	—	—
Odor	Not unpleasant	—	—	—	—	—	—	—	—	—
T-N (mg/L)	—	—	—	—	—	—	25	15	—	—
T-P (mg/L)	—	—	-	—	—	—	5	2	—	—
Intestinal nematodes (No./L)	—	1	—	ND	—	—	—	—	1	l(/10L)
pH	5.8–8.5	—	6.0–9.0	—	—	—	6.5–8.5	6.0–9.5	65–8.4	—
EC (μs/cm)	Food crops: 700 Processed food crops: 2,000	—	—	—	—	—	1400	3000	1000	—

ND: not detected; TC: total coliform; FC: fecal coliform; TSS: total suspended solids. [1] Standards for direct wastewater reuse. [2] The most stringent verification monitoring level, which refers to what has previously been referred to as effluent guideline levels, for each irrigation type and arithmetic mean value. [3] For vegetables eaten raw is not allowed and maximum value allowed. [4] Maximum monthly averages for unrestricted irrigation. [(a)] Vegetables whose edible parts are in close contact with the irrigated soil are not included and drip irrigation can be only employed. [(b)] No recommendation. [(c)] In accordance with wastewater treatment standards.

Table 5.11. Species of *Heterodera* associated with wastewater reuse.

Species	Name	Host crops
H. avenae	Cereal cyst-nematode	Oats, barley, wheat
H. carotae	Carrot cyst-nematode	Carrot
H. cruciferae	Cabbage cyst-nematode	Brassicas and other plants
H. goeltingiana	Pea cyst-nematode	Peas and beans
H. pallida	Pale potato cyst-nematode	Potato, tomato
H. rostochiensis	Potato root eelworm	Potato, tomato
H. schachtii	Beet cyst-nematode	Sugar beet and fodder beet

vegetables retain a greater proportion of soil on their surface there is a particular risk of spreading the disease. Cysts accumulate in the soil washed from potatoes, sugar beet, and carrots during processing, while the sludge from vegetable-processing factories must be carefully disposed of if *Heterodera* contamination of cyst-free soil is to be avoided. Potato growing is excluded from heavily infected soils, but the danger that sludge could introduce the pathogen to cyst-free soils is real. The EEC directive that concerns potato growing stipulates that potatoes grown for export can only be produced on certified cyst-free soil. Several species of cyst nematodes, including *H. rostochiensis* and *H. pallida*, can also parasitize the roots of tomatoes. If tomato plants are allowed to grow on sludge beds or sludge dumps, or even as weeds, after the application of sludge to the land, this would greatly increase the number of cysts in the sludge or soil. Sludge can be examined for cysts by using a simple flotation method. So long as the level of infection in the soil remains below 25 cysts per 100 g of dry soil, no significant losses in production will occur (Linfield, 1977).

References

Agrafioti, E. and Diamadopoulos, E. (2010) A strategic plan for reuse of treated municipal wastewater for crop irrigation on the Island of Crete. *Agricultural Water Management*, **105**, 57–64.

Allen, R.G., Pereira, L.S., Raes, D. and Smith, M. (1998) Crop evapotranspiration: Guidelines for computing crop water requirements. FAO Irrigation and Drainage Paper 56. Food and Agriculture Organization of the United Nations, Rome.

Angelakis, A.N., Durham, B., Marecos Do Monte, M.H.F., Salgot, M., Witgens, T. and Thoeye, C. (2008) Wastewater recycling and reuse in European countries: Trends and challenges. *Desalination*, **218**, 3–12.

Ayers, R.S. and Westcot, D.W. (1985) Water quality for agriculture. FAO Irrigation and Drainage Paper 29. Food and Agriculture Organization of the United Nations, Rome.

Ayres, R.M., Stott, R., Lee, D.L., Mara, D.D. and Silva, S.A. (1992) Contamination of lettuces with nematode eggs by spray irrigation with treated and untreated wastewater. *Water Science and Technology*, **26**(7/8), 1615–1623.

Becerracastro, C., Lopes, A.R., Vazmoreira, I., Silva, E.F., Manaia, C.M. and Nunes, O.C. (2015) Wastewater reuse in irrigation: A microbiological perspective on implications in soil fertility and human and environmental health. *Environment International*, **75**(4), 117–135. https://doi.org/10.1016/j.envint.2014.11.001

Boll, R. and Kayser, R. (1986) Reuse of wastewaters in agriculture by means of sprinkler irrigation on a farmland area of 300 ha: Large scale experience in Germany. *Water Science and Technology*, **18**(9), 163–173.

Bond, W.J. (1998) Effluent irrigation — An environmental challenge for soil science. *Australian Journal of Soil Research*, **36**(4), 543–555.

Bourrouet, A., Garcia, J., Mujeriego, R. and Penuelas, G. (2001) Faecal bacteria and bacteriophage inactivation in a full-scale UV disinfection system used for wastewater reclamation. *Water Science and Technology*, **43**(10), 187–194.

Brissaud, F. (2008) Criteria for water recycling and reuse in the Mediterranean countries. *Desalination*, **218**, 24–33.

Bruce, A.M. and Hawkes, H.A. (1983) Biological filters. In *Ecological Aspects of Used-Water Treatment, Vol. 3. The Processes and Their Ecology* (eds. C.R. Curds and H.A. Hawkes). Academic Press, London, pp. 1–111.

De Boer, I. and Linstedt, K.D. (1985) Advances in water reuse applications. *Water Research*, **19**, 1455–1461.

Dean, R.B. and Lund, E. (1981) *Water Reuse: Problems and Solutions.* Academic Press, London.

Doorenbos, J. and Pruitt, W.O. (1977) Crop requirements. FAO Irrigation and Drainage Paper 24. Food and Agricultural Organization of the United Nations, Rome.

Ellis, B.G. (1974). The soil as a chemical filter. In: *Conference on Recycling Treated Municipal Wastewater Through Forest and Cropland* (eds. W.E. Sopper, L.T. Kardos and R.E. Thomas). Report: EPA660/2-74-002. US Environmental Protection Agency, Cincinnati, OH, USA, pp. 47–72.

Environment Protection Agency Victoria. (2003) Guidelines for environmental management: Use of reclaimed water. Environment Protection Agency, Victoria. http://www.epa.vic.gov.au/~/media/Publications/464 2.pdf

Fattal, B., Yekutiel, P. and Shuval, H.I. (1986) Cholera outbreak in Jerusalem 1970, revisited: The evidence for transmission by wastewater irrigated vegetables. In *Environmental Epidemiology: Epidemiological Investigation of Community Environmental Health Problems* (ed. J.R. Goldsmith). CRC Press, Boca Raton, FL, pp. 49–59.

Fuller, R.J. and Glue, D.E. (1980). Sewage works as bird habitats in Britain. *Biological Conservation*, **17**, 165–181.

Fuller, R.J. and Glue, D.E. (1981). The impact on bird communities of the modernisation of sewage treatment works. *Effluent and Water Treatment Journal*, **21**, 27–31.

Geldreich, E.E., Nash, H.D., Reasoner, D.J. and Taylor, R.H. (1972) The necessity of controlling bacterial populations in potable supplies: Community water supply. *Journal of the American Water Works Association*, **64**(9), 596–602.

Gerba, C.P., Betancourt, W.Q. and Kitajima, M. (2017) How much reduction of virus is needed for recycled water: A continuous changing need for assessment? *Water Research*, **108**, 25–31. http://dx.doi.org/10.1016/j.watres.2016.11.020

Gibson, K.E. (2014) Viral pathogens in water: Occurrence, public health impact, and available control strategies. *Current Opinion in Virology*, **4**, 50–57. http://dx.doi.org/10.1016/j.coviro.2013.12.005

Ibrahim, T., Meehan, B. and Carpenter, P. (2007) TDS in sewage: Sources and trends in Melbourne's west. *Water*, **34**(5), 44–47.

Ito, T., Kitajima, M., Kato, T., Ishii, S., Segawa, T., Okabe, S. and Sano, D. (2017) Target virus $\log_{10}$ reduction values determined for two reclaimed wastewater irrigation scenarios in Japan based on tolerable annual disease burden. *Water Research*, **125**, 438–448.

Jeong, H., Kim, H. and Jang, T. (2016) Irrigation water quality standards for indirect wastewater resuse in agriculture: A contribution towards sustainable wastewater reuse in South Korea. *Water*, **8**, 169. doi:10.3390/w8040169

Johnson, R.D. (1973) Land treatment of wastewater. *The Military Engineer*, **65**, 375.

Kalavrouziotis, I.K., Kokkinos, P., Oron, G., Fatone, F., Bolzonella, D., Vatyliotou, M., Fatta-Kassinos, D., Koukoulakis, P.H. and Varnavas, S.P. (2015) Current status in wastewater treatment, reuse and research in some Mediterranean countries. *Desalination and Water Treatment*, **53**, 2015–2030.

Krishnazwami, S.K. (1971) Health aspects of land disposal of municipal wastewater efflunets. *Canadian Journal of Public Health*, **62**(1), 36–41.

Larkin, E.P., Tierney, J.T. and Sullivan, R. (1976). Persistence of virus on sewage irrigated vegetables. *Journal of the Environmental Engineering Division,* **102**(EE1), 29–55.

Li, P., Wang, X., Allinson, G., Li, X. and Xiong, X. (2009) Risk assessment of heavy metals in soils previously linked with industrial wastewater in Shenyang, China. *Journal of Hazardous Materials*, **161**, 516–521.

Li, X., Zhang, W., Liu, T., Chen, L., Chen, P. and Li, F. (2016) Changes in the composition and diversity of microbial communities during anaerobic nitrate reduction and Fe(ii) oxidation at circumneutral pH in paddy soil. *Soil Biology Biochemistry*, **94**, 70–79. https://doi.org/10.1016/j.soilbio.2015.11.013

Linfield, R. (1977) Potato cyst eelworm studies in the Anglian Water Authority. In *Research Seminar on Pathogens in Sewage Sludge*. Technical Note No. 7. Department of the Environment, Research and Development Division, London.

Liu, W.H., Zhao, J.Z., Ouyang, Z.Y., Soderlund, L. and Liu, G.H. (2005) Impacts of sewage irrigation on heavy metal distribution and contamination in Beijing, China. *Environment International*, **31**(6), 805–812.

Mara, D.D. and Pearson, H.W. (1998) Design manual for waste stabilization ponds in Mediterranean countries. Lagoon Technology International, Leeds, UK.

Marecos do Monte, M.H.F. (2007) Guidelines for good practice of water reuse for irrigation: Portuguese standard NP 4434. In *Wastewater Reuse-Risk Assessment, Decision-Making and Environmental Security* (ed. M. Zaidi). Springer, Dordrecht, pp. 253–265.

Miller, R.H. (1974). The soil as a biological filter. In: *Conference on Recycling Treated Municipal Wastewater Through Forest and Cropland* (eds. W.E. Sopper, L.T. Kardos and R.E. Thomas). Report: 660/2-74-003.

US Environmental Protection Agency, Cincinnati, OH, USA, pp. 73–94.

Müller, K., Magesan, G.N. and Bolan, N.S. (2007) A critical review of the influence of effluent irrigation on the fate of pesticides in soil. *Agriculture, Ecosystems and Environment*, **120**(2–4), 93–116.

Muyen, Z., Moore, G.A. and Wrigley, R.J. (2011) Soil salinity and sodicity effects of wastewater irrigation in South East Australia. *Agricultural Waste Management*, **99**, 33–41.

Ortega, E. and Iglesias, R. (2009) Reuse of treated municipal wastewater effluents in Spain: Regulations and most common technologies, including extensive treatments. *Desalination and Water Treatment*, **4**, 148–160.

Paranychianakis, N.V., Angelakis, A.N., Leverenz, H. and Tchobanoglous, G. (2006) Treatment of wastewater with Slow Rate systems: A review of treatment process and plant functions. *Critical Reviews in Environmental Science and Technology*. **36**, 187–259.

Paranychianakis, N.V., Salgot, M., Snyder, S.A. and Angelakis, A.N. (2015) Water reuse in EU states: Necessity for uniform criteria to mitigate human and environmental risks. *Critical Review of Environmental Science and Technology*, **45**, 1409–1468.

Pettygrove, G. and Asano, T. (1984). Irrigation with reclaimed wastewater: A guidance manual. Third Water Reuse Symposium. American Water Works Association, Denver, CO.

Rahman M.M., Hagare D. and Maheshwari B. (2016) Use of recycled water for irrigation of open spaces: Benefits and risks. In *Balanced Urban Development: Options and Strategies for Liveable Cities* (eds. B. Maheshwari, B. Thoradeniya and V.P. Singh). Springer, Cham. https://doi.org/10.1007/978-3-319-28112-4_17

Rowe, D.R. and Abdel-Magid, I.M. (1995) *Handbook of Wastewater Reclamation and Reuse*. CRC Press Inc., Boca Raton, FL.

Salgot, M., Priestley, G.K. and Folch, M. (2012) Golf course irrigation with reclaimed water in the Mediterranean: A risk assessment matter. *Water*, **4**, 389–429. https://www.mdpi.com/2073-4441/4/2/389/htm

Sano, D., Amarasiri, M., Hata, A., Watanabe, T. and Katayama, H. (2016) Risk management of viral infectious diseases in wastewater reclamation and reuse: Review. *Environmental International*, **91**, 220–229. http://dx.doi.org/10.1016/j.envint.2016.03.001

Shahid S.A., Zaman M. and Heng L. (2018a) Introduction to soil salinity, sodicity and diagnostics techniques. In *Guideline for Salinity Assessment,*

Mitigation and Adaptation Using Nuclear and Related Techniques (eds. M. Zaman, S. Shahid and L. Heng). Springer, Cham, pp. 1–42.

Shahid S.A., Zaman M. and Heng L. (2018b) Soil salinity: Historical perspectives and a world overview of the problem. In *Guideline for Salinity Assessment, Mitigation and Adaptation Using Nuclear and Related Techniques* (eds. M. Zaman, S. Shahid and L. Heng). Springer, Cham, pp. 43–53.

Sheikh, B., Jacques, R.S. and Cooper, R.C. (1984) Wastewater effluent reuse for irrigation of raw-eaten food crops: A five year study. Third Water Reuse Symposium. American Water Works Association, Denver, CO.

Shuval, H.I. and Fattal, B. (1980). Epidemiologic study of wastewater irrigation. In *Wastewater Aerosols and Disease* (eds. H. Pahren and W. Jakubowski). EPA 600/9–80–028. Health Effects Research Laboratory, US EPA, Cincinnati, OH, pp. 228–238.

Shuval, H.I., Lampert, Y. and Fattal, B. (1997) Development of a risk assessment approach for evaluating wastewater reuse standards for agriculture. *Water Science and Technology*, **35**(11–12), 15–20.

Singh, A., Sharma, R.K., Agrawal, M. and Marshall, F.M. (2010) Risk assessment of heavy metal toxicity through contaminated vegetables from waste water irrigated area of Varanasi, India. *Tropical Ecology*, **51**(2), 375–387. https://doi.org/10.1016/j.fct.2009.11.041

Steadman, J.R., Maier, C.R., Schwartz, H.F. and Kerr, E.D. (1975) Pollution of surface irrigation waters by plant pathogenic organisms. *Water Research Bulletin*, **11**, 796–804.

Stine, S.W., Song, I., Chol, C.Y. and Gerba, C.P. (2005) Effect of relative humidity on preharvest survival of bacterial and viral pathogens on the surface of cantaloupe, lettuce and bell peppers. *Journal of Food Production*, **68**(7), 1352–1358.

The State of Queensland. (2008) Water quality guidelines for recycled water schemes. The State of Queensland. https://www.dews.qld.gov.au/__data/assets/pdf_ file/0019/45172/water-quality-guidelines.pdf

Tierney, J.T., Sullivan, R. and Larkin, E.P. (1977) Persistence of poliovirus I in soil and on vegetables grown in soil previously flooded with inoculated sewage sludge or effluent. *Applied Environmental Microbiology*, **33**, 109–113.

USDA. (1954) *Diagnosis and Improvement of Saline and Alkaline Soils.* United States Department of Agriculture, Washington, DC.

USEPA. (2006) Land treatment of municipal wastewater effluents. Process design manual. EPA/625/R-06/016. US Environmental Protection Agency, Cincinnati, OH.

USEPA. (2012) Guidelines for water reuse. Report: 600/R-12/618. US Environmental Protection Agency, Washington, DC.

Verbyla, M.E., Symonds, E.M., Kafle, R.C., Cairns, M.R., Iriarte, M., Guzmán, A.V., Coronado, O., Breitbart, M., Ledo, C. and Mihelcic, J.R. (2016) Managing microbial risks from indirect wastewater reuse for irrigation in urbanizing watersheds. *Environmental Science and Technology,* **50**(13), 6803–6813. https://doi.org/10.1021/acs.est.5b05398

Wang, M., Chen, S., Chen, L., Wang, D. and Zhao, C. (2019) The responses of a soil bacterial community under saline stress are associated with Cd availability in long-term wastewater-irrigated fields. *Chemosphere*, **232**, 124372.

Wang, Q., Dong, Y., Cui, Y. and Liu, X. (2001) Instances of soil and crop heavy metal contamination in China. *Soil and Sediment Contamination*, **10**, 497–510.

WHO. (1989) Health guidelines for the use of wastewater in agriculture and aquaculture. Technical Report Series: **778**. World Health Organization, Geneva.

WHO. (2006) Guidelines for the safe use of wastewater, excreta and greywater — Volume 2: *Wastewater Use in Agriculture*. World Health Organization, Geneva. http://www.who.int/water_sanitation_health/wastewater/wwuvol2intro.pdf

Woods Ballard, B., Wilson, S., Udale-Clarke, H., Illman, S., Scott, T., Ashley, R. and Kellagher, R. (2015) The SuDS manual (C753). CIRIA, London.

Zaman, M., Shahid, S. and Heng, L. (2018) *Guideline for Salinity Assessment, Mitigation and Adaptation using Nuclear and Related Techniques.* Springer, Cham.

6

Subsurface Wastewater Infiltration Systems

6.1 Introduction

On-site subsurface wastewater infiltration systems (SWISs) are known more widely as either leach fields or percolation areas. There is estimated to be at least 250 million SIWSs worldwide, with the majority used for treating primary treated wastewater from individual houses. Their function is different from those used for stormwater management and sustainable urban drainage systems (Woods Ballard *et al.*, 2015). The majority of SWISs are employed for individual homes or small businesses to allow the primary settled wastewater from either a septic tank or Imhoff tank, or secondary settled wastewater, to seep through the soil and into the water table. This is achieved by a system of narrow trenches (0.6 to 1.5 m deep) containing perforated or open jointed distribution pipes, which are laid on a porous medium, 20 to 30 mm diameter gravel or clean-crushed stone, to allow the water to freely drain away (Fig. 6.1). The role of the medium is to maintain the structure of the trenches to provide partial treatment of the effluent by acting as a substrate for the growth of heterotrophs, ensure even distribution of the effluent to the soil interface, and provide for the temporary storage of effluent during peak flows from the tank. Problems associated with the use of gravel or crushed-stone media (18 to 64 mm in diameter) have been reported by May (1996). These include physical damage to the

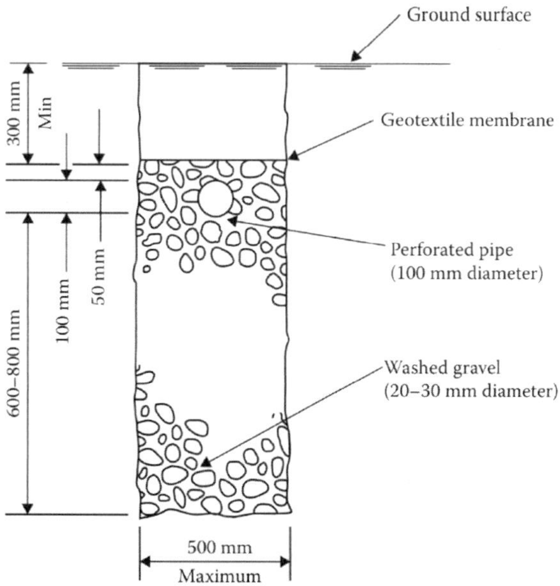

Fig. 6.1. Cross-section of a percolation area disposal trench.

infiltrative surface, compaction, negative effects of entrained fines, poor storage volume, native soil, or root intrusion into the trench, and the promotion of excessive biofilm development within the interstices of the medium or biomat formation at the trench interface with the soil reducing leaching rates. According to May, this leads to a significantly reduced long-term acceptance (infiltration) rate (LTAR) of natural soil. However, the use of such media is universal in subsurface infiltration systems, and although the United States Environmental Protection Agency (USEPA, 2003) allows the use of gravelless distribution systems where drainage pipes are laid in either open-bottom chambers, wrapped in geotextile material, or laid in polystyrene chips, washed stone medium remains the most common option. Trenches should also have a large length to width ratio with narrow infiltration surfaces. A major problem with irrigation systems is maintaining sufficient soil aeration around and

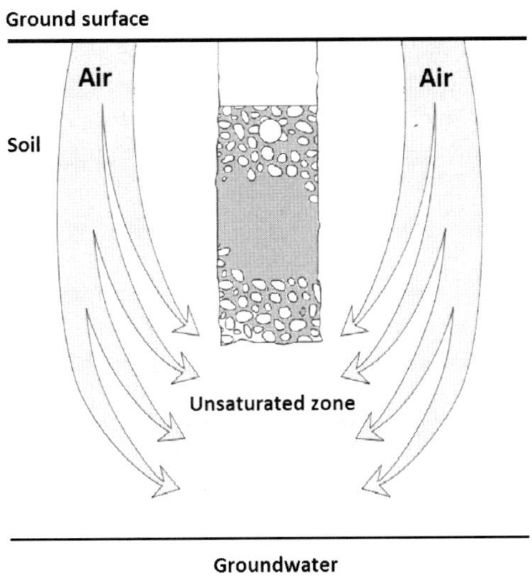

Fig. 6.2. Biological oxidation and nitrification within subsurface wastewater infiltration systems depend on oxygen diffusion from the soil surface (USEPA, 2003). Oxygen status within the pipework and trench depends on adequate ventilation of the distribution network as well as the depth of effluent held in the trench. However, some modern distribution systems are often not ventilated, resulting in anoxic conditions within the pipework and trench, with methane and hydrogen sulfide building up within the septic tank, creating a hazard. Reproduced with permission from the US Environmental Protection Agency.

especially below the trench. Oxygen diffuses from the soil surface of the percolation area to the soil adjacent to the infiltration surface of the trench (Fig. 6.2). The wider the trench, the more likely anaerobic conditions will develop due to the high oxygen demand of the biota oxidizing the organic matter in the effluent (Siegrist *et al.*, 1986). The effluent is normally fed by gravity to subsurface infiltration areas serving individual homes, with the flow being intermittent. Larger systems should be dosed periodically using either a siphon or pump (Laak, 1986).

Subsurface drip irrigation also falls under the definition of a SWIS but is used primarily for agricultural irrigation, usually of high-value crops such as fruit. Subsurface drip irrigation uses either surface or groundwater to drip feed individual plants at the roots, ensuring minimal evaporation losses and maximum uptake of water by the plant. In this way, optimal use is made of available water resources and also prevents weed development. Wastewaters are not normally used due to problems of solids and biofilm development within the pipework, leading to blockages at the outlets. All surface water has to be filtered to ensure that the outlets do not become blocked. Specialized drip tubes (5 to 10 mm in internal diameter) or tapes are buried and then fed by a low pressure (7–14 psi) using either mains pressure, simple pumps, or the hydraulic head from a reservoir. Soluble fertilizer can also be added to the water, which is highly cost-effective and reduces the risk of nutrient losses from soil-causing pollution. The main use of subsurface drip irrigation is in horticulture, especially greenhouses and polytunnels, but also in lawn irrigation in arid areas, although maintenance costs are high. A particular problem is the longevity of the tubes or tape when used in commercial situations, with cheaper materials having to be replaced within 18 to 36 months. The Biocycle® system, developed in Australia, is a compact integrated household wastewater treatment system comprising a single tank separated into different sections. A septic tank area is followed by a submerged aerated biofilter that feeds into a settlement tank area, which discharges into a holding tank via a passive disinfection system. The highly treated, disinfected, final effluent is then pumped out of the tank via a system of subsurface or surface tubes used for water hedging or ornamental flowerbeds. The United States Department of Agriculture (USDA, 2013; 2016) has reviewed the design and operation of these micro-irrigation systems.

6.2 Design of the SWIS

The area required for infiltration is based on initial soil permeability and groundwater conditions. In the United States (US), it is calculated as:

$$A = 1.5Q/k, \tag{6.1}$$

Table 6.1. US Environmental Protection Agency's recommended hydraulic loading rates for on-site subsurface wastewater infiltration systems (USEPA, 1980). Reproduced with permission from the US Environmental Protection Agency.

Range of soil type	Percolation rate (min. per 100 mm)	Hydraulic loading based on area of trench ($L\ m^{-2}d^{-1}$)
Gravel, coarse sand	<4	Not recommended
Coarse, medium sand	4–20	48
Fine sand, loam sand	21–60	32
Sandy loam, loam	61–120	24
Loam, porous silt loam	121–240	18
Silty loam, clay loam	241–480	8
Clays, colloidal clays	>480	Not recommended

where A is the infiltration area (m^2), Q is the average wastewater flow (m^3d^{-1}), and k is the permeability of the soil below the infiltration trench ($m^3m^{-2}d^{-1}$) (WPCF, 1990). The USEPA (1980) gives guidelines for hydraulic loading rates based on approximate percolation rates (Table 6.1). As soil infiltration rates can fall dramatically over time, there is some concern about the long-term validity of using percolation tests to determine hydraulic loading rates. However, in practice, percolation tests are useful in identifying unsuitable and problem soils, as well as giving useful guidance about the relative infiltration area required for particular soil conditions.

The British Code of Practice (BS6297:2007+A1:2008) uses a percolation test to determine the value V_p, which is the average time (in seconds) required for the water level to drop 1 mm. A percolation test hole is dug to a minimum of 0.3 m below the proposed invert level of the infiltration pipe, filled with water, and the time is then measured in seconds for the water to seep away completely (BSI, 1983; 2008). The mean V_p must be between 15 and 100 s for the construction of a percolation area and is used to calculate the floor area of the subsurface drainage trench A_t in m^2, where septic tank effluent is being used:

$$A_t = P \times V_p \times 0.25, \tag{6.2}$$

where P is the population equivalent. For effluents that have received secondary treatment followed by settlement, this area is reduced by 20%:

$$A_t = P \times V_p \times 0.20. \tag{6.3}$$

In the United Kingdom (UK), subsurface infiltration areas are permitted only where the geology and hydrogeology are suitable. Drainage trenches should have a uniform gradient no greater than 1:200, and be from 300 to 900 mm wide with a minimum of 2 m between parallel trenches (Table 6.2). Perforated or porous pipes (75 to 110 mm in diameter) are laid on a 150 to 250 mm layer of clean gravel or crushed stone (20 to 50 mm grade) and then covered by 50 to 150 mm of the same material. The trench is sealed by a strip of air-permeable sheeting to prevent the entry of silt and then filled with soil. Pipes should be a minimum of 500 mm below the surface. Where high percolation values are obtained, the soil is not suitable for the construction of an infiltration area. In this case, a complete treatment system must be employed or an infiltration area constructed above ground level. Where V_p is between 140 and 100 s (i.e., between 10 and 7 hrs to fall 250 mm), underdrains are required to prevent the

Table 6.2. Conversion of subsurface drainage trench area A_t to linear trench length using three different trench widths (BSI, 1983). Reproduced with permission from BSI Standards Limited.

Drainage field floor area (a_t) m^2	Linear trench length (in meters)		
	0.3 m width trench	0.6 m width trench	0.9 m width trench
20	66	33	22
30	100	50	33
40	133	67	44
50	167	83	56
60	200	100	66
70	233	117	78
80	266	134	88
90	300	150	100

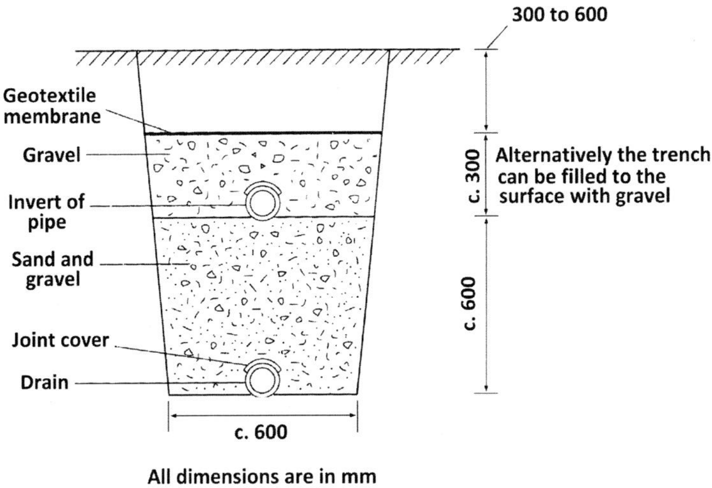

Fig. 6.3. Cross-section of a typical underdrain used in percolation areas where there is insufficient percolation (BSI, 1983). Reproduced with permission from BSI Standards Limited.

infiltration system from flooding. In this type of system, a second series of drainage pipes are laid on the bottom of the trench to convey the surplus water via outfall to a ditch or watercourse (Fig. 6.3). A typical layout of an infiltration area is shown in Fig. 6.4. The suitability of a site for subsurface wastewater irrigation can be initially assessed by looking at the native plants that are present, many of which can indicate either poor or good drainage potential. This helps in the selection of specific percolation test sites (Table 6.3). Full details of a site suitability assessment and the construction of infiltration areas are given by the Environmental Protection Agency (EPA, 2000), USEPA (2003) and British Standards Institute (BSI, 1983; 2008).

In most cases, a reserve infiltration area is also required, which is used in rotation with the main area and can be used in the event of a failure of the primary infiltration area. Surface runoff from paved areas or roofs is piped to a separate infiltration area or, where the water is

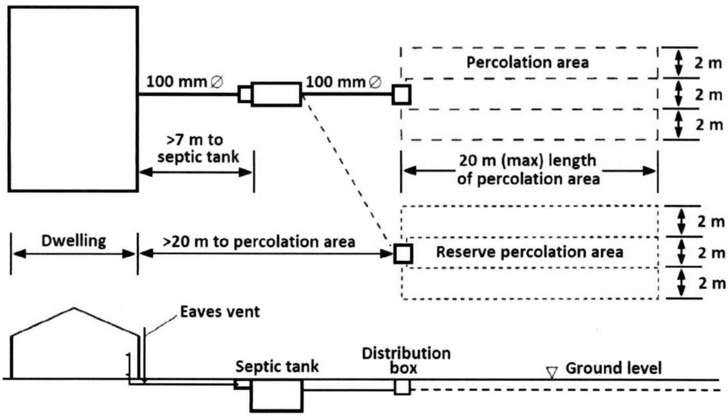

Fig. 6.4. Typical plan and section of a septic tank and percolation area (NSAI, 1991). Reproduced with permission from the National Standards Authority of Ireland, Dublin.

Table 6.3. Plants found in the British Isles that are indicative of the potential suitability of a site for the location of a percolation area. Common names are in parentheses.

Plants indicative of poor drainage
Alnus glutinosa (Common Alder)
Carex spp. (Sedges)
Cirsium palustre (Marsh thistle)
Epilobium palustre (Marshwillow herb)
Filipendula ulmaria (Meadowsweet)
Iris pseudacorus (Yellow flag iris)
Juncus articulatus (Jointed rush)
Juncus inflexus (Hard rush)
Juncus subulifolius (Rush)
Lychnis flos-cunculi (Ragged Robin)
Lythrum salicaria (Purple loosestrife)
Ranunculus reptans (Creeping buttercup)
Salix spp. (Willow)
Senecio aquaticus (Marsh ragwort)

(*Continued*)

Table 6.3. (*Continued*)

Succisia pratensis (Devil's-bit scabious)
Viburnum lantana (Wayfaring trice)
Plants indicative of good drainage all year
Achillea millefolium (Yarrow)
Anthriscus sylvestris (Cow parsley)
Bellis perennis (Daisy)
Cirsium arvense (Creeping thistle)
Daucus carotus (Wild carrot)
Galium verum (Lady's bedstraw)
Hyacinchoides non-scripta (Bluebell)
Juncus effuses (Soft rush)
Medicago lupulina (Black medick)
Papaver phoeas (Poppy)
Primula veris (Cowslip)
Pteris aquilina (Bracken)
Ranunculus bulbosus (Bulbous buttercup)
Senecio jacobaea (Ragwort)
Taraxacum officinale (Dandelion)

clean and not subject to contamination, to a soakaway (i.e., a soak or disposal pit). Soakaways must not be used for the disposal of septic tank effluents as this invariably leads to a point source of contaminants, resulting in severe groundwater pollution (Gondwe *et al.*, 1997). When the percolation rates are either too slow or rapid, the water table is too close to the base of the infiltration trench (i.e., normally <1.5 m), or the site is too steep, new infiltration areas can be constructed by raising the level of the existing land using a suitable soil (Fig. 6.5). Alternatively, intermittent sand filters can be employed. These are either buried or in open shallow beds (of coarse sand) that are 0.6 to 0.8 m in depth. The wastewater is applied periodically using a simple distribution system with the treated effluent collected by an underdrain system at the base of the filter (Metcalf and Eddy, 1991). Sand filters have been shown to be more efficient in removing

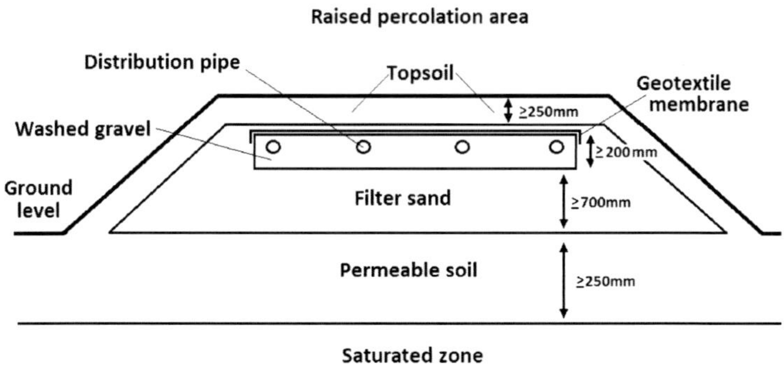

Fig. 6.5. Where there is inadequate infiltration for subsurface wastewater irrigation, a raised percolation area or mound can be employed.

pathogens and nutrients than other media, thereby offering a greater degree of protection to groundwaters (Harrison *et al.*, 2000). Peat is also used as an alternative filter medium with a number of commercial units available. It has a high adsorption capacity and water-holding characteristics that make it an ideal treatment medium for a wide range of wastewaters. The design of peat filters is similar to intermittent sand filters (Kennedy and van Geel, 2000; van Geel and Parker, 2003) with an example and performance data given by Gray and Coady (2015).

6.3 Biomat Formation and Treatment

When a percolation area is newly constructed, the effluent from the treatment unit rapidly infiltrates the soil, with very little of the overall length of the trench utilized. Over time, a layer of organic debris and microorganisms forms over the base of the trenches at the interface with the soil. This is the biomat that forms part of the physicochemical and biological treatment that occurs in the SWIS (Kreissel, 1982). Fine solids are physically strained out of the effluent, while the heterotrophic microorganisms metabolize the organic fraction. So long as the biomat is kept moist, protozoans, rotifers, nematodes, and other mesofauna will colonize it. Over time, the biomat slowly covers the

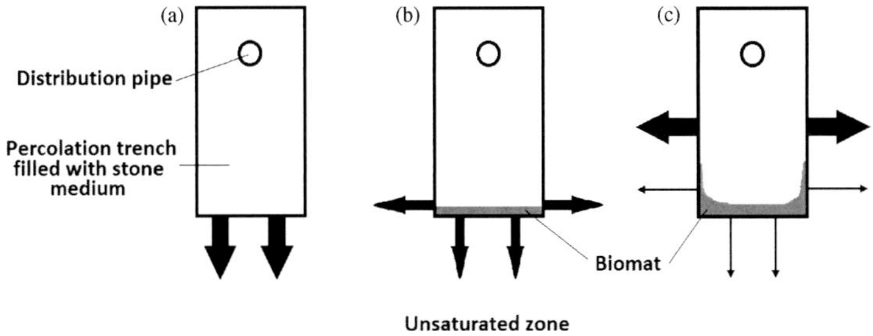

Fig. 6.6. Direction and relative volume of effluent discharged from a percolation trench: (a) without biomat formation, the majority of effluent is discharged through the trench base, (b) with biomat development over the trench base, there is reduced downwards percolation with lateral flow mainly at corners, and (c) as biomat begins to build up on trench walls, the majority of the flow is through the upper sidewalls.

entire base of the trench system, ensuring an even distribution of effluent throughout the percolation area. By ensuring that the entire trench area is utilized, it ensures longer retention times in the underlying unsaturated zone for further biological oxidation to occur.

However, the biomat can continue to increase in thickness and force the effluent to infiltrate into the soil through the sidewalls of the trench (Fig. 6.6). This was demonstrated experimentally by Beal *et al.* (2008) who showed that the effluent preferentially flows out of the trench above the biomat zone, forcing it to migrate laterally outwards and then downwards into the unsaturated zone, with the area immediately under the biomat allowing little infiltration, resulting in the soil remaining relatively dry (Fig. 6.7). As this occurs, the depth of effluent held in the trench increases so that the effluent becomes anoxic and eventually anaerobic. This reduces the heterotrophic biological treatment provided by the biofilm associated with the shallow layer of medium, although when used for effluents from nitrifying biological treatment systems, denitrification can occur. Finally, the trench becomes filled with effluent and overflows either onto the field surface, eventually polluting surface water or finding a rapid pathway to the groundwater, also causing pollution.

Treatment in septic tank percolation areas and the unsaturated soil layer into which the effluent is distributed are difficult to

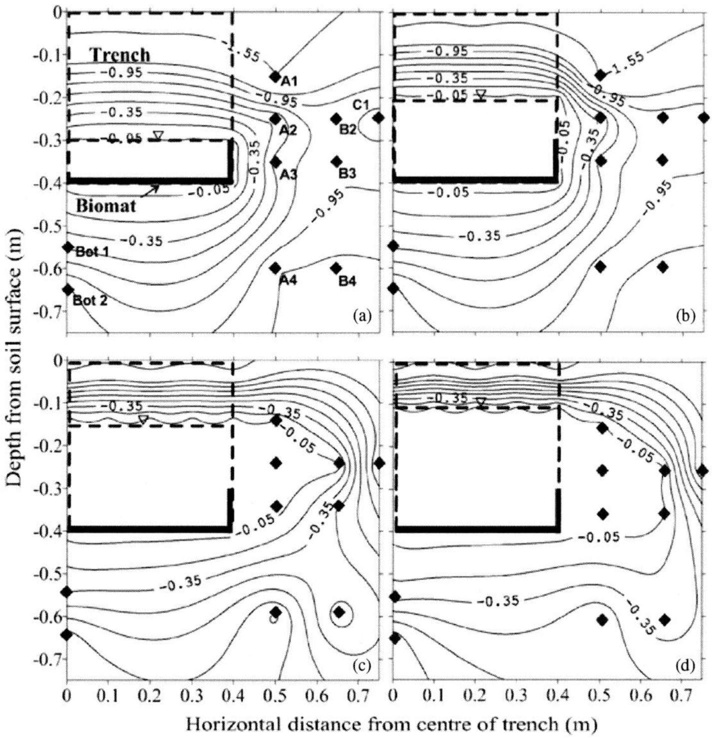

Fig. 6.7. Effluent movement from a trench with a biomat into the adjacent unsaturated soil as measured by the soil matric potential (ψ_{m}). Measurements were taken at different trench water heights: (a) 0.10 m, (b) 0.20 m, (c) 0.25 m, and (d) 0.30 m. Diamonds show the positions of tensiometers used to measure ψ_m (Beal *et al.*, 2008). Reproduced with permission from the American Society of Civil Engineers.

quantify but requires the distribution network to be kept aerobic by adequate ventilation. Treatment of the settled wastewater begins as the effluent passes through the distribution network where a biofilm develops on the internal walls of the pipework, similar to that found in a low-rate trickling filter (Gray, 2021). Initially, a similar biofilm develops on the medium within the trench, as the effluent flows over the medium to the base of the trench and then through the biomat into the soil below. When treating primary settled wastewater, biochemical oxygen demand (BOD) removals have been estimated at 5 to 10% within the vented distribution pipes, 20 to 40% within the stone medium (depending on depth), and a further 10 to 30% by the biomat. More treatment occurs as the effluent percolates through the unsaturated soil layer towards the groundwater. Metabolism will be aerobic when the medium is loaded intermittently, but anaerobic if it is continuously submerged. Overall, biological activity within the infiltration system depends on a number of factors: quality and quantity of wastewater applied, the hydraulic gradient, whether the system is adequately rested, oxygen availability within the trench and at the soil interface, and, most importantly, the temperature. To date, it has not been possible to accurately model or predict the degree of treatment provided by these systems. This is due to the majority of studies on SWISs based on newly installed systems and studied for 12 to 24 months. It is probable that it will take a percolation area and its associated biomat many years to reach some form of equilibrium. In colder climates, biological activity may be so limited that infiltration areas become primarily a system for hydraulic dispersal of effluents.

Biomat formation encourages an even distribution of effluent into the soil by ensuring all the trench is utilized, which promotes longer retention times in the soil for biological oxidation to occur. Hydraulic failure of SWISs occur when the loading to the infiltration trenches exceeds the infiltration rate of the soil. This is often caused by changes in the hydraulic behavior of the soil caused by the biomat formation. The biomat is a layer of organic and inorganic debris, which is very effective at filtering out pathogens and adsorbing other contaminants. When the infiltration rate through the biomat falls, it becomes covered by a permanent layer of effluent, creating an anaerobic environment that causes a gray gelatinous bacterial layer to develop on top, further impeding infiltration through the biomat.

Equilibrium develops between biomat accumulation and biomat reduction through microbial metabolism and grazing by soil invertebrates. Therefore, except where the soil has very high or low permeability (i.e., coarse gravel or clay, respectively), it is the hydraulic characteristics of the biomat that controls the LTAR of the system and not the permeability of the soil. If infiltration areas are rested and allowed to dry out, the biomat will be largely removed by the action of the grazers. The remaining biomat will eventually dry and crack, exposing the soil below and restoring its original permeability. For continuously loaded systems, resting is especially important if acceptable infiltration rates are to be maintained.

Although subsurface infiltration systems provide some degree of treatment of septic tank effluents, this is extremely variable, making groundwater contamination common. Chemical tracers have been widely used to identify potential pollution problems in ground and surface waters, either by using dyes, elevated chloride levels, fecal sterols, caffeine, and, more recently, micropollutants and DNA markers (Alhajjar *et al.,* 1990; Fahrenfeld *et al.,* 2016; Richards *et al.,* 2017). Elevated concentrations of metals, nutrients and pathogens have all been reported (De Borde *et al.,* 1998; Robertson *et al.*, 1998; Schwartz *et al.*, 1998; Khwaja *et al.*, 1999). Subsurface infiltration systems located in coarse sand soils are particularly at risk (Scandura and Sobsey, 1997). A study by Harman *et al.* (1996) on a 44-year-old septic tank system serving a school in Ontario, Canada, showed a residence time of 7 to 14 days for the septic tank effluent as it passed through the 1.6-m-deep, unsaturated zone below the infiltration trench. This allowed enough time for the complete oxidation of all the nitrogen present to nitrate. Within the groundwater, a 15-m-wide plume emanated from the site for over 110 m. The plume had detectable dissolved oxygen present, high nitrate (20–120 mg L^{-1} as nitrogen), chloride (42–209 mg L^{-1}), sodium (34–101 mg L^{-1}), calcium (120–249 mg L^{-1}), and above background concentrations of both sulfate and potassium. The phosphate concentrations present in plumes are controlled by mineral precipitation reactions that occur in the unsaturated zone, resulting in elevated concentrations in the soil. Once in the groundwater, phosphate plume migration is controlled by sorption processes (Robertson *et al.*, 1998).

References

Alhajjar, B.J., Chesters, G. and Harkin, J.M. (1990) Indicators of chemical pollution from septic systems. *Ground Water*, **28**, 559–568.

Beal, C.D., Rassam, W., Gardner, E.A., Kirchhof, G. and Menzies, N.W. (2008) Influence of hydraulic loading and effluent flux on surface surcharging in soil absorption systems. *Journal of Hydrologic Engineering*, **13**(8), 681–692.

BSI. (1983) Design and installation of small sewage treatment works and cesspools. Code of Practice, CP 6297. British Standards Institution, London.

BSI. (2008) Code of practice for the design and installation of drainage fields for use in wastewater treatment. CP 6297:2007 + A1: 2008. British Standards Institution, London.

De Borde, D.C., Woessner, W.W., Lauerman, B. and Ball, P.N. (1998) Virus occurrence and transport in a school septic system and unconfined aquifer. *Ground Water*, **36**(5), 825–834.

EPA. (2000) Wastewater treatment manuals: Treatment systems for single houses. Environmental Protection Agency, Wexford, Ireland.

Fahrenfeld, N.L., Del Monaco, N., Coates, J.T. and Elzerman, A.W. (2016) Fecal sterol and runoff analysis for nonpoint source tracking. *Journal of Environmental Quality*, **45**, 312–322.

Gondwe, E., Mwanuzi, F.L. and Mbwette, T.S.A. (1997) Septic tank soakpit systems in Dar es Salaam, Tanzania. *Journal of Environmental Engineering*, **123**(1), 93–95.

Gray, N.F. (2021) *Fixed-Film Reactors in Wastewater Treatment.* World Scientific Publishing, Singapore.

Gray, N.F. and Coady, J. (2015) Evaluation of full-scale vertical peat biofilters for treatment of septic tank effluents. *International Journal of Environmental Studies*, **72**(1), 154–165.

Harman, J., Robertson, W.D., Cherry, J.A. and Zanini, L. (1996) Impacts of a sand aquifer from an old septic system: Nitrate and phosphate. *Ground Water*, **34**(6), 1105–1114.

Harrison, R.B., Turner, N.S., Hoyle, J.A., Krejs, L.J., Tone, D.D., Henry, C.L., Isaken, P.J. and Xue, D. (2000) Treatment of septic tank effluent for faecal coliform and nitrogen in coarse-textured soils: Use of soil only and sand filter systems. *Water, Air and Soil Pollution*, **124**(1–2), 205–215.

Kennedy, P. and van Geel, P.J. (2000) Hydraulics of peat filters treating septic tank effluent. *Transport in Porous Media*, **41**(1), 47–60.

Khwaja, A.A., Lisa, M., Boustani, M., Jaffar, M. and Masud, M.K. (1999) An assessment study of septic tank based sewage disposal system on the quality of underground water. *Journal of the Chemical Society of Pakistan,* **21**(2), 141–145.

Kreissl, J.F. (1982) On-site wastewater disposal research in the United States. In *Alternative Wastewater Treatment* (eds. A.S. Eikum and R.W. Seabloom). D. Reidel, Boston, pp. 45–71.

Laak, R. (1986) *Wastewater Engineering Design For Unsewered Areas, 2nd ed.* Technomic Publishing Co., Lancaster, PA.

May, R. (1996) Problems associated with the use of gravel in septic tank leachfield systems. *Journal of Environmental Health,* **59**(2), 6–11.

Metcalf and Eddy, Inc. (1991) *Wastewater Engineering: Treatment, Disposal and Reuse, 3rd ed.* McGraw-Hill, New York.

NSAI. (1991) Septic tank systems recommendations for domestic effluent treatment and disposal from a single dwelling house. SR6, National Standards Authority of Ireland, Dublin.

Richards, S., Withers, P.J.A., Paterson, E., McRoberts, C.W. and Stutter, M. (2017) Potential tracers for tracking septic tank effluent discharges in watercourses. *Environmental Pollution*, **228**, 245–255.

Robertson, W.D., Schiff, S.L. and Ptacek, C.J. (1998) Review of phosphate mobility and persistence in 10 septic system plumes. *Ground Water,* **36**(6), 1000–1010.

Scandura, J.E. and Sobsey, M.D. (1997) Viral and bacterial contamination of groundwater from on-site sewage treatment systems. *Water Science and Technology,* **35**(11–12), 141–146.

Schwartz, J.J., Waterman, A.B., Lemley, A.T., Wagenet, L.P., Landre, P. and Allee, D.J. (1998) Homeowner perceptions and management of private water supplies and wastewater treatment. *Journal of Soil and Water Conservation,* **53**(4), 315–319.

Siegrist, R.L., Anderson, D.L. and Hargett, D.L. (1986) Large soil absorption systems for wastewaters from multiple-home developments. EPA/600/S2-86/023. US Environmental Protection Agency, Cincinnati, OH.

USDA. (2013) Microirrigation, Chapter 7. In *Part 623 Irrigation National Engineering Handbook.* Natural Resources Conservation Service, United States Department of Agriculture, Washington DC. https://www.wcc.nrcs.usda.gov/ftpref/wntsc/waterMgt/irrigation/NEH15/ch7.pdf

USDA. (2016) Irrigation system: Microirrigation. Conservation Practice Standard Code 441-CPS-1. Natural Resources Conservation Service,

United States Department of Agriculture, Washington DC. https://efotg.sc.egov.usda.gov/references/public/NY/nyps441.pdf

USEPA. (1980) Onsite wastewater treatment and disposal systems. EPA 625/1-80-012. Office of Water Program Operations. US Environmental Protection Agency, Washington DC.

USEPA. (2003) Onsite wastewater treatment systems manual. EPA 625/1-00/008, Office of Water Program Operations. US Environmental Protection Agency, Washington DC.

van Geel, P.J. and Parker, W.J. (2003) Estimating water budget for a peat filter treating septic tank effluent in the field. *Journal of Hydrology*, **271**, 52–64.

Woods Ballard, B., Wilson, S., Udale-Clarke, H., Illman, S., Scott, T., Ashley, R. and Kellagher, R. (2015) The SuDS manual (C753). CIRIA, London.

WPCF. (1990) Natural systems for wastewater treatment. Manual of practice: FD-16, Water Pollution Control Federation, Alexandria, VA.

7

Slow-rate Irrigation Processes

7.1 Introduction to Slow-rate Land Treatment

Irrigation is the most widely used and oldest land treatment system and involves the disposal of screened and primary settled wastewater to agricultural or recreational land. Slow-rate irrigation, or as it is more generally known, slow-rate land treatment (SRLT), is normally defined as the application of wastewater to a vegetated surface where the wastewater percolates into the soil where a significant level of treatment is possible as it passes through the plant-root-soil matrix. On a larger scale, SRLT is used to supplement rainfall and maximize crop production by utilizing the weak nutrient content of the wastewater as a growth promoter. It is also used for tertiary treatment of secondary-treated effluents, although the degree of pretreatment required is dependent on the treatment objectives and design of the irrigation system. The design of SRLT systems is based on several different processes: infiltration, percolation, lateral flow, and evapotranspiration. The main designs are: (i) percolation to groundwater, (ii) percolation with lateral movement to replenish surface waters, (iii) percolation to underdrains or wells for recovery/reuse, (iv) vegetative uptake of water and nutrients and enhanced evapotranspiration, and (v) crop irrigation and percolation (Fig. 7.1), each with specific treatment and reuse capabilities.

Slow-rate irrigation is unsuitable for shallow soils or those comprised primarily of clay where overland flow is normally used (Chap. 9). The ideal soil is either loam or silt, with good permeability

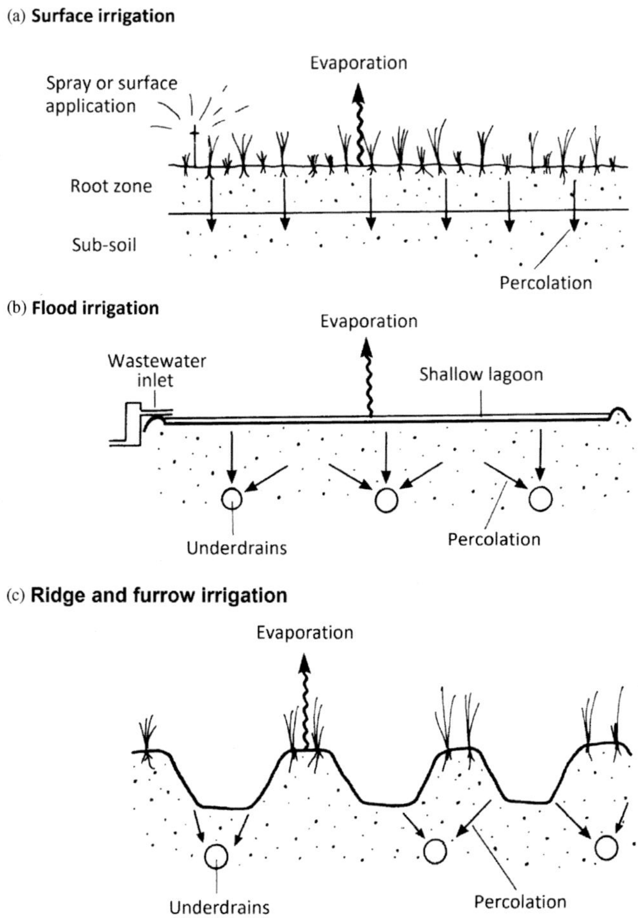

Fig. 7.1. Different types of surface irrigation land treatment.

and a minimum depth of 1.75 m of unsaturated soil or permeable rock to the groundwater. The rate of application of wastewater depends on soil permeability, evapotranspiration rates, and precipitation. Particle-size distribution of soil is important in the removal efficiency of wastewater constituents; smaller particles are most effective as they reduce the rate of percolation (Tare and Bokil, 1982). Soil profile and crop cover have been identified as major factors in nutrient removal. Although pasture grass species are normally used, trees (Sec. 7.4), legumes and corn (Sec. 7.3) are also utilized as a vegetative cover for such systems. Complete treatment can be obtained with almost complete biochemical oxygen demand (BOD) and suspended solids removal. The slow irrigation rate is important to ensure most of the phosphorus is immobilized and for the bulk of the nitrogen to be mineralized. The wastewater that has not evaporated or released into the atmosphere by evapotranspiration eventually enters the groundwater in the aquifer below, percolates through a bank to enter surface water, or is intercepted by underground drainage pipes (Fig 7.1). Irrigation is not continuous, therefore, storage is required, unless sufficient area is available to allow the land to be rested between application periods. This is to ensure that aerobic soil conditions are re-established, which are essential for the growth of non-wetland vegetation, maintenance of a permeable soil structure, and optimum nutrient removal. Application cycles are normally every 4 to 10 days with hydraulic loading rates varying between 0.15 and 1.6 cm d^{-1}, which is equivalent to an area of 6 to 67 ha per 1,000 m^3d^{-1} (USEPA, 1977; Metcalf and Eddy, 1991; WEF, 1992). Irrigation may not be possible during periods of high rainfall or when the land is frozen. The vegetation, coupled with the active soil layer, makes slow-rate irrigation systems an effective biological treatment process, as long as wastewater hydraulic loading rates are kept low. This makes them particularly suitable for the tertiary treatment of wastewaters.

7.2 Process Design

Slow-rate land treatment (SLRT) design falls into two broad categories — those designed to dispose of the greatest volume of wastewater per unit area and those designed to maximize the reuse of water and nutrients that comprise the wastewater, to irrigate and fertilize crops. The former designs are used in temperate areas or for industrial treatment while the latter are utilized in water-stressed areas as part of a water management plan.

Wastewater is distributed onto the land either by surface flooding or by spraying. Surface flooding is only possible where the land is level and may involve flooding specific areas for several days and then resting it for a couple of weeks, or using conventional furrows as were used on the original sewage farms (Fig. 7.1c). Where deep furrows are preferred, each furrow is fed by a siphon connected to an open supply channel or simply filled with wastewater several times a day (Fig. 7.2). In all surface flooding operations, it is necessary to contain the wastewater on the site by surrounding it with a dyke to prevent runoff. Spraying is done from either a fixed set of pipes or rotary sprayers, with different areas of the system rested intermittently, or by using a mobile rig that is moved manually from area to area. Spraying allows the unleveled ground to be used, although there is a health risk from aerosols (Bausum *et al.*, 1982), even though chlorination of treated effluents has been successfully used in the United States (US)

Fig. 7.2. Ridge and furrow irrigation in Australia. Image by HoraceG, reproduced under the Creative Commons License CC BY-SA 3.0.

to reduce downwind levels of microbial and coliphage aerosols. Unless severely overloaded organically, damage to the soil layer is unlikely, although groundwaters are at risk. Studies have shown that the total dissolved solids (TDS) concentration and nitrate concentration immediately below the site will be enhanced, although most pathogens, protozoan cysts, and bacteria will be filtered out by the soil. However, prolonged application of wastewater to the land may result in the translocation of some viruses through the soil into the groundwater or the treated water used for crop irrigation (Shaub *et al.*, 1982). Metals are immobilized onto the soil particles with the major metals in municipal wastewater, cadmium, copper, nickel, lead, and zinc rarely migrating deeper into the soil than 1.5 m (Brown *et al.*, 1983). However, some persistent pesticides have been detected in groundwater after SRLT (Koerner and Haws, 1979; Müller *et al.*, 2007).

The hydraulic loading is controlled by the percolation capacity of the soil.

$$L_h = ET_c - P + Pw, \tag{7.1}$$

where L_h is the hydraulic loading rate in cm d^{-1}, ET_c is the crop evapotranspiration rate in cm d^{-1}, P is the mean precipitation in cm d^{-1}, and Pw is the deep percolation rate in cm d^{-1} (USEPA, 2006).

The largest slow-rate irrigation wastewater treatment plant in operation was at Werribee Farm, now the Western Treatment Plant, in Melbourne, Australia, which treated the wastewater from 800,000 people (546,000 m^{3}d^{-1}). A series of paddocks were flooded in rotation with filtered wastewater to a depth of 10 cm over a period of 24 to 48 hours, which then percolated through the grass sward over a further 5 days. Treated effluent was collected by underlying drains that channeled the final effluent into Port Philip Bay. Each paddock was rested for 14 days, during which it was grazed with sheep or cattle before the cycle was repeated. To deal with expansion and to improve winter performance, the overland flow was added to the existing SFLT grass plots during the 1930s and then, subsequently lagoons. While no adverse effects had been recorded to the soil, vegetation or the livestock (Dinges, 1982), the high nitrogen inputs into the Bay and associated protected wetlands from the treated effluent resulted in severe eutrophication. To protect these fragile environments, a new lagoon system replaced the existing treatment plant, including the land filtration systems in 2004 (Chap. 16).

7.3 Crop Irrigation

7.3.1 *Performance*

While perennial grasses give optimum treatment, have a longer application season, are able to withstand higher hydraulic loadings, and achieve greater nitrogen removal when compared to agricultural crops, there is still a large scope to use wastewater irrigation to supplement surface and groundwater supplies for irrigation and to utilize the nutrients and trace elements required.

The use of wastewater for crop irrigation and as a nutrient supplement is more complex than permanent treatment grass plots due to the variable demand by the crop plants themselves, their management (i.e., planting, weeding, possible spraying, and harvesting that require dry periods), periods of high precipitation, and during very cold periods (mean monthly average temperatures $<0°C$) when infiltration is significantly reduced. This means that the storage of wastewater will be required, which in the case of seasonal crops may be up to four months. Details of how to calculate storage requirements are given by the United States Environmental Protection Agency (USEPA, 2006). Crop irrigation can also be operated alongside another irrigation system such as rapid infiltration basins, which can be used to treat the excess wastewater from crop irrigation (Chap. 8).

Rice paddy fields have been used for SRLT in China since the mid-1980s with hydraulic loading rates of 200 cm yr^{-1} (Ou *et al.*, 1992a). In an experimental plot of 7 ha, which was irrigated with primary treated wastewater (BOD, 74 mg L^{-1}; COD, 189 mg L^{-1}; TN, 21 mg L^{-1}; TP, 1.8 mg L^{-1}; and SS, 69 mg L^{-1}) from May to November only due to the cold winter temperatures, where almost half of the infiltration effluent was recovered and reused. In the winter, the wastewater was diverted to a rapid infiltration basin of 1 ha in area (Fig. 7.3). The SRLT system also incorporated a

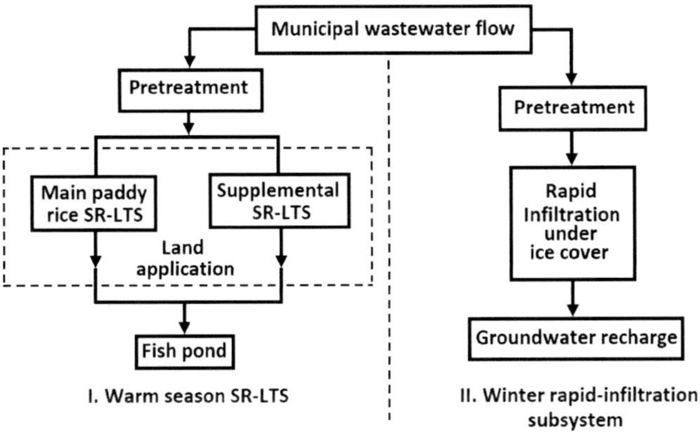

Fig. 7.3. Schematic layout of the Shenyang slow-rate land treatment (SR-LT) systems showing summer and winter operational designs.

fishpond. Over the four-year study period the average final effluent BOD concentration was 5 mg L^{-1} (95% removal); COD, 36 mg L^{-1} (84% removal); TP, 0.2 mg L^{-1} (90% removal); and TN, 6 mg L^{-1} (72% removal). No significant difference was found in the quality of the rice in terms of priority organic pollutants when compared to those grown using clean water (Ou *et al.*, 1992b).

Flood irrigation can lead to the soil becoming intermittently anaerobic, which can result in the release of N_2O, a greenhouse gas (Fig. 3.3). However, unless the wastewater is supplemented by artificial nitrogen fertilizer, which may be required in growing crops such as maize where 1% of the nitrogen applied was lost as N_2O, then N_2O emissions from grass plots are relatively low and do not pose a significant addition to emissions in arid areas (González-Méndez *et al.*, 2020). The formation of N_2O is a potentially major problem for irrigation systems using wastewater due to climate, the crop grown, soil, and operational conditions, so an assessment should be made at the pilot-scale trials of potential emissions and a mitigation strategy should be developed before full-scale operation begins (Xue *et al.*, 2012) (Sec. 3.2). Nitrogen dynamics in slow SRLT systems are explained by Hernández-Martínez *et al.* (2018) who stress the importance of matching nitrogen inputs to crop uptake to prevent groundwater and atmospheric pollution.

The relationship between crop yield and evapotranspiration has been studied by Bennett and Harms (2011) who have produced a number of improved methods to predict irrigation water requirements of crops and the effect of water stress on yields. In their paper, they calculated the relative yield reductions arising from different water stress scenarios using the relationship:

$$1-\frac{Ya}{Ym}=ky\left(\frac{1-ETc}{ETm}\right), \tag{7.2}$$

where Ya is the predicted crop yield (Mg ha^{-1} which is equivalent to t ha^{-1}), Ym is the maximum potential yield (Mg ha^{-1}), ET_c is the seasonal crop evapotranspiration (mm), ET_m is the maximum seasonal crop evapotranspiration (mm), and ky is the crop-specific yield response factor (Stewart *et al.*, 1977).

The relative yield is then predicted as:

$$\frac{Ya}{Ym}=1-ky+ky\left(\frac{ETc}{ETm}\right). \tag{7.3}$$

In Alberta, Canada, examples of the yield response factors (ky) are alfalfa, 1.05; barley, 1.15; corn silage, 1.25; beans, 1.15; grass for hay, 1.05; spring wheat, 1.15; potatoes, 1.10; and sugar beet, 1.10 (Doorenbos and Kassam, 1979).

7.3.2 *Salinity intolerance*

Apart from adequate water, the effect of salinization is a critical factor in the success of growing crops irrigated with wastewater. Salinization and sodicity have been explored in Sec. 5.4, but here, the effect of salts from wastewater on growing crops is explored in greater detail.

TDS is a measure of the total concentration of salts in irrigation water as well as in the soil of the growing plot and is used to evaluate salinity. It is most easily calculated using electrical conductivity (EC) multiplied by a conversion factor for irrigation water. For EC values of irrigation water between 0.1 and 5.0 mS cm^{-1}, then:

$$\text{TDS (ppm)} = \text{EC (mS cm}^{-1}) \times 640. \quad (7.4)$$

For EC values > 5.0 mS cm^{-1}, then:

$$\text{TDS (ppm)} + \text{EC (mS cm}^{-1}) \times 800. \quad (7.5)$$

The units of EC can be confusing with 1 deciSiemens per meter (dS m^{-1}) equivalent to the older unit of 1 milliMhos per centimeter (mmhos cm^{-1}), which is also equivalent to the more commonly used measure of the conductivity of 1milliSiemens per centimeter (mS cm^{-1}).

$$\text{dS m}^{-1} \equiv \text{mmhos cm}^{-1} \equiv \text{mS cm}^{-1}. \quad (7.6)$$

Crops have very different tolerances to salinity, which can result in significant reductions in yields and even crop failures. Where the water table is <2 m below ground level, groundwater can be drawn up into the root zone by capillary action, although many plants have very deep roots. So where the groundwater has a high degree of mineralization, the root zone will become increasingly salinized until only salt-tolerant crops can be grown.

Salinity problems in irrigation plots are generally caused by salts in the wastewater applied but can be controlled by leaching, that is, flushing the soluble salts below the root zone. This is done by applying more water than is just required by evapotranspiration (Et_c) so that enough water percolates through the root zone to remove the salts. The amount of salts leached must exceed or at least equal the amount added by the wastewater; if not, the salinity will increase in the root zone.

Leaching is calculated using the EC of the irrigation water (EC_w) and the tolerance of specific crops to soil salinity (EC_e), both in mS cm^{-1}. Table 7.1 gives the salt tolerance (EC_e) for a wide range of crops and vegetative cover plants for different expected yield losses (Doorenbos and Pruitt, 1977; Maas, 1984; Maas and Grattan, 1999). An EC_e value to give a minimum yield of 90% or more is

Table 7.1. Crop tolerance and yield potential of major crops as influenced by irrigation water salinity (EC_w) and soil salinity (EC_e)[1,2] (Doorenbos and Pruitt, 1977; Maas and Grattan, 1999). Reproduced with permission from the FAO, Rome, Italy, and John Wiley Inc., New York.

	100%		90%		75%		50%		0% "Maximum"[3]	
Field crops	EC_e	EC_w	EC_e	EC_w	EC_e	EC_w	EC_e	EC_w	EC_e	EC_w
Barley *(Hordeum vulgare)*[4]	8.0	5.3	10	6.7	13	8.7	18	12	28	19
Cotton *(Gossypium hirsutum)*	7.7	5.1	9.6	6.4	13	8.4	17	12	27	18
Sugarbeet *(Beta vulgaris)*[5]	7.0	4.7	8.7	5.8	11	7.5	15	10	24	16
Sorghum *(Sorghum bicolor)*	6.8	4.5	7.4	5.0	8.4	5.6	9.9	6.7	13	8.7
Wheat *(Triticum aestivum)*[4,6]	6.0	4.0	7.4	4.9	9.5	6.3	13	8.7	20	13
Wheat, durum *(Triticum turgidum)*	5.7	3.8	7.6	5.0	10	6.9	15	10	24	16
Soybean *(Glycine max)*	5.0	3.3	5.5	3.7	6.3	4.2	7.5	5.0	10	6.7
Cowpea *(Vigna unguiculata)*	4.9	3.3	5.7	3.8	7.0	4.7	9.1	6.0	13	8.8
Groundnut (Peanut) *(Arachis hypogaea)*	3.2	2.1	3.5	2.4	4.1	2.7	4.9	3.3	6.6	4.4
Rice (paddy) *(Oriza sativa)*	3.0	2.0	3.8	2.6	5.1	3.4	7.2	4.8	11	7.6
Sugarcane *(Saccharum officinarum)*	1.7	1.1	3.4	2.3	5.9	4.0	10	6.8	19	12
Corn (maize) *(Zea mays)*	1.7	1.1	2.5	1.7	3.8	2.5	5.9	3.9	10	6.7
Flax *(Linum usitatissimum)*	1.7	1.1	2.5	1.7	3.8	2.5	5.9	3.9	10	6.7
Broadbean *(Vicia faba)*	1.5	1.1	2.6	1.8	4.2	2.0	6.8	4.5	12	8.0
Bean *(Phaseolus vulgaris)*	1.0	0.7	1.5	1.0	2.3	1.5	3.6	2.4	6.3	4.2
Vegetable crops										
Squash, zucchini (courgette) *(Cucurbita pepo melopepo)*	4.7	3.1	5.8	3.8	7.4	4.9	10	6.7	15	10
Beet, red *(Beta vulgaris)*[5]	4.0	2.7	5.1	3.4	6.8	4.5	9.6	6.4	15	10
Squash, scallop *(Cucurbita pepo melopepo)*	3.2	2.1	3.8	2.6	4.8	3.2	6.3	4.2	9.4	6.3

Broccoli *(Brassica oleracea botrytis)*	2.8	1.9	3.9	2.6	5.5	3.7	8.2	5.5	14	9.1
Tomato *(Lycopersicon esculentum)*	2.5	1.7	3.5	2.3	5.0	3.4	7.6	5.0	13	8.4
Cucumber *(Cucumis sativus)*	2.5	1.7	3.3	2.2	4.4	2.9	6.3	4.2	10	6.8
Spinach *(Spinacia oleracea)*	2.0	1.3	3.3	2.2	5.3	3.5	8.6	5.7	15	10
Celery *(Apium graveolens)*	1.8	1.2	3.4	2.3	5.8	3.9	9.9	6.6	18	12
Cabbage *(Brassica oleracea capitata)*	1.8	1.2	2.8	1.9	4.4	2.9	7.0	4.6	12	8.1
Potato *(Solanum tuberosum)*	1.7	1.1	2.5	1.7	3.8	2.5	5.9	3.9	10	6.7
Corn, sweet (maize) *(Zea mays)*	1.7	1.1	2.5	1.7	3.8	2.5	5.9	3.9	10	6.7
Sweet potato *(Ipomoea batatas)*	1.5	1.0	2.4	1.6	3.8	2.5	6.0	4.0	11	7.1
Pepper *(Capsicum annuum)*	1.5	1.0	2.2	1.5	3.3	2.2	5.1	3.4	8.6	5.8
Lettuce *(Lactuca sativa)*	1.3	0.9	2.1	1.4	3.2	2.1	5.1	3.4	9.0	6.0
Radish *(Raphanus sativus)*	1.2	0.8	2.0	1.3	3.1	2.1	5.0	3.4	8.9	5.9
Onion *(Allium cepa)*	1.2	0.8	1.8	1.2	2.8	1.8	4.3	2.9	7.4	5.0
Carrot *(Daucus carota)*	1.0	0.7	1.7	1.1	2.8	1.9	4.6	3.0	8.1	5.4
Bean *(Phaseolus vulgaris)*	1.0	0.7	1.5	1.0	2.3	1.5	3.6	2.4	6.3	4.2
Turnip *(Brassica rapa)*	0.9	0.6	2.0	1.3	3.7	2.5	6.5	4.3	12	8.0
Wheatgrass, tall *(Agropyron elongatum)*	7.5	5.0	9.9	6.6	13	9.0	19	13	31	21
Wheatgrass, fairway crested *(Agropyron cristatum)*	7.5	5.0	9.0	6.0	11	7.4	15	9.8	22	15
Bermuda grass *(Cynodon dactylon)*[7]	6.9	4.6	8.5	5.6	11	7.2	15	9.8	23	15
Barley (forage) *(Hordeum vulgare)*[4]	6.0	4.0	7.4	4.9	9.5	6.4	13	8.7	20	13
Ryegrass, perennial *(Lolium perenne)*	5.6	3.7	6.9	4.6	8.9	5.9	12	8.1	19	13
Trefoil, narrowleaf birdsfoot[8] *(Lotus corniculatus tenuifolium)*	5.0	3.3	6.0	4.0	7.5	5.0	10	6.7	15	10
Harding grass *(Phalaris tuberosa)*	4.6	3.1	5.9	3.9	7.9	5.3	11	7.4	18	12
Fescue, tall *(Festuca elatior)*	3.9	2.6	5.5	3.6	7.8	5.2	12	7.8	20	13
Wheatgrass, standard crested *(Agropyron sibiricum)*	3.5	2.3	6.0	4.0	9.8	6.5	16	11	28	19

(Continued)

Table 7.1. (*Continued*)

									0%	
	100%		90%		75%		50%		"Maximum"[3]	
Field crops	EC_e	EC_w	EC_e	EC_w	EC_e	EC_w	EC_e	EC_w	EC_e	EC_w
Vetch, common (*Vicia angustifolia*)	3.0	2.0	3.9	2.6	5.3	3.5	7.6	5.0	12	8.1
Sudan grass (*Sorghum sudanense*)	2.8	1.9	5.1	3.4	8.6	5.7	14	9.6	26	17
Wildrye, beardless (*Elymus triticoides*)	2.7	1.8	4.4	2.9	6.9	4.6	11	7.4	19	13
Cowpea (forage) (*Vigna unguiculata*)	2.5	1.7	3.4	2.3	4.8	3.2	7.1	4.8	12	7.8
Trefoil, big (*Lotus uliginosus*)	2.3	1.5	2.8	1.9	3.6	2.4	4.9	3.3	7.6	5.0
Sesbania (*Sesbania exaltata*)	2.3	1.5	3.7	2.5	5.9	3.9	9.4	6.3	17	11
Sphaerophysa (*Sphaerophysa salsula*)	2.2	1.5	3.6	2.4	5.8	3.8	9.3	6.2	16	11
Alfalfa (*Medicago sativa*)	2.0	1.3	3.4	2.2	5.4	3.6	8.8	5.9	16	10
Lovegrass (*Eragrostis sp.*)[9]	2.0	1.3	3.2	2.1	5.0	3.3	8.0	5.3	14	9.3
Corn (forage) (maize) (*Zea mays*)	1.8	1.2	3.2	2.1	5.2	3.5	8.6	5.7	15	10
Clover, berseem (*Trifolium alexandrinum*)	1.5	1.0	3.2	2.2	5.9	3.9	10	6.8	19	13
Orchard grass (*Dactylis glomerata*)	1.5	1.0	3.1	2.1	5.5	3.7	9.6	6.4	18	12
Foxtail, meadow (*Alopecurus pratensis*)	1.5	1.0	2.5	1.7	4.1	2.7	6.7	4.5	12	7.9
Clover, red (*Trifolium pratense*)	1.5	1.0	2.3	1.6	3.6	2.4	5.7	3.8	9.8	6.6
Clover, alsike (*Trifolium hybridum*)	1.5	1.0	2.3	1.6	3.6	2.4	5.7	3.8	9.8	6.6
Clover, ladino (*Trifolium repens*)	1.5	1.0	2.3	1.6	3.6	2.4	5.7	3.8	9.8	6.6
Clover, strawberry (*Trifolium fragiferum*)	1.5	1.0	2.3	1.6	3.6	2.4	5.7	3.8	9.8	6.6
Fruit crops[10]										
Date palm (*Phoenix dactylifera*)	4.0	2.7	6.8	4.5	11	7.3	18	12	32	21
Grapefruit (*citrus paradisi*)[11]	1.8	1.2	2.4	1.6	3.4	2.2	4.9	3.3	8.0	5.4

Orange *(Citrus sinensis)*	1.7	1.1	2.3	1.6	3.3	2.2	4.8	3.2	8.0	5.3
Peach *(Prunus persica)*	1.7	1.1	2.2	1.5	2.9	1.9	4.1	2.7	6.5	4.3
Apricot *(Prunus armeniaca)*[11]	1.6	1.1	2.0	1.3	2.6	1.8	3.7	2.5	5.8	3.8
Grape *(Vitus sp.)*[11]	1.5	1.0	2.5	1.7	4.1	2.7	6.7	4.5	12	7.9
Almond *(Prunus dulcis)*[11]	1.5	1.0	2.0	1.4	2.8	1.9	4.1	2.8	6.8	4.5
Plum, prune *(Prunus domestica)*[11]	1.5	1.0	2.1	1.4	2.9	1.9	4.3	2.9	7.1	4.7
Blackberry *(Rubus sp.)*	1.5	1.0	2.0	1.3	2.6	1.8	3.8	2.5	6.0	4.0
Boysenberry *(Rubus ursinus)*	1.5	1.0	2.0	1.3	2.6	1.8	3.8	2.5	6.0	4.0
Strawberry *(Fragaria sp.)*	1.0	0.7	1.3	0.9	1.8	1.2	2.5	1.7	4	2.7

[1]These data should only serve as a guide to relative tolerances among crops. Absolute tolerances vary depending upon climate, soil conditions and cultural practices. In gypsiferous soils, plants will tolerate about 2 dS/m higher soil salinity (EC_e) than indicated but the water salinity (EC_w) will remain the same as shown in this table.

[2] EC_e means average root zone salinity as measured by electrical conductivity of the saturation extract of the soil, reported in deciSiemens per meter (dS/m) at 25°C. EC_w means electrical conductivity of the irrigation water in deciSiemens per meter (dS/m). The relationship between soil salinity and water salinity ($EC_e = 1.5\ EC_w$) assumes a 15–20% leaching fraction and a 40-30-20-10% water use pattern for the upper to lower quarters of the root zone.

[3] The zero yield potential or maximum EC_e indicates the theoretical soil salinity (EC_e) at which crop growth ceases.

[4] Barley and wheat are less tolerant during germination and seeding stage; EC_e should not exceed 4–5 dS/m in the upper soil during this period.

[5] Beets are more sensitive during germination; EC_e should not exceed 3 dS/m in the seeding area for garden beets and sugar beets.

[6] Semi-dwarf, short cultivars may be less tolerant.

[7] Tolerance given is an average of several varieties; Suwannee and Coastal Bermuda grass are about 20% more tolerant, while Common and Greenfield Bermuda grass are about 20 % less tolerant.

[8] Broadleaf Birdsfoot Trefoil seems less tolerant than Narrowleaf Birdsfoot Trefoil.

[9] Tolerance given is an average for Boer, Wilman, Sand and Weeping Lovegrass; Lehman Lovegrass seems about 50% more tolerant.

[10] These data are applicable when rootstocks are used that do not accumulate Na^+ and Cl^- rapidly or when these ions do not predominate in the soil. If either ions do, refer to the toxicity discussion in Sec. 4.

[11] Tolerance evaluation is based on tree growth and not on yield.

recommended to protect soils from salinization in the long term. The leaching requirement (*LR*), which is the same as the leaching fraction (the portion of irrigation water that infiltrates below the root zone that is required to control salts to within the EC_e for the crop using surface irrigation), is expressed as a fraction and is calculated as:

$$LR = \frac{ECw}{5(ECe) - ECw}. \tag{7.7}$$

The net irrigation requirement (*NIR*), which is the plant water requirement (ET_c) and the *LR*, is:

$$NIR = \frac{ETc}{1 - LR}, \tag{7.8}$$

where ET_c is the water demand of the crop (mm yr^{-1}) and calculated using Eq. (5.1), and is fully explained in Sec. 5.3. With seasonal vegetables, the salinity may be naturally leached out of the root zone during winter by precipitation. So, leaching may only be required towards the end of the growing season as the soil conductivity rises to inhibitory levels.

For example, when using irrigation water with a conductivity of 1.4 mS cm^{-1} for growing soybeans, the EC_e for a 90% yield is 5.5 mS cm^{-1} (Table 7.1). Using Eq. (7.6), the leaching requirement is calculated as:

$$LR = \frac{1.4}{5(5.5) - 1.4} = 0.054.$$

The ET_c for soybeans is 750 mm per season, so the required net irrigation rate is calculated using Eq. (7.7) as:

$$NIR = \frac{750}{1 - 0.054} = 793 \textit{ mm per season}.$$

The efficiency of irrigation varies with each plot and is known as the crop irrigation efficiency (CIE), which is calculated as:

$$CIE = \frac{Irrigation\, needed}{Irrigation\, supplied}, \tag{7.9}$$

which in this example can be approximated as 80%, so that the actual gross irrigation, when adjusted, is calculated as:

$$Gross\, irrigation = \frac{NIR}{CIE} = \frac{793}{0.80} = 991\ mm\ per\ season.$$

Approximate ranges for seasonal ET_c values are given by Doorenbos and Pruitt (1977). Common examples are: 600–1,500 mm for alfalfa, 250–500 mm for beans, 800–1,200 mm for coffee, 550–950 mm for cotton, 700–1,050 mm for deciduous trees, 450–900 mm for flax, 300–450 mm for barley and wheat, 400–750 mm for maize, 350–625 mm for potatoes, 500–950 mm for rice, 300–650 mm for sorghum, 450–825 mm for soybeans, and 250–500 mm for vegetables.

Plants have different tolerances to salinity (Table 7.2), largely due to their ability to make osmotic adjustments in order to extract sufficient water from affected soils. However, in order to achieve optimal yields, careful selection of the crops selected must be based on their tolerance or the ability to leach out excess salts. There is sufficient tolerance to allow crops or vegetative cover to be grown on all but the most saline soils, allowing a wide range of wastewater irrigation salinities (EC_w) to be used (Table 7.1). From this table, a full yield of all the crops listed is achievable where the irrigation water has an EC of <0.7 mS cm^{-1}. The relative tolerance ratings for specific crops are given in Table 7.2, with the effect on crop yield summarized in Fig. 7.4 (Maas, 1984).

Apart from abandoning irrigation sites for crop growth due to salts accumulation, or leaching the entire irrigation plot, some success has been achieved in restoring the yields of avocados in wastewater-irrigated orchards in Israel — wastewater was diluted by blending with uncontaminated water (in a 1:1 ratio) and using direct drip application. Using trenches filled with an adsorptive material (volcanic tuff) for irrigation using normal wastewater also improved plant health and yields (Nemera *et al.*, 2020).

Table 7.2. Salinity tolerance level of a range of crops and vegetative cover plants (Doorenbos and Pruitt, 1977; Maas, 1984; Paranychianakis *et al.,* 2006). Reproduced with permission from the FAO, Rome, Italy, and the Taylor & Francis Group.

Salinity tolerant	
Fibre, seed and sugar crops	
Barley	*Hordeum vulgare*
Cotton	*Gossypium hirsutum*
Jojoba	*Simmondsia chinensis*
Sugarbeet	*Beta vulgaris*
Grasses and forage crops	
Alkali grass, Nuttall	*Puccinellia airoides*
Alkali sacaton	*Sporobolus airoides*
Bermuda grass	*Cynodon dactylon*
Kallar grass	*Diplachne fusca*
Saltgrass, desert	*Distichlis stricta*
Wheatgrass, fairway crested	*Agropyron cristatum*
Wheatgrass, tall	*Agropyron elongatum*
Wildrye, Altai	*Elymus angustus*
Wildrye, Russian	*Elymus junceus*
Vegetable crops	
Asparagus	*Asparagus officinalis*
Fruit and nut crops	
Date palm	*Phoenix dactylifera*
Salinity moderately tolerant	
Fiber, seed and sugar crops	
Cowpea	*Vigna unguiculata*
Oats	*Avena sativa*
Rye	*Secale cereale*
Safflower	*Carthamus tinctorius*
Sorghum	*Sorghum bicolor*
Soybean	*Glycine max*
Triticale	*X Triticosecale*
Wheat	*Triticum aestivum*
Wheat, Durum	*Triticum turgidum*
Grasses and forage crops	
Barley (forage)	*Hordeum vulgare*

Table 7.2. (*Continued*)

Brome, mountain	*Bromus marginatus*
Canary grass, reed	*Phalaris arundinacea*
Clover, Hubam	*Melilotus alba*
Clover, sweet	*Melilotus*
Fescue, meadow	*Festuca pratensis*
Fescue, tall	*Festuca elatior*
Harding grass	*Phalaris tuberosa*
Panic grass, blue	*Panicum antidotale*
Rape	*Brassica napus*
Rescue grass	*Bromus unioloides*
Rhodes grass	*Chloris gayana*
Ryegrass, Italian	*Lolium italicum multiflorum*
Ryegrass, perennial	*Lolium perenne*
Sudan grass	*Sorghum sudanense*
Trefoil, narrowleaf birdsfoot	*Lotus corniculatus tenuifolium*
Trefoil, broadleaf birdsfoot	*Lotus corniculatus arvenis*
Wheat (forage)	*Triticum aestivum*
Wheatgrass, standard crested	*Agropyron sibiricum*
Wheatgrass, intermediate	*Agropyron intermedium*
Wheatgrass, slender	*Agropyron trachycaulum*
Wheatgrass, western	*Agropyron smithii*
Wildrye, beardless	*Elymus triticoides*
Wildrye, Canadian	*Elymus canadeneis*
Vegetable crops	
Artichoke	*Helianthus tuberosus*
Beet, red	*Beta vulgaris*
Squash, zucchini	*Cucurbita pepo melopepo*
Fruit and nut crops	
Fig	*Ficus carica*
Jujube	*Ziziphus jujuba*
Olive	*Olea europaea*
Papaya	*Carica papaya*
Pineapple	*Ananas comosus*
Pomegranate	*Punica granatum*

Table 7.2. (*Continued*)

(*Continued*)

Salinity moderately sensitive	
Fiber, seed and sugar crops	
Broadbean	*Vicia faba*
Castorbean	*Ricinus communis*
Maize	*Zea mays*
Flax	*Linum usitatissimum*
Millet, foxtail	*Setaria italica*
Groundnut/Peanut	*Arachis hypogaea*
Rice, paddy	*Oryza sativa*
Sugarcane	*Saccharum officinarum*
Sunflower	*Helianthus annuus*
Grasses and forage crops	
Alfalfa	*Medicago sativa*
Bentgrass	*Agrostis stolonifera palustris*
Bluestem, Angleton	*Dichanthium aristatum*
Brome, smooth	*Bromus inermis*
Buffelgrass	*Cenchrus ciliaris*
Burnet	*Poterium sanguisorba*
Clover, alsike	*Trifolium hydridum*
Clover, Berseem	*Trifolium alexandrinum*
Clover, ladino	*Trifolium repens*
Clover, red	*Trifolium pratense*
Clover, strawberry	*Trifolium fragiferum*
Clover, white Dutch	*Trifolium repens*
Corn (forage) (maize)	*Zea mays*
Cowpea (forage)	*Vigna unguiculata*
Dallis grass	*Paspalum dilatatum*
Foxtail, meadow	*Alopecurus pratensis*
Grama, blue	*Bouteloua gracilis*
Lovegrass	*Eragrostis* sp.
Milkvetch, Cicer	*Astragalus cicer*
Oatgrass, tall	*Arrhenatherum danthonia*
Oats (forage)	*Avena sativa*
Orchard grass	*Dactylis glomerata*
Rye (forage)	*Secale cereale*
Sesbania	*Sesbania exaltata*

Table 7.2. (*Continued*)

Siratro	*Macroptilium atropurpureum*
Sphaerophysa	*Sphaerophysa salsula*
Timothy	*Phleum pratense*
Trefoil, big	*Lotus uliginosus*
Vetch, common	*Vicia angustifolia*
Vegetable crops	
Broccoli	*Brassica oleracea botrytis*
Brussels sprouts	*B. oleracea gemmifera*
Cabbage	*B. oleracea capitata*
Cauliflower	*B. oleracea botrytis*
Celery	*Apium graveolens*
Corn, sweet	*Zea mays*
Cucumber	*Cucumis sativus*
Eggplant	*Solanum melongena esculentum*
Kale	*Brassica oleracea acephala*
Kohlrabi	*B. oleracea gongylode*
Lettuce	*Latuca sativa*
Muskmelon	*Cucumis melo*
Pepper	*Capsicum annuum*
Potato	*Solanum tuberosum*
Pumpkin	*Cucurbita peop pepo*
Radish	*Raphanus sativus*
Spinach	*Spinacia oleracea*
Squash, scallop	*Cucurbita pepo melopepo*
Sweet potato	*Ipomoea batatas*
Tomato	*Lycopersicon lycopersicum*
Turnip	*Brassica rapa*
Watermelon	*Citrullus lanatus*
Fruit and nut crops	
Grape	*Vitis* sp.
Salinity sensitive	
Fibre, seed and sugar crops	
Bean	*Phaseolus vulgaris*
Guayule	*Parthenium argentatum*
Sesame	*Sesamum indicum*

(*Continued*)

Table 7.2. (*Continued*)

Vegetable crops	
Bean	*Phaseolus vulgaris*
Carrot	*Daucus carota*
Okra	*Abelmoschus esculentus*
Onion	*Allium cepa*
Parsnip	*Pastinaca sativa*
Fruit and nut crops	
Almond	*Prunus dulcis*
Apple	*Malus sylvestris*
Apricot	*Prunus armeniaca*
Avocado	*Persea americana*
Blackberry	*Rubus sp.*
Boysenberry	*Rubus ursinus*
Cherimoya	*Annona cherimola*
Cherry, sweet	*Prunus avium*
Cherry, sand	*Prunus besseyi*
Currant	*Ribes* sp.
Gooseberry	*Ribes* sp.
Grapefruit	*Citrus paradisi*
Lemon	*Citrus limon*
Lime	*Citrus aurantiifolia*
Loquat	*Eriobotrya japonica*
Mango	*Mangifera indica*
Orange	*Citrus sinensis*
Passion fruit	*Passiflora edulis*
Peach	*Prunus persica*
Pear	*Pyrus communis*
Persimmon	*Diospyros virginiana*
Plum: Prume	*Prunus domestica*
Pummelo	*Citrus maxima*
Raspberry	*Rubus idaeus*
Rose apple	*Syzygium jambos*
Sapote, white	*Casimiroa edulis*
Strawberry	*Fragaria* sp.
Tangerine	*Citrus reticulata*

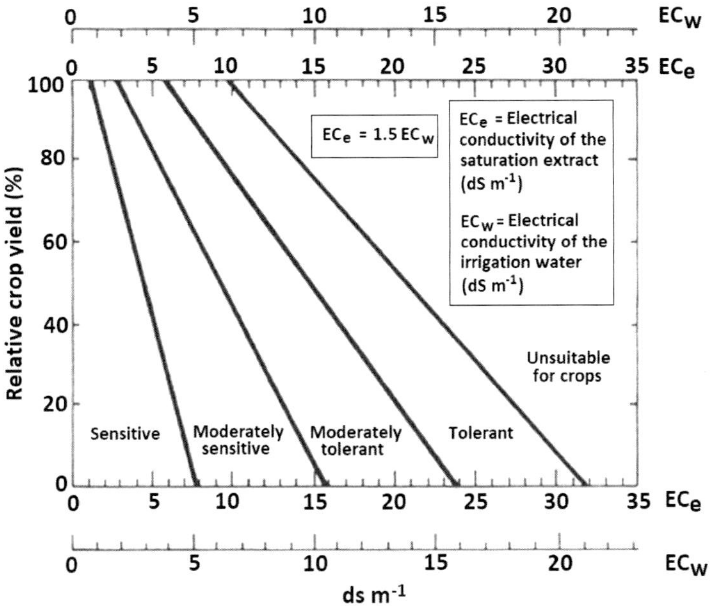

Fig. 7.4. Relative crop yields for relative salt tolerance ratings based on soil salinity (EC_e) and water salinity (EC_w) in dS m^{-1} (≡ mS cm^{-1}) (Doorenbos and Pruitt, 1977; Maas and Grattan, 1999). Reproduced with permission from the FAO, Rome, Italy, and John Wiley Inc., New York.

7.4 Forestry

The use of short rotation coppicing using species of willow (e.g. *Salix vimimalis, S. schwerinii* and *S. dasyclados*) and poplar (*Populus* spp.) is increasingly popular in Europe for the treatment of secondary or tertiary treated wastewaters, mainly as a cost-effective alternative for nitrogen removal, and also for reusing nutrients and water (i.e., fertigation) (Dickman *et al.*, 1996; Rosenqvist *et al.*, 1997). Elsewhere, Eucalyptus are also used (e.g. *Eucalyptus botryoides, E. camaldulensis, E. cyanophylla, E. globulus, E. grandis, E. ovata*) for treating a wide range of wastewaters, including landfill leachate. Unlike other SRLT systems, short rotation coppicing using wastewater produces biomass for renewable energy, while also increasing carbon storage in the soil, both of which make a significant contribution to reducing all the overall carbon dioxide emissions of treatment. Another advantage of the system is that there is no risk of pathogen or chemical contamination as opposed to food or fodder crops (Perttu, 1999). Willow trees are ideal trees as they utilize significant nutrients, have a high filtering capacity, tolerate wet conditions, have high evapotranspiration rates and good wood yields, and promote denitrification in the root zone. Wastewater is an ideal source of nutrients and trace elements needed for high growth rates although willow selectively bioaccumulates a range of heavy metals, especially cadmium (Aronsson and Perttu, 2001).

In 2002, there was approximately 18,000 ha of land in Sweden being used for short rotation willow coppice (SRWC) and this has continued to grow throughout the country and elsewhere in Europe. Most treatment areas in Sweden are planted using cuttings (which readily root) and spaced in double rows 0.75 m apart with each double row 1.5 m apart. Cuttings are planted 0.5 m apart within the rows, giving an approximate planting density of 18,000 trees per ha. Coppicing is done by a harvester, which also chips the stems for use as biomass energy, every 3 to 5 years depending on growth. In commercial systems, fertilization rates of 80 to 100 kg N $ha^{-1}yr^{-1}$ and 4.5 kg P kg N $ha^{-1}yr^{-1}$ are required, which can be achieved by utilizing the nitrogen and phosphorus in wastewater, which reduces operational costs by 25 to 30% (Dimitriou and Rosenqvist, 2011). The standard hydraulic loading of wastewater to Swedish SRWC plantations is approximately 600 mm yr^{-1}, with average timber yields in Northern Europe between 6.2 and 11.0 t DM ha^{-1} yr^{-1}. Wastewater irrigation can increase wood yields by up to 40% compared to non-wastewater treated sites (Borjesson and Berndes, 2006). In Sweden, treated effluent is applied by drip irrigation systems to reduce potential pathogen transfer, which makes even distribution difficult. Even the distribution of wastewater is important, otherwise, it can lead to uneven growth, resulting in neither maximum treatment potential nor timber production being achieved. As nitrogen retention in

SRWC systems fed by wastewater is correlated with biomass productivity, maximum plant growth is required if nutrients are to be effectively removed to prevent ground or surface water pollution (Aronsson *et al.*, 2002).

Pilot-scale SRWCs were trialed in Canada to treat primary-treated wastewater from small communities and households (Lachapelle *et al.*, 2019). Two wastewater hydraulic loadings (L1 at 10 mm d^{-1} and L2 at 16 mm d^{-1}) were compared with a control using groundwater (L0 at 14 mm d^{-1}) for 111 days. Removal rates were excellent, a 91% removal of chemical oxygen demand (COD) for both wastewater plots L1 and L2 was achieved; and 94 and 87% for L1 and L2 total nitrogen (TN) removal rates, respectively. As with other studies, the removal of total phosphorus (TP) was very high (98%), although this coincided with a similar increase in soil phosphorus, which suggests that once the soil has become saturated, uptake will be reduced significantly. The fate of both TN and TP are shown graphically for these experimental plots in Fig. 7.5. Organic loading was not seen to be a limiting factor in regard to loading, but rather nitrogen removal. These plots were continuously loaded at a constant rate and it became evident that there was a large seasonal variation between the nutrients applied and what was needed by the plants, leading to potential groundwater contamination. The authors suggest controlling the irrigation rate according to both plant nutrient needs and the evapotranspiration rate, which both decline during plant dormancy when temporary storage of wastewater will be required. The storage of wastewater is also required during colder periods when the ground is frozen. Storage has design implications in terms of the extra loading generated due to storage during active growth periods.

Many clones have been developed for use with the SRWC system. For example, Gudrun (*Salix burjatica* × *S. dasyclados*) and Tora (*S. schwerinii* × *S. verminali*) both have high productivity with lower fertilization but high irrigation requirements. Holm and Heinsoo (2013) compared a range of willow clones in Estonia, including Gudrun and Tora, for nutrient removal from treated wastewater using a planting density of 14,800 plants per hectare with irrigation pipes every 9 m and loaded at 160 mm yr^{-1}. The plots were irrigated during

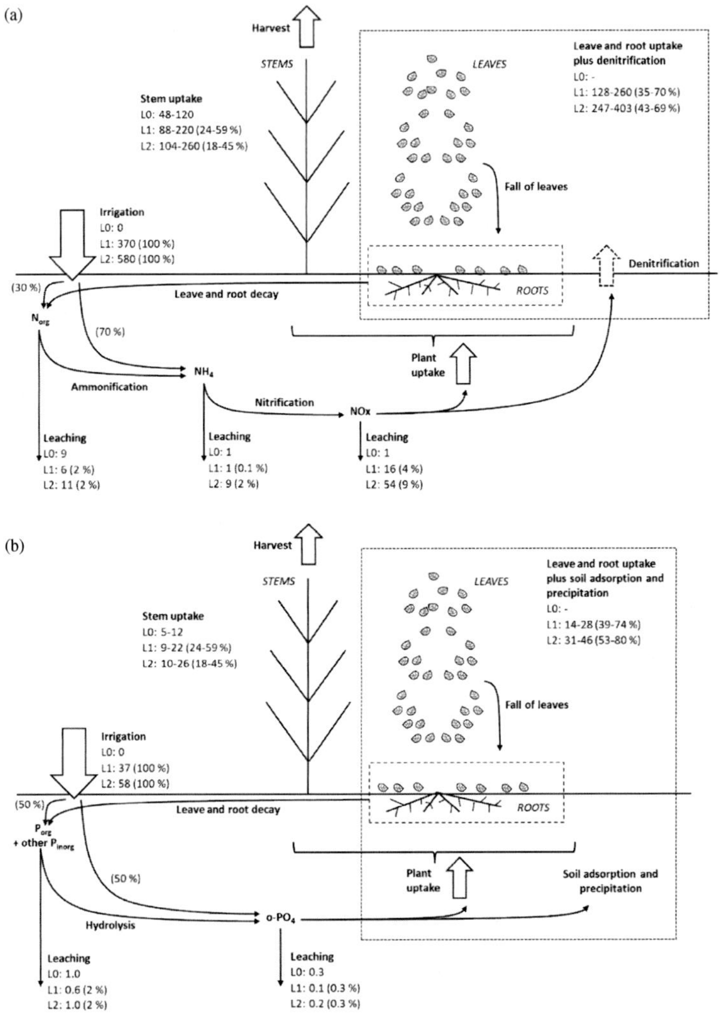

Fig. 7.5. The fate of nitrogen and phosphorous in a short rotation willow coppice system loaded with primary-treated wastewater at two hydraulic loadings (L1 at 10 mm d^{-1} and L2 at 16 mm d^{-1}) compared with a control using groundwater (L0 at 14 mm d^{-1}) (Lachapelle *et al.*, 2019). Reproduced with permission from Elsevier Publishing.

the growing season from May to November and only at night. Nutrient removal was 58% TN and 70% TP as measured by lysimeters located 40 cm below the plantation, which was sufficient to protect the groundwater. They also found that using wastewater increased the number of shoots produced and the plant's dry weight and wood yield. Employing higher planting density or denser irrigation networks did not increase productivity.

Vegetative filter systems such as the SRWC can treat wastewaters to the same degree of purity as conventional biological secondary wastewater treatment, with even better TN and TP removal at very low costs. Conventional short rotation coppice plantations are less profitable than those irrigated, with wastewater making SRWC not only a sustainable treatment option but viable biomass for an energy system. Coupling biomass production for energy and phytoremediation is a perfect example of the concept of a circular economy. Cozzi *et al.* (2015) carried out an economic feasibility study of using SRWC for wastewater irrigation in Northern Italy and found them to be beneficial in providing local bioenergy and other economic advantages. There are problems in adopting the technology apart from climatic conditions, and these include finding suitable sites in terms of soil and groundwater suitability and sufficient land availability close enough to existing wastewater treatment plants. The distance to potential plantation sites was the major limiting factor in Northern Italy, with only 25 out of 163 wastewater treatment plants evaluated as being suitable in terms of cost-effectiveness. This is explored in more detail in Chap. 2.

Irrigation with wastewater is not only limited to short rotation coppice plantations. In the US, wastewater is used on large-scale commercial forestry, such as a 1,460-ha coniferous forest in Clayton County, Georgia, using 75,700 m^3d^{-1}, and a 3,232-ha forest located in Dalton, Georgia (McKim, 1982). Forested systems have longer wastewater application seasons than agricultural crops, and have high hydraulic loadings but relatively poor nitrogen removal, although this is dependent not only on tree species, but the age of the trees and, importantly, the soil conditions (USEPA 1974; McKim 1982). Careful selection of forestry species is important with growth requirements, climatic suitability, soil salinity (Table 7.3), and tolerance to flooding (Table 7.4), all being important selection criteria.

Table 7.3. Salt tolerance rating of plant species commonly used in slow-irrigation land treatment systems (Paranychianakis *et al.*, 2006). Reproduced with permission from the Taylor & Francis Group.

Tolerance to salt	Scientific name	Common name
High	*Pinus radiata*	Monterey pine
	Eucalyptus spathulata	Swamp mallet
	Eucalyptus occidentalis	Swamp yate
	Eucalyptus camaldulensis	Red gum
	Acacia stenophylla	Shoestring acacia
	Acacia saligna	Orange wattle
	Populus euphratica	Euphrates poplar
	Melaleuca lanceolata	Moonah
	Melaleuca bracteata	Black teatree
	Cynodon spp.	Bermuda grass
	Beta vulgaris	Sugar beet
Moderate	*Acacia auriculiformis*	Darwin black wattle
	Acacia salicina	Willow wattle
	Eucalyptus grandis	Rose gum
	Eucalyptus robusta	Swamp mahogany
	Eucalyptus astringens	Brown mallet
	Eucalyptus botryoides	Southern mahogany
	Melaleuca leucadendra	Cajuput
	Melaleuca armillaris	Green globe
	Sorghum verticilliflorum	Sudan grass
	Chloris gayana	Rhodes grass
	Festuca spp.	Tall fescue
	Medicago sativa	Alfalfa
	Zea mays	Corn
Low	*Acacia pendula*	Weeping myall
	Acacia victoriae	Bramble wattle
	Eucalyptus globulus	Tasmanian bluegum
	Eucalyptus intermedia	Bloodwood
	Populus x euramericana	Hybrid poplar
	Platanus spp.	Sycamore
	Populus spp.	Poplar
	Sorgum spp.	Sorgum
	Trifolium spp.	Clover

Table 7.4. Flooding tolerance of plants commonly used in slow-irrigation land treatment systems (Paranychianakis *et al.*, 2006). Reproduced with permission from the Taylor & Francis Group.

Tolerance to flooding	Scientific name	Common name
High	*Eucalyptus ovata*	Swamp gum
	Eucalyptus robusta	Swamp mahogany
	Eucalyptus occidentalis	Swamp yate
	Populus deltoides	Eastern cottonwood
	Salix nigra	Black willow
	Populus balsamifera	Balsam poplar
Moderate	*Eucalyptus camaldulensis*	Red gum
	Plantanus occidentalis	American sycamore
Low	*Eucalyptus grandis*	Rose gum
	Eucalyptus globulus	Tasmanian bluegum
	Pinus radiata	Monterey pine
	Populus grandidentata	Big tooth aspen
	Populus tremuloides	Quaking aspen

In conclusion, the green spaces occupied by slow-rate irrigation systems allow them to be used for wastewater treatment in the most sensitive landscapes, preserve open space and promote insects in particular, and provide a varied array of habitats. At a time when natural wetlands are at risk, a number of specialist plants thrive in treatment areas. Slow rate irrigation provides sustainable and low-energy treatment systems that capture and reuse useful nutrients and water to produce crops, making them a key tool in water conservation management in water-stressed and arid zones. Importantly, the reuse of treated water for irrigation preserves water supplies and protects surface and ground waters from over-abstraction. Of course, these advantages vary according to local conditions and the degree of water stress.

However, public health issues should be considered when handling partially treated wastewaters and in the design of irrigated plantations. Carlander *et al.* (2009) carried out quantitative microbial risk assessments (QMRAs) at a number of European SRWC sites. The main risks are to workers from the accidental ingestion of wastewater, and to workers and also those living close by the plantations from exposure to aerosols and the ingestion of pathogen-contaminated groundwater. They are examples of pathogen transfer to workers handling the wastewater as well as groundwater contamination. They recommended site-specific QMRAs for all sites to estimate the required level of pathogen reduction in wastewaters before use and to be part of the design of new plantations.

References

Aronsson, P. and Perttu, K. (2001) Willow vegetation filters for wastewater treatment and soil remediation combined with biomass production. *The Forestry Chronicle*, 77(2), 293–299.

Aronsson, P., Heinsoo, K., Perttu, K. and Hesselgren, K. (2002) Spatial variation in above-ground growth in unevenly wastewater-irrigated willow *Salix viminalis* plantations. *Ecological Engineering*, **19**, 281–287.

Bausum, H.T., Schaub, S.A., Kenyon, K.F. and Small, M.J. (1982) Comparison of coliphage and bacterial aerosols at a wastewater spray irrigation site. *Applied Environmental Microbiology*, **43**, 28.

Bennett, D.R. and Harms, T.E. (2011) Crop yield and water requirement relationships for major irrigated crops in Southern Alberta. *Canadian Water Resources Journal*, **36**(2), 159–170.

Borjesson, P. and Berndes, G. (2006) The prospects for willow plantations for wastewater treatment in Sweden. *Biomass Bioenergy*, **30**, 428–438.

Brown, K.W., Thomas, J.C. and Slowey, J.F. (1983) The movement of metals applied to soils in sewage effluent. *Water, Air and Soil Pollution*, **19**, 43–54.

Carlander, A., Shonning, C. and Strenstrom, T.A. (2009) Energy forest irrigated with wastewater: A comparative microbial risk assessment. *Journal of Water and Health*, 7, 413–433.

Cozzi, M., Viccaro, M., Di Napoli, F., Fagarazzi, C., Tirinnanzi, A. and Romano, S. (2015) A spatial analysis model to assess the feasibility of short rotation forestry fertigated with urban wastewater: Basilicata region case study. *Agricultural Water Management,* **159**, 185–196.

Dickman, D.I., Nguyen, P.V. and Pregitzer, K.S. (1996) Effects of irrigation and coppicing on above-ground growth, physiology and fine-root dynamics of two field-grown hybrid poplar clones. *Forest Ecology and Management*, **80**, 163–174.

Dimitriou, I. and Rosenqvist, H. (2011) Sewage sludge and wastewater fertilization of short rotation coppice (SRC) for increased energy production — Biological and economic potential. *Biomass Bioenergy*, **35**, 835–842.

Dinges, R. (1982) *Natural Systems For Water Pollution Control.* Van Nostrand Reinhold, New York.

Doorenbos, J. and Kassam, A.H. (1979) Yield response to water. FAO Irrigation and Drainage Paper 33, Food and Agricultural Organization of the United Nations, Rome.

Doorenbos, J. and Pruitt, W.O. (1977) Crop water requirements. FAO Irrigation and Drainage paper: 24. Food and Agriculture Organization, Rome. http://www.fao.org/3/f2430e/f2430e.pdf

González-Méndez, B., Ruiz-Suárez, L. and Siebe, C. (2020) N_2O emission factors from a wastewater irrigated land in a semiarid environment in Mexico. *Science of the Total Environment*, **709**, 136177.

Hernández-Martínez, J., Prado, B., Cayetano-Salazar, M., Bischoff, W.-A. and Siebe, C. (2018) Ammonium-nitrate dynamics in the critical zone during single irrigation events with untreated sewage effluents. *Journal of Soils and Sediments*, **18**, 467–480.

Holm, B. and Heinsoo, K. (2013) Municipal wastewater application to short rotation coppice of willows — Treatment efficiency and clone response in Estonian case study. *Biomass and Bioenergy*, **57**, 126–135.

Koerner, E.L. and Haws, D.A. (1979) Long-term effects of land application of domestic wastewater, Roswell, New Mexico Slow-Rate irrigation site. EPA 600.2-79—047. US Environmental Protection Agency, Cincinnati, OH.

Lachapelle, X., Labrecque, M. and Comeau, Y. (2019) Treatment and valorization of a primary municipal wastewater by a short rotation willow coppice vegetation filter. *Ecological Engineering*, **130**, 32–44.

Maas E.V. (1984) Salt tolerance of plants. In *The Handbook of Plant Science in Agriculture* (ed. B.R. Christie). CRC Press, Boca Raton, Florida.

Maas, E.V. and Grattan, S.R. (1999) Crop yields as affected by salinity. In *Agricultural Drainage* (eds. R.W. Skaggs and J.A. Schilfgaarde), American Society of Agronomy, Madison, WI, USA, pp. 55–108.

McKim, H.L. (1982) Wastewater application in forest ecosystems. Report 82-19. USA CRREL, Hanover, NH, USA. https://erdc-library.erdc.dren.mil/jspui/bitstream/11681/9364/1/CR-82-19.pdf

Metcalf and Eddy, Inc. (1991) *Wastewater Engineering: Treatment, Disposal and Reuse, 3rd ed.* McGraw-Hill, New York, USA.

Müller, K., Magesan, G.N. and Bolan, N.S. (2007) A critical review of the influence of effluent irrigation on the fate of pesticides in soil. *Agriculture, Ecosystems and Environment*, **120**(2–4), 93–116.

Nemera, D.B., Bar-Tala, A., Levya, G.J., Lukyanova, V., Tarchitzkyb, J., Paudela, I. and Cohen, S. (2020) Mitigating negative effects of long-term treated wastewater application via soil and irrigation manipulations: Sap flow and water relations of avocado trees (*Persea americana* Mill.). *Agricultural Water Management*, **237**, 106178.

Ou, Z., Gao, Z. and Sun, T. (1992a) Paddy rice slow-rate land treatment systems: I. Design of hydraulic loading rates. *Water Research*, **26**(11), 1479–1486.

Ou, Z., Chang, S., Gao, Z., Sun, T., Qi, E., Ma, X., Li, P., Zhang, L., Yang, G., Han, S., Song, Y., Zhang, H., Ren, L. and Ou, X. (1992b) Paddy rice slow-rate land treatment systems: II. Hydraulic balances and results of 4 years operation. *Water Research*, **26**(11), 1487–1494.

Paranychianakis, N.V., Angelakis, A.N., Leverenz, H. and Tchobanoglous, G. (2006). Treatment of wastewater with Slow Rate systems: A review of treatment process and plant functions. *Critical Reviews in Environmental Science and Technology*. **36**, 187–259.

Perttu, K.L. (1999) Environmental and hygenic aspects of willow coppice in Sweden. *Biomass Bioenergy*, **16**, 291–297.

Rosenqvist, H., Aronsson, P., Hasselgren, K. and Perttu, K. (1997) Economics of using municipal wastewater irrigation of willow coppice crops. *Biomass and Bioenergy*, **12**, 1–8.

Shaub, S.A., Bausum, H.T. and Taylor, G.W. (1982) Fate of viruses in waste-water applied to slow-infiltration land treatment systems. *Applied Environmental Microbiology*, **44**, 383–394.

Stewart, J.I., Danielson, R.E., Hanks, R.J., Jackson, E.B., Hagan, R.M., Pruitt, W.O., Franklin, W.T. and Riley, J.P. (1977) Optimizing crop production through control of water and salinity levels in the soil. Report, PRWG 151-1. Utah Water Laboratory, Logan, Utah.

Tare, V. and Bokil, S.D. (1982) Wastewater treatment by soils: Role of particle size distribution. *Journal of Environmental Quality*, **11**, 596–602.

USEPA. (1974) Renovation of secondary effluent for reuse as a water resource. EPA 660/2-74-016. US EPA CERI, Cincinnati, OH.

USEPA. (1977) Process design manual for land treatment of municipal wastewater. EPA 625/1-77-008. US Environmental Protection Agency, Cincinnati, OH.

USEPA. (2006) Land treatment of municipal wastewater effluents. Process design manual, EPA/625/R-06/016. US Environmental Protection Agency, Cincinnati, OH.

WEF. (1992) Design of municipal wastewater treatment plants. Manual of practice No. 76. Water Environment Federation, Alexandria, VA.

Xue, Y., Yang, P., Luo, Y., Li, Y., Ren, S., Su, Y. and Niu, Y. (2012) Characteristics and driven factors of nitrous oxide and carbon dioxide emissions in soil irrigated with treated wastewater. *Journal of Integrative Agriculture*, **11**(8), 1354–1364.

8
Rapid Infiltration Processes

High-rate irrigation and infiltration–percolation systems are very similar, except that the former is used on natural sandy soils, which are irrigated at very high rates using sprayers, whereas the latter uses specially constructed basins in suitably permeable soils so that the treated wastewater can soak into the soil to the groundwater. Rapid infiltration basins are used for high-quality treated effluents primarily for aquifer recharge. Where used for tertiary wastewater treatment, infiltration–percolation basins are referred to as soil aquifer treatment (SAT) systems. Rapid infiltration processes are all passive treatment systems able to treat large volumes of wastewater relying on gravity feed, making them very sustainable in terms of energy use and greenhouse gas emissions. Pumping is only required where water is reused. They require no chemicals, are simple to operate, and are especially suited for small-medium-sized wastewater treatment plants. They do, however, require significant land area compared to other treatment and recharge systems.

8.1 High-rate Irrigation

High-rate irrigation uses hydraulic loading rates of between 1.6 and 25 cm d^{-1}, which is equivalent to an area of 0.4 to 6 ha per 1,000 m^3d^{-1} (Metcalf and Eddy, 1991; WEF, 1992). Operation is continuous and due to the small area of land used, factors such as evapotranspiration and precipitation are unimportant. While irrigation sites are normally bare, they can have vegetation, but this requires regular plowing to prevent the accumulation of organic matter on the surface which could restrict long-term percolation efficiency. Even small amounts of clay, silt or humic material can severely limit the capacity of the soil to remove metals, phosphorus and other wastewater constituents. Thus, the groundwater is much more vulnerable and receives a significant volume of mineral contamination. At these high loading rates, there is only partial removal occurring with pathogens, especially viruses, reaching the groundwater, although both primary and secondary wastewaters have been successfully treated (Carlson *et al.*, 1982). Although widely used in the United States (US) as a secondary treatment process, high-rate irrigation is generally used for aquifer recharge using high-quality secondary treated effluents.

8.2 Irrigation–percolation

8.2.1 *Rapid infiltration basins (RIBs)*

Irrigation–percolation is also restricted to sandy and sandy-loam soils, which have high permeability, allowing the applied wastewater to pass quickly through the soil to the groundwater. Lagoons or basins are specially constructed to enhance rapid percolation and are usually referred to as rapid infiltration basins (RIBs) (Fig. 8.1).

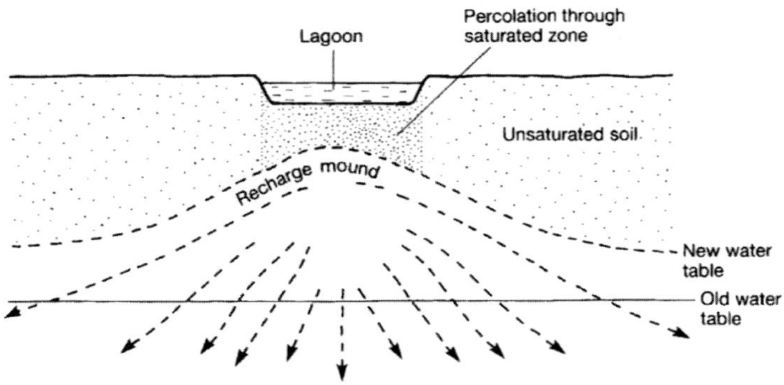

Fig. 8.1. Land treatment by infiltration–percolation using a rapid infiltration basin. The high volumes of water entering the unsaturated (vadose) create a recharge mound that interfaces with the existing water table.

RIBs have been used throughout North America and other arid regions including Australia. They have been successfully employed throughout the southern states of the US for over a hundred years, originally for effluent disposal, but now primarily for groundwater recharge using high-quality advanced treated effluents. There is an urgent need for water recycling in many southern states of the US and other arid regions, and aquifer recharge has many significant benefits, including long-term water storage, protection of surface waters from over-abstraction, free distribution of water via the aquifer, and also a small tertiary treatment capability, especially for compounds not fully removed during the wastewater treatment, such as nitrogen, phosphorous, micro-pollutants including pharmaceutical and personal care products (PPCPs), and pathogens (Sec. 8.2.2).

RIBs are shallow basins overlying a deep layer of highly porous soil that extends to the aquifer below, so soil permeability and the depth of soil above the water table are important criteria for site selection. The base of the basin is kept free from vegetation and is regularly scarified to ensure maximum permeability is maintained. Sometimes, the surface is contoured into a series of low dykes to ensure equal loading per unit surface area of basin. The RIB is designed with a slight slope to help encourage the infiltration water to flow away from the basin to prevent flooding at the site. The area of land required for rapid infiltration basins used for recharge can be estimated as:

$$A = \frac{(1.9)(Q)}{(L)(P)}, \tag{8.1}$$

where A is the area required for treatment (ha), Q is the expected daily flow (m^3d^{-1}), L is the annual designed percolation rate (m yr^{-1}), and P is the period of operation (weeks per year). P is expected to be 52 for normal operation, working on a year-round basis.

Loading is done in cycles with the basin flooded, then left to allow for complete infiltration, and then left to dry using the Wet to Dry ratio of 1:2. The top 1.5 to 3.0 m of the underlying soil should be unsaturated before the next loading cycle begins, which can take up to 3 weeks. The design should allow for a high rate of infiltration followed by rapid percolation, which occurs both vertically and laterally.

Long narrow basins are preferred with the longer side perpendicular to the direction of groundwater flow, which ensures rapid dispersal of infiltrated water and the prevention of water table mounding (Sec. 8.2.3). Treatment of the infiltration water occurs in the soil layer due to filtration, adsorption, ion exchange, precipitation, and microbial action. Therefore, the rate of percolation must allow sufficient time for the required level of adsorption of nutrients and contaminants as well as microbial activity to occur. The minimum soil permeability required is 1.5 cm h^{-1} (Fig. 8.2). High hydraulic loading rates create high flow velocities through the soil, preventing sufficient contact with the soil particles for effective nutrient removal, and for that reason, hydraulic loading rarely exceeds 50 mm per hour. The annual hydraulic loading rate range from 6 to 90 m yr^{-1} is equivalent to a surface loading rate of 6 to 92 $m^3m^{-2}yr^{-1}$. Whilst there is no minimum number of RIBs, there must always be at least two separate basins to allow for each to dry out between alternate use. Depending

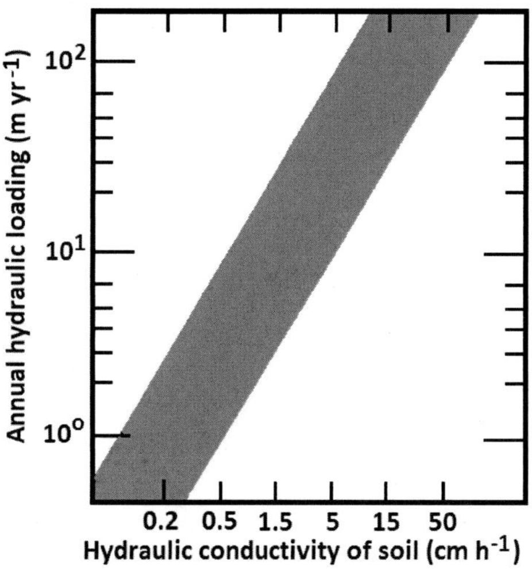

Fig. 8.2. The acceptable design range (shaded area) for hydraulic loading (m yr^{-1}) to a rapid infiltration basin (RIB) based on the hydraulic conductivity (cm h^{-1}) of the underlying soil.

on the hydraulic loading, the area required will be between 0.0003 and 0.0055 ha per m^3d^{-1} with individual basins varying in size from 0.2 to 2.5 ha for small to medium systems, rising to 5 to 8 ha each for large systems. Larger basins require higher freeboards of at least 1.0 m due to larger wind-induced wave action.

Groundwater quality is protected by employing monitoring wells, usually one up gradient and at least two down gradient following the path of the groundwater from the RIB.

The type of soil or medium employed in the RIB controls the infiltration rate as well as overall permeability, with coarse sand or sandy gravel preferred, which, in turn, controls the degree of purification that the wastewater receives from the various physicochemical and microbial mechanisms before it joins the underlying groundwater. All RIBs suffer from an accumulation of organic material on the sand surface, which must be removed every 12 to 18 months. For this reason, a good secondary settlement at the wastewater treatment plant is required, so they are generally not suitable for effluents from facultative oxidation or maturation ponds where algal is present, as this accumulates at the bottom of the RIB, impeding infiltration, unless they are first removed by microfiltration. Coupled with different operational procedures, treatment efficiencies can vary widely, for example, using model simulations, Akhaven *et al.* (2013) found denitrification to vary between 2 and 49% and have proposed a number of operational actions to improve treatment efficiency. The Landis Sewerage Authority in New Jersey recharges the local groundwater with 3,100 m^3d^{-1} of advanced treated wastewater effluent. Other examples are given in (USEPA, 2003). One of the largest RIB systems in the world is being used for the Water Conserv II project in Orlando, Florida, where advanced treated wastewater from two large municipal treatment plants is pumped to a central distribution center where the high-quality effluent is either used for irrigation of 1,110 ha of citrus crops or discharged into RIBs for aquifer recharge. There are multiple groups of RIBs, each comprising a single or up to five basins, each approximately 110 m long and 45 m wide, built over a natural unsaturated sand layer of between 9 and 60 m in depth. Full details can be found in (Water Conserv II, 2020) (Fig. 8.3).

Fig. 8.3. Rapid infiltration basins used for groundwater recharge of wastewater that has received advanced treatment at Orlando, Florida. Reproduced with permission from the US Geological Survey.

In terms of aquifer recharge, RIBs serve the same purpose as vadose injection wells or direct injection wells (USEPA, 2006), except they provide a level of extra protection to groundwater quality as they offer a limited degree of treatment. Full details of site selection and the design of RIBs have been given by the United States Environmental Protection Agency (USEPA, 1984; 2003; 2006), although useful comments on the design are also available from local state agencies in the US (NDEP, 2017).

8.2.2 *Soil aquifer treatment (SAT) systems*

No matter how clean the water used to recharge an aquifer is, there will be a change in the chemical nature of the groundwater it is discharged into. The objective is to minimize the negative impact of recharge on the quality by ensuring that nutrients (nitrogen and phosphorus), soluble organic matter, pathogens, and trace contaminants such as metals and complex organic compounds, including PPCPs, do not reach the water table. So RIBs should only be used for the hydraulic dispersal of very clean, treated wastewaters that should ideally conform to the drinking water regulations. Where the treated wastewaters required further tertiary treatment, then a specially designed SAT system should be used, but only where the aquifer and the unsaturated (vadose) zone above it are deemed suitable. Expected long-term removal efficiencies in a well-designed and operated SAT are between 50 and 60% total nitrogen (TN) and 80 and 99% total phosphorus (TP). While SAT systems utilize a rapid infiltration basin, a number of modifications are made to the design, including underground drainage systems (Fig. 8.4b) and the use of specialist media to encourage higher microbial activity and adsorption. The impact on groundwater can be minimized by locating RIBS and SAT systems close to the surface water (Fig. 8.4c).

Biosorption activated media (BAM) was originally developed for use in sustainable urban drainage systems, especially swales, to improve nutrient removal. There is growing interest in its use in RIBs. The media is commercially produced and comes in many different combinations such as sands, recycled materials, such as granulated tires (i.e., tire crumb), and clay, providing a wide range of permeability and treatment capabilities. Compared to sand, BAM can remove 30% more nitrate (Sand 47%; BAM 78%) and 62% more total phosphorus (Sand 12%; BAM 28%) using a 0.6-m-deep layer of medium (O'Reilly *et al.*, 2012). Wen *et al.* (2020) were able to directly compare the removal of nitrogen from a RIB treating reclaimed wastewater using two different filtration media by dividing a full-scale basin into two equal sections. The local sandy soil (2 to 4% clay) was placed 60 cm deep at the bottom of one section and the other with an equal depth of BAM. The BAM used was a mixture of granulated tires, sand and clay with a specific surface area of 0.71 m^2g^{-1}, a porosity of 40%, and a hydraulic conductivity of 0.026 cm s^{-1}, compared to 0.32 m^2g^{-1}, 34% and 0.020 cm s^{-1} for the sandy medium used as the control. The nitrogen removal in the BAM was 83%, double that of the control medium at 40% (Cormier and Duranceau, 2019).

The removal of nitrogen is dependent on the presence of specialist microorganisms (Fig. 8.5) that form a biofilm on the active medium layer of RIBs. This biofilm is similar to that found in many other

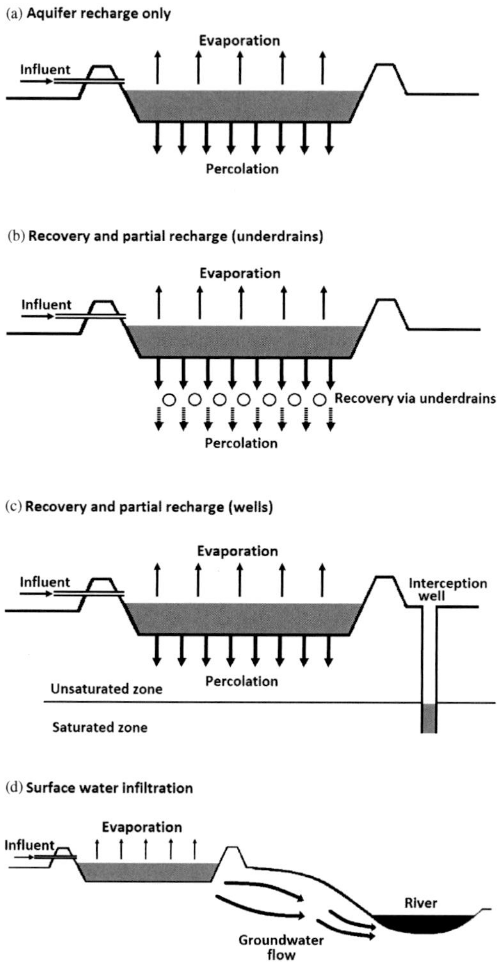

Fig. 8.4. Basic layout for rapid sand filtration: (a) for aquifer recharge directly into unsaturated zone, (b) recovery of treated effluent below the basin using a drainage pipe network for disposal or reuse elsewhere, with a portion entering groundwater, (c) recovery of treated effluent by interception wells that are located in the effluent plume down flow of the basin, which is reused elsewhere usually for irrigation, with a portion entering groundwater, and (d) where hydrology permits the treated effluent to flow directly into the river through porous soil, similar to bankside infiltration, except that the unsaturated zone is used under the basin and towards the river with some loss to groundwater.

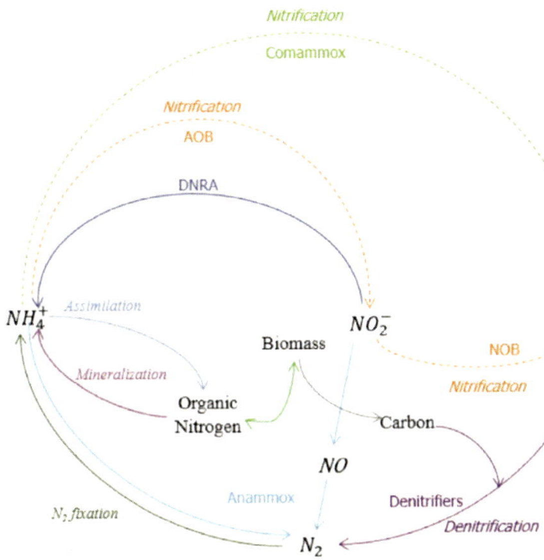

Fig. 8.5. The removal of nitrogen comprises many different microbial pathways involving many different microorganisms, including ammonia-oxidizing bacteria (AOB), nitrite-oxidizing bacteria (NOB), ammonia-oxidizing archaea (AOA), anaerobic ammonium oxidation (anammox) bacteria, denitrifiers, complete ammonia oxidation (comammox) bacteria, and dissimilatory nitrate reduction to ammonia (DNRA) (Wen *et al.*, 2020). Reproduced with permission from Elsevier Publishing.

fixed-film systems (Gray, 2021), with an oxygen gradient through the biofilm resulting in an aerobic zone at the surface and an anoxic zone where denitrification occurs closer to the medium on which it grows. The homogeneous structure of the BAM with its large interstices and high specific surface area ensures good, even distribution of treated wastewater within the medium layer and a more stable supply of nutrients and oxygen. This provides an ideal environment for good biofilm development, and especially for the denitrifiers that are more abundant than in the biofilm growing on the conventional sandy media (Fig. 8.6). In contrast, the smaller voids, lower surface area, and the heterogeneous structure of the sandy medium results in slower infiltration rates and less effective treatment, with the

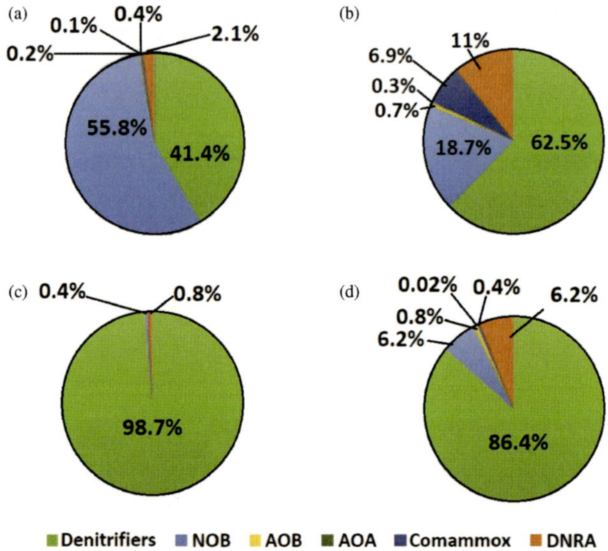

Fig. 8.6. Breakdown of the percentage composition denitrifiers — nitrite-oxidizing bacteria (NOB), ammonia-oxidizing bacteria (AOB), ammonia-oxidizing archaea (AOA), complete ammonia oxidation (comammox) and dissimilatory nitrate reduction to ammonia (DNRA) — in the biofilm collected from the full-scale SAT basins (Wen *et al.*, 2020) using the control sand medium: (a) initial sampling and (b) sampling 12 months later; and for the biological active medium: (c) initial sampling and (d) sampling 12 months later. Reproduced with permission from Elsevier Publishing.

short-circuiting of the treated wastewater via channels through a medium also a problem caused by finer particles blocking the smaller interstices leading to potential pollution of the groundwater (Wen *et al.,* 2020). Groundwater quality is protected by setting total nitrogen limits, which dictate the required level of treatment the wastewater must receive before hydraulic disposal via RIBs. For example, in Nevada, this has been set at <7 mg N L^{-1}, but when the total nitrogen has risen to 10 mg N L^{-1}, the recharge must cease (NDEP, 2017).

Because SAT systems are used as an active treatment system for wastewaters, it may not always be desirable to allow the treated water to enter the groundwater either completely or in part, as with a RIB. There are a number of alternative arrangements that can be used, apart from a 100% direct aquifer recharge: (i) use of underlying drainage pipes to recover a significant proportion of the treated water for reuse or discharge to the surface water, (ii) partial recovery of treated water for reuse using extraction wells positioned at the down flow of the basin, and (iii) recharge of the adjacent surface waters (Fig. 8.4).

One of the best known SAT systems is the Dan Region Wastewater Reclamation project in Tel-Aviv, Israel, originally set up in 1977. It acts as a tertiary treatment process for secondary municipal wastewater for reuse in agricultural irrigation. It effectively further reduces the biochemical oxygen demand (BOD), nutrients, suspended solids, metals, organic contaminants, and pathogens. It currently treats 130×10^6 m^3 of wastewater treated by the activated sludge process. The wastewater infiltrates via four large irrigation basins (15 to 30 m) through sandy soil into the groundwater where it flows to recovery wells situated 300 m and 1,500 m away. Water is recovered and subsequently pumped away and used for agricultural irrigation. Hydraulic loading of secondary-treated settled wastewater is between 80 and 150 m yr^{-1}, with the operational cycle consisting of loading for 24 hours, followed by infiltration over a 48 to 72-hour period. The performance of the SAT system has been extensively researched, achieving good conversion of organic nitrogen and ammonium to nitrate with an overall removal of TN of between 49 and 83% by denitrification and absorption (Idelovitch *et al.,* 2003). However, when the concentration of TN in the feed exceeds 10 mg N L^{-1}, both organic nitrogen and ammonium end up in the groundwater (Mienis and Arye, 2018).

There is still a significant amount of organic matter in settled wastewater, which can cause the SAT process some problems. The organic matter loading can be as much as a hundred-fold higher than applied to a RIB. As with other surface irrigation systems, the majority of the organic matter present in the wastewater is retained in the top 15 cm layer of the soil, thus altering its permeability. The build-up of organic matter can increase the soil's water repellence or hydrophobicity as the sand particles become coated with organic material, which is associated with a retardation of infiltration rates. However, in practice, hydrophobicity is restricted to the upper surface layer of the soil in SAT and this appears to be limited due to the regular wetting drying cycle that controls the development of hydrophobicity by a number of different physicochemical actions (Aryle *et al.*, 2011). Regular plowing can break up any surface organic matter crust and help maintain overall organic matter content and infiltration efficiency (Nadav *et al.*, 2012).

The design, performance and operation of soil-aquifer treatment systems have been reviewed by Sharma and Kennedy (2017). Further details on water reuse are given in the USEPA Guidelines (USEPA, 2012).

8.2.3 *Problems with RIB and SAT processes*

One of the major problems with both RIB and SAT processes is that groundwater mounding can occur beneath the basin, resulting in potential flooding (Fig. 8.1) and an increased risk of aquifer contamination. This makes a constant infiltration rate difficult to maintain as the rate of infiltration becomes a function of the ratio between the bottom surface area of the basin and the length of its edges. Thus, high length to width ratios, or smaller basin areas, tend to increase infiltration rates. In practice, multiple basins are used and are rotated, ensuring sufficient time for each basin to dry and the recharge mound to recede. While the basin is being rested, the permeability of the soil can be improved by harrowing or rotavating. Where a layer of sand has been used to line the basin, the top surface layer will need to be renewed periodically, depending on the organic loading, and replaced with clean sand. Where partially treated, wastewater is used; basins are flooded for only a couple of days before being rested for 1 to 2 weeks. As with high-rate irrigation, the limiting factor for both processes is the degree of groundwater contamination. Groundwater mounding can be minimized if potential sites for RIBs or SAT systems are fully evaluated using an appropriate groundwater model (Yihdego, 2017).

Due to over-pumping, aquifers have become depleted in many countries, which has resulted in attempts to recharge aquifers from other, usually less pure, water sources such as rivers. There has been growing interest in using primary and secondary wastewater to replenish groundwater and relying on the soil and aquifer to provide a high degree of purification (Idelovitch, 1978; Idelovitch and Michail, 1980; Roberts and McCarthy, 1978). However, it has proved very difficult to predict or model the movement of the recharge effluent within the aquifer so that, in practice, the water quality has often become impaired. There may be a lag period of several years between the commencement of recharge and contamination arriving in observation boreholes, with a much longer period required for the recovery of water quality once recharge has ceased. As recharge commences, a mixture of recharge effluent and natural groundwater will eventually be pumped from the supply boreholes, therefore, regular monitoring of water quality is essential to ensure potable standards are maintained. For example, an infiltration system used in Milton, Wisconsin, US, over a 30-year-period receiving primary-treated wastewater was discharged into a set of four basins, which were continuously operated in series. The soil was glacial sand and gravel with inclusions of silt and clay. Test boreholes showed that the quality of the groundwater below the site had become contaminated, with a BOD of 4.2 mg L^{-1}, ammonia-N of 13.7 mg L^{-1}, nitrate of 1.5 mg L^{-1}, total phosphorus of 4.0 mg L^{-1}, and fecal coliforms reaching 258 per 100 mL (Anon, 1979). Groundwater contamination by trace organics has also been associated with rapid infiltration systems (Tomson *et al.*, 1981). Where there is a risk of contaminating water supplies, RIBs should only be used for infiltration using fully treated effluents from advanced treatment systems that also incorporate nutrient removal.

Pathogen removal is critical where treated wastewater is reused. While cysts and oocysts are readily retained, bacteria and viruses can penetrate deep into the vadose zone beneath an SAT infiltration basin. Viruses are particularly mobile and can reach groundwater where they can remain viable for long periods (months to years) (Kauppinen *et al.*, 2018). Due to the unpredictability of removing viruses from any wastewater treatment system, the current trend has been to look for reliable viral indicators such as Reovirus and Adenovirus. Alternative non-human pathogenic viral indicators have been proposed, including the Pepper Mild Mottle Virus (PMMoV) or the bacteriophage (crAssphage), which are both common in human feces (Symonds *et al.*, 2018). Both were found to be good indicators of virus removal in the SAT treatment of wastewater (Morrison *et al.*, 2020).

Because of the potential long-term risks to important groundwater resources where rapid infiltration basins are used for groundwater recharge, it is advisable only to use wastewater that has been treated to near drinking water quality, and to regard the basin system solely as a means of recharging the aquifer, much in the same way that injection wells are used. Therefore, any benefits from the potential treatment capacity of the soil as the water percolates to the aquifer should be regarded as incidental, with the main nutrient removal and contaminant removal dealt with by integrating suitable treatment process steps at the wastewater treatment plant prior to recharge. Where tertiary treatment is required, then a SAT system should be employed but only used when the aquifer is not used for domestic supply. Where that is the case, then the groundwater should only be supplied to consumers where advanced water treatment is in place prior to supply.

References

Akhaven, M., Imhoff, P.T., Andres, A.S. and Finsterle, S. (2013) Model evaluation of denitrification under rapid infiltration basin systems. *Journal of Contaminant Hydrology*, **152**, 18–34.

Anon. (1979). Advanced wastewater treatment: A low-risk proving-ground for experiment and invention. *World Water*, **6**, 29–41.

Arye, G., Tarchitzky, J. and Chen, Y. (2011) Treated wastewater effects on water repellency and soil hydraulic properties of soil aquifer treatment infiltration basins. *Journal of Hydrology*, **397**, 136–145.

Carlson, R.R., Linstept, K.D., Bennett, E.R. and Hartman, R.B. (1982) Rapid infiltration treatment of primary and secondary effluents. *Journal of the Water Pollution Control Federation*, **54**, 270–280.

Cormier, J.A. and Duranceau, S. (2019) Comparison of a modified and traditional rapid infiltration basin for treatment of nutrients in wastewater effluent. *Water Environment Research*, 1–14.

Gray, N.F. (2021) *Fixed-Film Reactors in Wastewater Treatment.* World Scientific Publishing, Singapore.

Idelovitch, E. (1978) Wastewater reuse by biological-chemical treatment and groundwater recharge. *Journal of the Water Pollution Control Federation*, **50**, 2723–2740.

Idelovitch, E. and Michail, M. (1980) Treatment effects and pollution dangers of secondary effluent percolation for groundwater. *Progress in Water Technology*, **12**, 949–966.

Idelovitch, E., Icekson-Tal, N., Avranham, O. and Michail, M. (2003) The long-term performance of a soil aquifer treatment (SAT) for effluent reuse. *Water Science and Technology Water Supply*, **3**(4), 239–246.

Kauppinen, A., Pitkanen, T. and Miettinen, I.T. (2018) Persistent norovirus contamination of groundwater supplies in two waterborne outbreaks. *Food and Environmental Virology*, **10**, 39–50.

Metcalf and Eddy, Inc. (1991) *Wastewater Engineering: Treatment, Disposal and Reuse, 3rd ed.* McGraw-Hill, New York.

Mienis, O. and Arye, G. (2018) Long-term nitrogen behavior under treated wastewater infiltration basins in a soil-aquifer treatment (SAT) system. *Water Research*, **134**, 192–199.

Morrison, C.M., Betancourt, W.Q., Quintanar, D.R., Lopez, G.U., Pepper, I.L. and Gerba, C.P. (2020) Potential indicators of virus transport and removal during soil aquifer treatment of treated wastewater effluent. *Water Research*, **177**, 115812.

Nadav, I., Tarchitzky, J. and Chen, Y. (2012) Soil cultivation for enhanced wastewater infiltration in soil aquifer treatment (SAT). *Journal of Hydrology*, **470–471**, 75–81.

NDEP. (2017) Guidance document for the design of rapid infiltration basins. Technical Publication: WTS-3A. Nevada Division of Environmental Protection, Carson City, NV.

O'Reilly, A.M., Wanielista, M.P., Chang, N., Xuan, Z. and Harris, W.G. (2012) Nutrient removal using biosorption activated media: Preliminary biogeochemical assessment of an innovative stormwater infiltration basin. *Science of the Total Environment*, **432**, 227–242.

Roberts, P.V. and McCarty, P.L. (1978). Direct injection of reclaimed water into the aquifer. *Journal of the Environmental Engineering Division, American Society of Civil Engineers*, **104**, 933–949.

Sharma, S.J. and Kennedy, M.D. (2017) Soil aquifer treatment for wastewater treatment and reuse. *International Biodeterioration and Biodegradation*, **119**, 671–677.

Symonds, E.M., Nguyen, K.H., Harwood, V.J. and Breitbart, M. (2018) Pepper mild mottle virus: A plant pathogen with a greater purpose in (waste)water treatment development and public health management. *Water Research*, **144**, 1–12.

Tomson, M.B., Dauchy, J., Hutchins, S., Curran, C., Cook, C.J. and Ward, C.H. (1981) Groundwater contamination by trace organics from a rapid infiltration site. *Water Research*, **15**, 1109–1116.

USEPA. (1984) Process design manual for land treatment of municipal wastewater: Supplement on rapid infiltration and overland flow. EPA/625/1-81-013a. US Environmental Protection Agency, Cincinnati, OH.

USEPA. (2003) Wastewater technology fact sheet: Rapid infiltration land treatment. EPA 832-F-03-025. Office of Water, US Environmental Protection Agency, Cincinnati, OH.

USEPA. (2006) Process design manual: Land treatment of wastewater effluents. EPA/625/R-06/16. US Environmental Protection Agency, Washington, DC.

USEPA. (2012) Guidelines for water reuse. EPA/600/R-12/618. US Environmental Protection Agency, Washington, DC.

Water Conserv II. (2020) Water Conserve II project. Winter Garden, FL. http://www.waterconservii.com/

WEF. (1992) Design of municipal wastewater treatment plants. Manual of Practice No. 76. Water Environment Federation, Alexandria, VA.

Wen, D., Valencia, A., Ordeonez, D., Chang, N.-B. and Wanielista, M. (2020) Comparative nitrogen removal via microbial ecology between soil and green sorption media in a rapid infiltration basin for co-disposal of stormwater and wastewater. *Environmental Research*, **184**, 109338.

Yihdego, Y. (2017) Simulation of groundwater mounding due to irrigation practice: Case of wastewater reuse engineering design. *Hydrology*, **4**, 19.

9

Overland Flow

9.1 Introduction

Overland flow is an ideal natural treatment system with many advantages, including low energy and environmentally sympathetic construction, low capital and operational costs, simple operation, uses no chemicals, has short retention times, and minimal carbon footprint. They are widely used today for the tertiary treatment of both municipal and industrial wastewaters but are also still used as a secondary biological process. Overland flow, which is based on grass filtration that has been used in Europe since the 19th century, was developed in the United States (US) in 1954 for the treatment of wastewaters on poor permeability soils comprising clay or clay-loam soils. The most suitable soils are those with a permeability of <0.5 cm h^{-1}. Where permeability is above the recommended threshold, the soil should be compacted or a 0.5-m layer of clay or similar material laid down and compacted over the existing soil to protect the groundwater.

When wastewater is applied, the bulk of the water flows over the surface rather than infiltrating the soil, resulting in the effluent having to be collected at the bottom of the slope for reuse or discharge to a watercourse. Flatland is preferred so that it can be carefully regraded to give a slope of between 2 and 6%, although naturally sloping land can also be utilized. Low slopes are more likely to pond, causing depletion of soil oxygen, excessive surface microbial growths, anaerobic conditions leading to odor problems, and the eventual death of the vegetation. In contrast, excessive slopes may lead to soil erosion and a high inert suspended solids concentration in the final effluent. Where natural slopes exceed 8%, they should be terraced to allow each section to be regraded to prevent erosion. Each sloped section is between 30 and 100 m in length, the width being governed by the total area of land required. While a total length of 30 to 45 m provides sufficient retention time for the adequate treatment of most wastewater types, full nitrification will require longer slope lengths (>45 m).

Where overland flow is used for secondary treatment, the wastewater is screened and the settleable solids are removed by primary settlement. The influent is then applied using overflow pipes located at the top of the slope, or when sprayers are employed to the top third of the gently sloping land, and then allowed to slowly flow by gravity down the slope to the collection area at the bottom (Fig. 9.1). It is

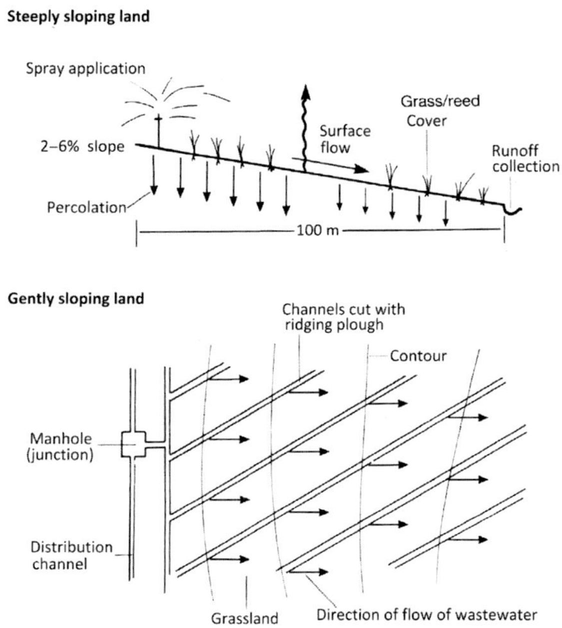

Fig. 9.1. Land treatment by overland flow using sprinkler and perforated pipe wastewater distribution systems.

important to ensure that an even flow is maintained over the vegetation, creating a continuous sheet of wastewater, in order to obtain maximum treatment efficiency. With domestic wastewaters, the slope is divided into blocks, each with three sections and loaded in rotation with 8 hours of loading followed by 16 hours of rest. The use of grass plots in series has been beneficial in terms of enhanced effluent quality but less efficient in overall removal per unit area of land.

9.2 The Treatment Process

The ground is planted with a suitable species of grass so that purification occurs by filtration and bacterial decomposition as the wastewater moves through the vegetation. The grass also slows the rate of flow of wastewater as it passes down the slope, preventing short-circuiting and minimizes the potential for erosion, especially during periods of heavy rainfall. The grass is also important as it stores both nutrients and contaminants while also increasing the rate of evapotranspiration, which reduces the overall volume of treated effluent requiring disposal.

The selection of grass species is important as they have to withstand constant saturation, which can also result in periods of anaerobic conditions in the root zone, and drying when oxygen is replenished. Treatment areas should be seeded with a mixture of at least five species of grass to ensure the best surface conditions for optimum treatment (Droste, 1997). Recommended grass species are given in Table 9.1, although consideration should be given to overall grass management and local climatic conditions. In a comparative study of four regional grass species in Taiwan — Paragrass (*Brachairia mutica*), Nilegrass (*Acroceras macrum*), Cattail (*Typha orientalis*), and Vetiver (*Vetiveria zizanioides*) — used to treat polluted stream water, Wen *et al.* (2007) found that nutrients were much higher in grass tissue from overland flow treatment systems compared to control pasture sites with nitrate accumulating in the grass during winter (up to 1,680 mg kg^{-1}), but significantly reducing in spring as active growth restarts. Performance in terms of biochemical oxygen demand (BOD), chemical oxygen demand (COD), total suspended solids

Table 9.1. Grasses recommended for seeding overland flow land treatment areas in the US with those in italics being the most widely used.

Cool season species	Warm season grasses
Reed canary	*Common Bermuda*
Tall fescue	Coastal Bermuda
Rye grass	Dallis grass
Red top	Bahia
Kentucky bluegrass	
Orchard grass	

(TSS), and coliform removal was similar for all species with no significant difference in removal efficiency between plots where the grass was either harvested or not. The removal efficiency of nutrients varied between grass species, with both the Paragrass and Nilegrass being far more efficient in removing both nitrogen and phosphorous. Both these species produced the highest average yields at 72 tons $ha^{-1}yr^{-1}$ (20 g $m^{-2}d^{-1}$) and 43 tons $ha^{-1}yr^{-1}$ (12 g $m^{-2}d^{-1}$), respectively, almost double that expected on normal pasture land in Taiwan. Although the soil had low permeability, Cu and Zn still accumulated, with the electrical conductivity increasing due to the NaCl, and the pH was lowered but this did not affect the grass growth. While Cu and Zn also accumulated in the grass, it was below the recommended limit for feeding to livestock. The addition of lime has been found to enhance the removal of phosphorus from overland flow systems (Khalid *et al.*, 1982). Practice has shown that it is important to ensure that the grass does not reach maturity and dieback, releasing nutrients and decomposition products into the water stream. Regular cutting ensures new grass to better control wastewater movement and to ensure the maximum uptake of nutrients. Mowing requires the slopes to be rested for several weeks to completely dry out, so as to ensure that no damage is done to the surface or grass while being harvested. In the US, slopes are cut for hay or silage. In Europe, concerns have been expressed about the potential quality of the fodder in terms of contaminant bioaccumulation and potential pathogen transfer. This is strongly influenced by the composition of the wastewater being treated. Increasingly, the grass is mown as for silage but used for the production of biogas by anaerobic digestion rather than as an animal feed.

Treatment is by both adsorption and filtration, but most importantly by microbial oxidation by biofilm formation on the surface of soil particles and plant material, including the roots, stems and leaves of the grass. The roots of the grass create a thin permeable layer, which provides limited filtration but good protection from drying for heterotrophic and autotrophic (nitrifying) bacteria during the rest period. The oxygen transfer rate through the water flowing over the surface to the biofilm below is dependent on the organic loading. This film of water should be about 5 mm in depth to ensure good oxygen exchange. When settled, wastewater is being treated, if the organic loading or the period of loading exceeds the recommended levels; then, thick heterotrophic growth may form and cover the surface and vegetation with a whitish-grey or pink slime (Gray, 1985). This is particularly a problem with continuous loading. These growths can also be sloughed off, carried down the slope, and end up in the collection drains and, subsequently, released to surface water where they decompose, resulting in the reduction of oxygen concentration. These slimes are particularly efficient at the bioaccumulation of contaminants, which are released when they decay. Residual BOD and suspended solids in the final effluent are also due to the release of natural decaying organic material and solids from the soil-plant system. The community structure of ammonia-oxidizing bacteria is complex (Fig. 8.4) and has been explored in detail in overland flow systems treating ammonia-rich landfill leachate in Sweden by Sundberg *et al.* (2007). They also discovered that ammonia-oxidizing bacteria adapted to temperature change and remained active throughout the year.

9.3 Operational Management and Performance

These systems are very dependent on the weather, so the hydraulic loading is normally kept to between 0.2 and 10 cm d^{-1}, which is equivalent to an area of 1 to 10 ha per 1,000 m^3d^{-1} (USEPA, 1977; Metcalf and Eddy, 1991; WEF, 1992). The recommended application rates (AR), which is the flow rate of wastewater applied per unit width of the slope, and hydraulic loadings are 0.07 to 0.12 $m^3m^{-1}h^{-1}$ and 2.0 to 7.0 cm d^{-1} for primary-treated wastewaters, 0.09 to 0.15 $m^3m^{-1}h^{-1}$ and 2.5 to 9.0 cm d^{-1} for effluents from wastewater treatment ponds, and 0.11 to 0.17 $m^3m^{-1}h^{-1}$ and 3.0 to 10.0 cm d^{-1} for secondary-treated effluents (USEPA, 1984). More specifically, the USEPA recommends application rates based on BOD removal using three categories: *Excellent final effluent quality* required (10 mg L^{-1} BOD and 15 mg L^{-1} TSS) in cold climates, the ARs are 0.08 to 0.12 $m^3m^{-1}h^{-1}$ for settled sewage and 0.19 to 0.24 $m^3m^{-1}h^{-1}$ for secondary-treated sewage; ***moderate final effluent quality*** required (≤20 mg L^{-1} BOD and TSS) in medium climates, the ARs are 0.19 to 0.29 $m^3m^{-1}h^{-1}$ for settled sewage and 0.24 to 0.39 $m^3m^{-1}h^{-1}$ for secondary-treated sewage; and the least stringent final effluent quality required (≤30 mg L^{-1} BOD and TSS) in warm climates, the ARs are 0.30 to 0.45 $m^3m^{-1}h^{-1}$ for settled sewage and 0.39 to 0.48 $m^3m^{-1}h^{-1}$ for secondary-treated sewage (USEPA, 2006).

Organic loading should not exceed 100 kg BOD $ha^{-1}d^{-1}$. The actual organic loading is calculated as:

$$L_{BOD} = 0.1(L_w)(C_0), \tag{9.1}$$

where $\boldsymbol{L_{BOD}}$ is the BOD loading rate (kg BOD $ha^{-1}d^{-1}$), **0.1** is a conversion factor, $\boldsymbol{L_w}$ is the hydraulic loading rate (mm d^{-1}), and $\boldsymbol{C_0}$ is the influent BOD_5 (mg L^{-1} or g m^{-3}).

The application of wastewater is not normally continuous but for 6 to 12 hours each day — the stronger the influent, the shorter the application period. This allows the retained organic matter to be microbially oxidized and the nutrients to be adsorbed onto the soil particles, utilized by the plant growth, or be nitrified. Slopes can be operated daily or rested completely for 1 or 2 days per week, which is common with industrial wastewaters due to work practices. Some systems are operated on a 24-hour basis, usually those that are used for advanced treatment or the treatment of polluted stream water.

Application rates in excess of 12 hours per day will reduce nitrification and increase anoxic or anaerobic conditions in the root zone. This favors denitrification, which also requires high BOD/nitrogen ratios.

The relationship between hydraulic loading and the application rate is expressed as:

$$L_w = \frac{qPF}{Z}, \tag{9.2}$$

where $\boldsymbol{L_w}$ is the hydraulic loading rate (m d^{-1}), $\boldsymbol{q}$ is the application rate per unit width of slope (m^{3}m^{-1}min^{-1}), $\boldsymbol{P}$ is the application period (h d^{-1}), $\boldsymbol{F}$ is the conversion factor 60 (min h^{-1}), and $\boldsymbol{Z}$ is the length of the slope (m).

The treatment area required is calculated as:

$$A = \frac{QZ}{qPF}, \tag{9.3}$$

where $\boldsymbol{A}$ is the grassed area required for treatment (ha), $\boldsymbol{Q}$ is the wastewater flowrate (m^{3}d^{-1}), $\boldsymbol{Z}$ is the slope length (m), $\boldsymbol{q}$ is the application rate per unit width of slope (m^{3}m^{-1}min^{-1}), $\boldsymbol{P}$ is the application period (h d^{-1}), and $\boldsymbol{F}$ is the conversion factor 10,000.

Performance is a function of the length of the slope over which the wastewater flows (Scott and Fulton, 1979), although the retention time, controlled by the slope and hydraulic loading, is also a factor (Jenkins *et al.*, 1985; Witherow and Bledsoe, 1986).

The performance of these systems is good for primary and secondary wastewaters, with final effluents of BOD (10 mg L^{-1}), TSS (15 mg L^{-1}), ammonium-nitrogen (1 mg L^{-1}), and 5 mg TN L^{-1} expected (WEF, 2001). However, unlike BOD or suspended solids, the removal of ammonia decreases as the wastewater application rate increases (Wightman *et al.*, 1983). The actual performance is dependent on specific site characteristics (e.g. nature of the wastewater and the degree of pretreatment, soil structure, vegetative cover type, and local climatic conditions), and for that reason, pilot-scale trails are always advisable before a full plant construction. For example, Taebi and Droste (2008) compared the treatment of three types of wastewater by overland flow in Isfahan, Iran, by using pilot plots, each 4.5 m wide and 40 m long with a 4% slope. Primary-treated municipal wastewater (COD, 428 mg L^{-1}), activated sludge secondary effluent (COD, 130 mg L^{-1}), and lagoon-treated textile wastewater (COD, 900 mg L^{-1}) (Table 9.2a) were treated under identical conditions for five months after an initial acclimatization period of three months. The results showed that while performance varied according to the type of influent used, the percentage of removal decreased with increasing hydraulic loading in each case (Table 9.2b). Li *et al.* (2014) explored the effect of the wetting-drying period on ammonium removal. Each experimental plot was dosed continuously for 24 hours, followed by different dry periods of 24 hours (W:D ratio of 1.0), 48 hours (0.5), 72 hours (0.3), 96 hours (0.25), and 120 hours (0.2). The control plot was continuously loaded each day with no intervening dry period. The ammonia removal efficiency improved significantly as the W:D ratio declined (i.e., longer dry periods were used) (Table 9.3) while, in contrast, the total nitrogen removal decreased as the W:D declined. There is a conflict here, in that strict aerobic conditions within the active microbial zone are required for nitrifying microorganisms, while the opposite is required for denitrifiers. The authors recommended a W:D ratio of 0.25 for optimum ammonium and total nitrogen removal rates. While nitrogen gas is the main end product of denitrification, nitrous oxide is also released from overland flow land treatment systems at a maximum rate of 1.2% of the ammonium applied, which affects the overall greenhouse gas emissions associated with the process (Smith *et al.,* 1981).

Overland flow is very effective at removing trace organic and other contaminants, such as metals from wastewater being a function of the application rate and residence time. Jenkins *et al.* (1985) examined the removal efficiency of 13 trace organics in an overland flow system; these were chloroform at a mean concentration of 33 μg L^{-1}. 19 μg L^{-1} of toluene, 58 μg L^{-1} of chlorobenzene, 68 μg L^{-1} of bromoform, 11 μg L^{-1} of dibromochloromethane, 50 μg L^{-1} of *m*-nitrotoluene, 63 μg L^{-1} of diethyl phalate, 1,242 μg L^{-1} of PCB, 63 μg L^{-1}

Table 9.2. The comparative performance of pilot-scale overland flow land treatment systems treating three different wastewaters. (a) The influent strength applied to the plots, (b) removal efficiencies at the three different application rates (i.e., the volumetric loading per unit width of plot expressed in $m^3m^{-1}h^{-1}$), which in the study is equivalent to hydraulic loading rates of 2.63, 4.38 and 6.13 cm d^{-1} (Taebi and Droste, 2008). Reproduced with permission from Elsevier Publishing.

(a)

Influent	Statistical parameter	$TBOD_5$ (mgL^{-1})	TCOD (mgL^{-1})	TSS (mgL^{-1})	TN (mgL^{-1})	TP (mgL^{-1})	Turbidity (NTU)
Primary effluent	Mean	149.0	428.0	125.0	82.3	20.4	56.9
	Std.dev.	17.0	52.3	20.0	9.7	5.5	8.9
	Range	115–172	332–503	95–159	68–99	10–32	36–72
Activated sludge secondary effluent	Mean	50.3	130.3	100.3	35.0	10.2	35.3
	Std.dev.	10.1	17.5	14.4	4.6	2.8	8.7
	Range	37–75	115–185	67.132	27–45	6–15	28–65
Lagoon effluent of textile wastewater	Mean	180.4	899.7	140.0	25.1	7.7	38.4
	Std.dev.	19.5	141.6	16.7	3.4	0.7	7.4
	Range	150–213	740–1209	105–161	19–31	6.5–9.1	26–49

(b)

Influent	AR ($m^3m^{-1}h^{-1}$)	$TBOD_5$ (%)	TCOD (%)	TSS (%)	TN (%)	TP (%)	Turbidity (%)
Primary effluent	0.15	84.0(4.0)	59.5(4.8)	71.9(4.4)	48.7(6.3)	40.6(5.2)	72.9(4.1)
	0.25	74.5(3.3)	54.7(3.7)	67.5(4.1)	36.5(4.5)	36.7(4.8)	66.3(3.3)
	0.35	63.3(2.5)	49.6(3.2)	57.8(3.6)	31.8(4.0)	29.2(5.4)	63.3(3.2)
Activated sludge secondary effluent	0.15	58.7(4.7)	58.5(4.0)	71.6(3.2)	53.3(5.0)	50.7(5.4)	54.3(5.2)
	0.25	52.4(4.4)	53.9(4.8)	68.7(4.1)	40.1(5.8)	45.0(5.6)	51.4(3.7)
	0.35	46.7(4.5)	45.3(2.6)	58.1(3.3)	38.9(3.8)	21.2(6.0)	43.8(5.2)
Lagoon effluent of textile wastewater	0.15	74.0(3.8)	65.4(4.1)	80.1(2.2)	47.6(5.3)	41.4(5.3)	60.2(3.7)
	0.25	64.6(2.4)	57.7(2.3)	67.6(5.2)	43.2(3.2)	38.7(5.3)	52.7(2.9)
	0.35	57.4(2.8)	52.0(3.1)	62.0(3.1)	33.1(3.1)	44.0(4.2)	51.1(2.5)

Table 9.3. The effect of the wetting:drying (WD) ratio on nitrogen removal in overland flow land treatment pilot plots (Li *et al.*, 2014). Reproduced with permission from Elsevier Publishing.

Pollutants	Continuous flow mode	WD 1.0	WD 0.5	WD 0.3	WD 0.25	WD 0.20
Effluent (mg L^{-1})						
NH3-N	10.4±0.2	9.0±0.3	8.1±0.2	7.0±0.4	4.0±1.0	4.3±0.5
TN	17.6±0.5	10.5±0.4	11.8±0.9	14.2±0.1	14.8±0.2	16.1±0.4
Removal efficiency (%)						
NH3-N	52.3±1.1	58.9±0.6	61.7±0.5	68.1±1.1	78.2±1.0	79.8±0.5
TN	41.2±0.5	62.3±1.0	60.5±1.2	51.3±0.4	49.8±0.3	46.4±1.0

of naphthalene, 45 μg L^{-1} of phenanthrene, 39 μg L^{-1} of pentachlorophenol, 79 μg L^{-1} of 2,4-Dinitrophrnol, and 113 μg L^{-1} of nitrobenzene. Apart from PCB and pentachlorophenol, organic contaminants did not accumulate in the soil due to sorption, biodegradation and volatilization. Removal rates were affected by temperature, for example, 94 to 100% in the summer, falling to 77 to 99% in the winter at 0.4 cm h^{-1}; this fell to 76 to 100% in the summer at the higher application rate of 0.8 cm h^{-1} (Table 9.4). Foco *et al.* (2018) studied the uptake of metals from municipal wastewater at two loading rates — 0.2 and 0.3 $m^3m^{-1}h^{-1}$. Using pilot-scale plots (4.25 m × 40 m) planted with *Cynodon* spp. (Tifton 85), good removals were recorded of Cd, Cr, Cu, Pb, Fe, Ni, and Zn, with all the metals achieving the required Brazilian discharge limits except for Cr, Cu and Pb. Higher application rates did not affect the metal uptake by the biomass.

Cold weather is a problem for overland flow treatment, especially when ice formation occurs over the surface. However, at soil and wastewater temperatures of >8°C, BOD removal is largely independent of temperature. Problems also arise when overland flow systems are used as a secondary-treatment phase for lagoon-treated wastewaters, as the algae in the applied wastewater are not removed by the land system, resulting in high suspended solids concentrations in the final effluent (Abernathy, 1983; Witherow and Bledsoe, 1983).

Table 9.4. Average removal rates (% by mass) of 13 trace organic compounds by a pilot-scale overland flow land treatment system at different application rates and seasons (Jenkins *et al.*, 1985). Reproduced with permission from Elsevier Publishing.

	Average removal			
Substance	**0.4 cm h^{-1} 1 Jun–4 Aug**	**0.4 cm h^{-1} 28 Oct–11 Dec**	**0.8 cm h^{-1} 5 Aug–16 Oct**	**1.2 cm h^{-1} 19–21 Oct**
Chloroform	97	95	96	—
Toluene	>99	—	>99	—
Chlorobenzene	>99	97	>99	—
Bromoform	98	94	89	73
Dibromochloromethane	>98	97	94	84
m-Nitrotoluene	>99	86	87	57
Diethylphthalate	96	77	76	15
PCB 1242	98	92	95	64
Naphthalene	>99	99	97	92
Phenanthrene	>99	98	99	—
Pentachlorophenol	99	94	97	44
2,4-Dinitrophenol	94	79	92	9
Nitrobenzene	—	81	—	—

Although the overland flow has lost some of its popularity in Europe, it still has a role as a sustainable tertiary or advanced treatment system for both domestic and industrial wastewaters, where the final effluent can be discharged surface water. Overland flow has also been employed to restore polluted streams, where a portion of the flow is intercepted and treated before being returned to surface waters. It is also integrated with other natural treatment systems where it is used as the final tertiary step. In Turkey, a combined free water surface constructed wetland planted with *Phragmites communis* followed by an overland flow land treatment system planted with Italian ryegrass successfully improved the quality of a small polluted stream (Tunçsiper, 2019).

Grass filtration is widely used for sustainable urban drainage solutions and in the treatment and disposal of road runoff. Overland flow systems have been modified to include more engineered surfaces using artificial soils, allowing water to flow through the soil rather than over it with improved organic matter and nitrogen removal. For example, replacing clay soil with a 25-cm thick layer of artificial substrate resulted in enhanced removal rates of 82% COD, 83% ammonium, 60% total nitrogen, and 89% total phosphorus using optimal loadings of 0.01 to 0.03 $m^3m^{-2}d^{-1}$ and 0.15 to 0.45 g BOD $m^{-2}d^{-1}$ (Li *et al.*, 2019).

In the US and Australia, overland flow land treatment remains an effective biological attached biofilm treatment system for the treatment of settled wastewater and a wide range of industrial wastewaters (USEPA, 2006). The current design of overland flow systems has been largely based on original USEPA designs (USEPA, 1977; 1984; 2006) and further modeled by Witherow and Bledsoe (1984; 1986). Hegg and Turner (1983) have reviewed their use with animal wastes and Smith and Schroeder (1983) for domestic wastewaters. Further design and operational details have also been published by the Water Pollution Control Federation (WPCF, 1990) and the Water Environment Federation (WEF, 1992).

References

Abernathy, A.R. (1983) Overland flow treatment of municipal sewage at Easley, SC. EPA-600/2-83-015, NTIS PB83-170985. US Environmental Protection Agency, Ada, OK.

Droste, R.L. (1997) *Theory and Practice of Water and Wastewater Treatment.* Wiley, New York. NY.

Foco, M.L.R., de Souza, E.S. and Nour, E.A. (2018) Tratament de esgoto sanitária por escoamento superficial no solo: Remoção de metais (Wastewater treatment by overland flow: removal of metals). *Engenharia Sanitária e Ambiental*, **23**(4), 823–832.

Gray, N.F. (1985) Heterotrophic slimes in flowing waters. *Biological Reviews of the Cambridge Philosophical Society*, **60**, 499–548.

Hegg, R.O. and Turner, A.K. (1983) Overland flow as a method of treatment for animal wastes — A review. *Animal Wastes*, **8**, 167.

Jenkins, T.F., Leggett, D.C., Parker. L.V. and Oliphant, J.L. (1985) Toxic organics removal in overland flow land treatment. *Water Research*, **19**(6), 707–718.

Khalid, R.A., Patrick, W.H. and Mixon, M.N. (1982) Phosphorus removal processes from overland flow treatment of simulated wastewater. *Journal of the Water Pollution Control Federation*, **54**, 61–69.

Li, H., Li, Y., Xu, X., Wang, P. and Zhou, Y. (2014) Effects of intermittent operation mode on nitrogen removal by an overland flow system: A pilot study. *Ecological Engineering*, **69**, 192–195.

Li, H.B., Li. Y.H., Xu, X.Y., Chen, X., Yang, L. and Su, F. (2019) How does the loading rate affect the performance of an overland flow system treating domestic sewage? *Desalination and Water Treatment*, **138**, 12–18.

Metcalf and Eddy, Inc. (1991) *Wastewater Engineering: Treatment, Disposal and Reuse, 3rd ed.* McGraw-Hill, New York.

Scott, T.M. and Fulton, P.M. (1979) Removal of pollutants in the overland flow (grass filtration) system. *Progress in Water Technology*, **11**, 301–313.

Smith, C.J., Chen, R.L. and Patrick, W.H. (1981) Nitrous oxide emissions from simulated overland flow wastewater treatment systems. *Soil Biology and Biochemistry*, **13**, 275–278.

Smith, R.G. and Schroeder, E.D. (1983) Physical design of overland flow systems. *Journal of the Water Pollution Control Federation*, **55**, 255–259.

Sundberg, C., Stendahl, J.S.K., Tonderski, K. and Lindgren, P.-E. (2007) Overland flow systems for treatment of landfill leachates — Potential nitrification and structure of the ammonia-oxidizing bacterial community during a growing season. *Soil Biology and Biochemistry*, **39**, 127–138.

Taebi, A. and Droste, R.L. (2008) Performance of an overland flow system for advanced treatment of wastewater plant effluent. *Journal of Environmental Management*, **88**, 688–696.

Tunçsiper, B. (2019) Combined natural wastewater treatment systems for removal of organic matter and phosphorus from polluted streams. *Journal of Cleaner Production*, **228**, 1368–1376.

USEPA. (1977) Process design manual for land treatment of municipal wastewater. EPA 625/1-77-008. US Environmental Protection Agency, Ada, OK.

USEPA. (1984) Process design manual for land treatment of municipal wastewater: Supplement on rapid infiltration and overland flow. EPA/625/1-81-013a. US Environmental Protection Agency, Cincinnati, OH.

USEPA. (2006) Land treatment of municipal wastewater effluents. Process design manual: EPA/625/R-06/016). US Environmental Protection Agency, Cincinnati, OH.

WEF. (1992) Design of municipal wastewater treatment plants. Manual of Practice No. 76. Water Environment Federation, Alexandria, VA.

WEF. (2001) Natural systems for wastewater treatment. Manual of Practice. Water Environment Federation, Alexandria, VA.

Wen, C.G., Chen, T.H., Hsu, F.H., Lu, C.H., Lin, J.B., Chang, C.H., Chang, S.P. and Lee, C.S. (2007) A high loading overland flow system: Impacts of soil characteristics, grass constituents, yields and nutrient removal. *Chemosphere*, **67**, 1588–1600.

Wightman, D., George, P. B., Zirschky, J. H. and Filip, P.S. (1983) High-rate overland flow. *Water Research*, **17**, 1679–1690.

Witherow, J.L. and Bledsoe, B.E. (1983) Algae removal by overland flow process. *Journal of the Water Pollution Control Federation*, **55**, 1256–1262.

Witherow, J.L. and Bledsoe, B.E. (1984) Design model for the overland flow process. Robert S. Kerr Environmental Research laboratory, US Environmental protection Agency, Ada, OK, USA.

Witherow, J.L. and Bledsoe, B.E. (1986) Design model for the overland flow process. *Journal of the Water Pollution Control Federation*, **58**, 381–386.

WPCF. (1990) Natural systems for wastewater treatment. Manual of Practice: FD-16. Water Pollution Control Federation, Alexandria, VA, USA.

Part III
Constructed Wetland Systems

10

The Use of Microalgae and Macrophytes in Wastewater Treatment

10.1 Introduction

The controlled culture of aquatic plants (microalgae and macrophytes) is now widely used in wastewater treatment, both as secondary and tertiary treatment systems, a process often referred to as phycoremediation. Not only do they accumulate nutrients and other soluble compounds from the water, but they also act as a food source for a variety of other organisms such as zooplankton, small crustaceans such as the brine shrimp (*Artemia salina*), and a wide variety of fish. The excess plant biomass itself can be harvested and used for various purposes. The plants used in wastewater treatment systems are all lake and wetland species and can be loosely classified as submerged algae and macrophytes, floating macrophytes, and emergent vegetation. The advantage of using plants is that the capital investment in the treatment works is low, and as little equipment, energy or chemicals are required, the operational costs are also very low. The cost of harvesting either the plant or the secondary animal biomass is reduced by its value as a source of energy or protein (Sec. 16.2). Much work has been done in evaluating plants for wastewater treatment in terms of treatment efficiency, biomass production, and value of the resultant biomass. The main selection criteria for such plants are: (i) ease of harvesting, (ii) low water content, (iii) high protein content, (iv) low fiber and lignin content, (v) high mineral absorption capability, (vi) extended growing and harvesting periods, (vii) non-toxic to either humans or livestock, (viii) capable of being processed into a useful byproduct, and (ix) have few natural pests. The selection of plants for wastewater treatment should, if possible, satisfy all of these criteria (Culley and Epps, 1973).

Equation (10.1) summarizes the stoichiometry of photosynthesis and shows the ratio of N:P that controls whether either nitrogen or phosphorus is limiting to growth in the wastewater. This is a molar ratio, so in practice, this corresponds to approximately 7:1 with only the inorganic fraction utilized by plants (i.e., PO_4^{3-}, HPO_4^{2-}, H_2PO^{4-}, and NO_3^- NH_4^+). For submerged macrophytes, the carbon source is normally dissolved carbon dioxide (CO_2), although carbonates can also be used under certain circumstances by some macrophytes.

$$106\ CO_2 + 16\ NO_3^- + HPO_4^{2-} + 18\ H^+ + \\ 122\ H_2O \leftrightarrow C_{106}H_{263}O_{110}N_{16}P + 138\ O_2. \qquad (10.1)$$

Constructed wetlands provide the optimum treatment conditions found in natural wetlands but can be built at almost any location, following contours and blending into the landscape. They can be used for secondary and/or tertiary treatment of sewage, and treating stormwater and other wastewaters, including acid mine drainage, landfill leachate, agricultural and industrial wastewaters (Hammer, 1989; Mulamoottil *et al.*, 1998; Kadlec *et al.*, 2000). They are also successfully employed for dewatering and stabilizing sewage sludge. In the United States (US), constructed wetland systems are classified into two basic designs: free water surface (FWS), or surface flow (SF) systems and subsurface flow (SSF) systems (Fig. 10.1) (Chap. 13). Free water surface treatment wetlands are densely vegetated and characterized by a significant depth of permanent water above the media surface. They can be further categorized by the type of vegetation used: (i) submerged macrophytes (Secs. 10.2 and 10.3), (ii) free-floating macrophytes (Sec. 10.4), (iii) floating leaves but bottom rooted macrophytes such as water lilies (*Nymphaea* spp.), lotus (*Nelumbo* spp.), and cow lily (*Nuphar* spp.), (iv) macrophytes forming mats, and (v) emergent vegetation (Chap. 12). Free water systems may not offer many advantages over waste stabilization ponds

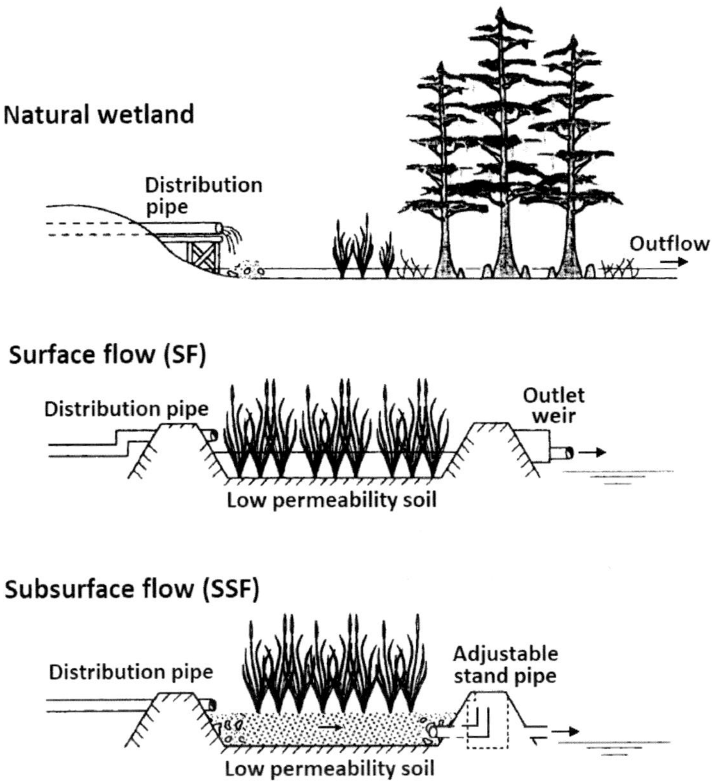

Fig. 10.1. Three basic wetland systems are employed in wastewater treatment: Natural wetland and constructed wetlands employing surface and subsurface flow (Kadlec and Knight, 1996). Reproduced with permission from the Taylor & Francis Group.

(Chap. 16), and as mosquito breeding, especially *Mansonia* spp., can be a serious problem in FWS systems, then waste stabilization ponds should always be considered as an alternative (Ringuelet, 1983).

The climatic conditions in Europe limit the type of constructed wetlands that can be employed successfully throughout the year. For that reason, European-constructed wetlands are almost exclusively SSF systems using emergent plant species commonly referred to as reed beds (Sec. 13.3). A bed of soil or gravel that is 0.6 to 1.0 m in depth is used as a substrate for the growth of rooted emergent species. Settled wastewater flows through the substrate by gravity, either horizontally or vertically, where it comes into contact with a mixture of facultative microorganisms living in association with the substrate and plant roots.

Two groups of aquatic plants that are not widely used for wastewater treatment are the microalgae and submerged macrophytes, both of which are grown primarily for their biomass, but have a secondary function in removing nutrients and, where present, contaminants. Like all plant systems, they do not require costly aeration but instead release oxygen while at the same time fixing carbon from the atmosphere, thereby reducing greenhouse gases. These are considered below along with floating macrophytes.

10.2 Microalgae

In facultative ponds, algae use the inorganic compounds released by aerobic and facultative bacteria for growth using sunlight for energy. They release oxygen into the solution, which, in turn, is utilized by the bacteria completing the symbiotic cycle (Fig. 10.2). The algae in facultative ponds are restricted to the euphotic zone, which is often only a few centimeters deep, depending on the organic loading and whether it is day or night (Sec. 16.2.2). High-rate algal ponds, another form of a stabilization pond, are not designed for optimum purification of wastewater but algal production (Chap. 11). Microalgae are known to produce thousands of useful byproducts as well as being able to bioaccumulate pollutants and nutrients from water (Sec. 11.3).

Microalgae are found as individuals, groups or chains of plant cells that do not possess roots, stems or leaves, although they carry out photosynthesis like all other plants. They are predominantly found in fresh or marine waters and sediments and comprise a huge diversity of species. In wastewater treatment applications, mono-species cultures of microalgae are generally used, such as cultures of the green algae *Chlorella* spp. or *Scenedesmus* spp., which have a protein content of

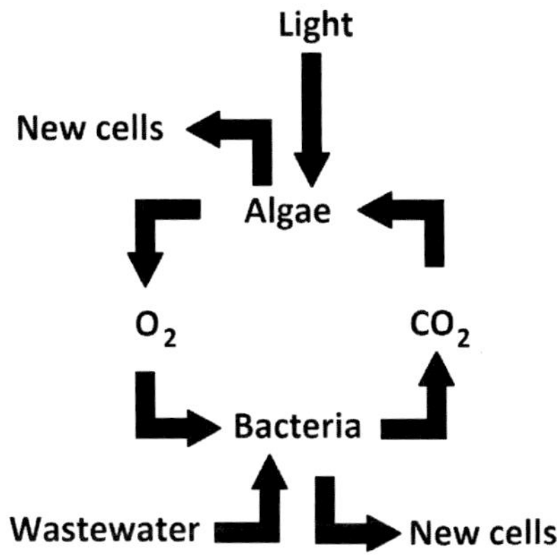

Fig. 10.2. Basic biological interactions in a facultative pond (Hawkes, 1983). Reproduced with permission from Academic Press.

50% (dry weight [DW]), compared to 60 to 70% for the blue-green alga *Spirulina*. Light availability is the most critical factor controlling algal growth and as it is generally uneconomical to use artificial light, ponds must be located in areas where there is maximum sunshine. Shallow lagoons are normally used for algal production because as the density of cells increases, the light penetration is reduced. However, the cells nearest the surface can become damaged by constant exposure to sunlight, so slightly deeper lagoons are used, which are gently mixed to maintain the algae in suspension, ensuring intermittent transport of the cells into the euphotic zone. Microalgal cells are comprised of protein, lipids, carbohydrates as well as fatty acids vitamins, pigments and enzymes, all of which can be utilized for the manufacture of useful byproducts (Cuellar-Bermudez *et al.*, 2015).

Microalgae have an important role in wastewater treatment, especially in the removal of nitrogen and phosphorus, which are key drivers in surface water eutrophication, where they offer a cost-effective and sustainable alternative to chemical or advanced biological (bacterial) nutrient removal systems (Hoh *et al.*, 2016; Solovchenko *et al.*, 2016). Wastewaters offer a cheap and continuous supply of these nutrients as well as other trace elements for microalgal

Table 10.1. Comparison of digestibility and biogas production of various plant biomass (Braun, 1982). Reproduced with permission from Springer.

Material	Hydraulic residence time (d)	Volumetric loading (kg TS $m^{-3}d^{-1}$)	Gas yield ($m^3kg^{-1}TS$)
Sea grass	20	2	0.1
Seaweed	20	2	0.17
Water hyacinth	10	4	0.1
Algae[a]	10	4	0.25
Algae[b]	3–30	1.4–11.2	0.26–0.5
Silage	15	6.7	0.45

[a] Freshwater algae.
[b] Sewage lagoon algae.

cultivation (Wang *et al.*, 2010), with those from food-processing industries that contain fewer pathogens normally used for their commercial cultivation. Large-scale algal ponds are being used in several countries as an effective means of sewage treatment (Larsdotter, 2006) as well as producing valuable algal biomass that can be used for many different purposes such as biogas production (Table 10.1), biodiesel, bioethanol, and bio-oil production, feed for aquaculture and livestock, single-cell protein production, health supplements, and as a fertilizer (Oswald, 1972; Shelef *et al.*, 1978; Taiganides *et al.*, 1979; Chisti, 2007; Cai *et al.*, 2013; Cuellar-Bermudez *et al.*, 2015; Solovchenko *et al.*, 2016) (Sec. 11.2). Algal ponds are also used in the removal and recovery of heavy metals and other contaminants from effluents (Cuellar-Bermudez *et al.*, 2017). The ability of certain algae to absorb and accumulate metals suggests the possibility of removing the low concentrations of heavy metals found in normal domestic wastewaters (Filip *et al.*, 1979; Harding and Whitton, 1981; Nakajima *et al.*, 1981). However, Becker (1983) indicates that significant removal of metals by algae may only be possible where long hydraulic retention times (HRTs) are used, for example, for the pretreatment of certain metal wastes. In ordinary algal ponds with relatively short HRTs, metal removal is only marginal. This is reviewed by Zeraatkar *et al.* (2016).

Algae have higher biomass production rates per hectare than plants, making them an important tool for sequestering the greenhouse gas, CO_2 (Ho *et al.*, 2011), and also for producing biofuels. Microalgae are a valuable raw ingredient for the production of a range of carbon-neutral fuels including biodiesel, bioethanol and bio-oil (Christi, 2007; Brennan and Owende, 2010). This makes microalgal treatment systems extremely sustainable and carbon neutral.

Freshwater microalgae have been used for the treatment of a number of industrial wastewaters, especially those from distilleries and breweries (Johnson *et al.*, 2019; Amenorfenyo *et al.*, 2019). Species isolated from marine environments are excellent at treating saline wastewaters including those from food processing and tanning. Dyes are produced worldwide with their process wastewaters, which usually comprised azo compounds, presenting a unique treatment challenge (Kaushik and Malik, 2009). These wastewaters are known to have toxic, carcinogenic and mutagenic properties, so they are potentially highly damaging to both the environment as well as humans (Ali, 2010; Priya *et al.*, 2011). Phytoremediation using cyanobacteria and other microalgae is seen as a potential breakthrough in the decolorization and detoxification of wastewaters contaminated with dyes (El-Sheekh, *et al.*, 2009; Sarkar and Dey, 2021). Treatment is achieved by several mechanisms including biosorption (which is dependent on the composition of the cell wall, especially the presence of functional groups such as carboxylate, hydroxyl, phosphate, etc.), biocoagulation/bioflocculation by extracellular biopolymers released by the algae, and finally by biodegradation/bioconversion, which is discussed in detail by Sarkar and Dey (2021). The authors also discuss new emerging treatment techniques such as immobilized algae cells embedded on nanoparticles, biochar and activated carbon.

Microalgal treatment is discussed in detail in Chap. 11.

10.3 Submerged Macrophytes

Submerged macrophytes play an important role in the self-purification of freshwaters, in particular, rivers where they are able to expand into the water body, creating a large leaf surface area that is able to maximize the adsorption of nutrients as well as contaminants such as soluble metals. They also support a diverse community of other microfauna and flora in the form of a biofilm or periphyton layer. Experiments that showed the effects of removing submerged macrophytes growing in stormwater retention ponds in Winnipeg, Canada, which were deemed unsightly by residents, highlighted the important role they play in improving water quality, with cyanobacterial blooms resulting after their removal (Badiou *et al.*, 2019). However, it has proved far more difficult to utilize submerged macrophytes in purpose-built wastewater treatment reactors, although they do form a minor role in free water surface constructed wetlands.

Submerged plants or macrophytes are only used for tertiary treatment of wastewaters as they require high clarity to ensure adequate light penetration and an aerobic environment for plant respiration. They are also used for treating eutrophied natural waters. Only a few of the many submerged aquatic plants found in natural waters can tolerate enriched waters, for example, *Potamogeton berchtoldii, P. filiformis, P. foliosus, P. pectinatus, P. zosterformis, Ceratophyllum demersum, Elodea canadensis, E. nuttalli, Ludwigia repens, Ulva lactuca, Najas marina*, and *Myriophyllum* spp., and hence have the potential for phycoremediation (Bishop and Eighmy, 1989; Nuttall, 1985; Gumbricht, 1993a). The macroalgae *Hydrodictyon* spp. and the ubiquitous *Cladophora* spp. are also associated with wastewaters. They remove soluble nutrients and some other compounds from the water but also provide a solid substrate for the development of an active periphyton slime, including heterotrophic bacteria. A drawback to their use for wastewater treatment is that while submerged plants grow well in the summer, they dieback in the winter. For example, in Michigan, US, the submerged macrophyte biomass degenerated by 98 to 99% in the winter (McNabb, 1976). Plants will compete with each other in ponds, so where mixed cultures are used, *E. canadensis, C. demersum*, or *P. foliosus* will eventually dominate the others present.

In order that the euphotic zone extends to the bottom of the pond, the depth should not exceed 2 m. Four experimental ponds that were 1.8 m deep and covering a total area of 16 ha were used to treat 1,900 m^3d^{-1} of final effluent from an activated sludge plant serving Michigan State University (Bahr *et al.*, 1977). The ponds contained *E. canadensis*, *C. demersum*, *P. foliosus*, and the epiphytic algae *Cladophora*. Harvesting of the filamentous alga *Cladophora* and the macrophytes is done four times a year during the growing season, resulting in biomass production of 400 g (DW) m^{-2}. After two years of operation, the nitrate concentration was reduced from 15 to 0.01 mg L^{-1} and the concentration of phosphorus from 4 to 0.03 mg L^{-1} by using all four ponds in series. The macrophytes can be used for fish food, feeding livestock, or for composting. *Elodea canadensis* contains 16.3 to 16.9% of ash, 2.89 to 3.86% of nitrogen, and 0.09 to 1.18% of phosphorus on a dry weight basis (Jewel, 1971) (Table 10.2).

Submerged macrophytes create ideal conditions for a wide variety of chemicophysical and biological reactions involved with the removal of nutrients from wastewater. Macrophytes can be harvested using a commercial weed harvester, although herbivorous fish can be used to crop excess vegetation (Sutton, 1974). Overstocking with fish will lead to overgrazing, therefore, care must be taken with stocking

Table 10.2. Chemical composition of selected microalgae (Rose, 1979). Reproduced with permission from Academic Press.

Component (% DW)	*Chorella* sp. 7–11–05	*Scenedesmus* sp.	*Chlorella–Scenedesmus* (10:1, sewage-grown)	*Spirulina maxima*
Protein*	55.5	53.0	41.8	55.5
Fat	7.5	13.0	7.2	12.7
Carbohydrate	17.8	13.5	27.4	17.4
Ash	8.25	6.5	19.1	7.4
Moisture	7.0	6.0	4.5	7.0
Crude fibre	3.1	8.0	—	—
Ascorbic acid	1.46	—	3.96	1.03
β-Carotene	5.02	—	6.02	2.25
Pantothenic acid	1.12	1.2	0.46	—
Pyridoxin	0.3	—	0.11	0.043
Thiamin	0.77	0.17	0.115	0.138
Riboflavin	—	0.42	0.269	0.285

* Protein contents were calculated by multiplying the value for the total nitrogen content by 6.25. True protein contents are about two-thirds of the values quoted.

levels. The fish can then be harvested annually and used for a much wider range of uses than the plant material (Dinges, 1982). Regular harvesting promotes macrophyte growth as well as maximizes nutrient removal. Significant biomass can be produced, for example, 5 to 90 t $ha^{-1}yr^{-1}$ for the *Mycriophyllum* spp., 10 to 12 t $ha^{-1}yr^{-1}$ for *Enteromorpha*, and 13 to 47 t $ha^{-1}yr^{-1}$ for the *Elodea* spp. (Gumbricht, 1993a). Where the biomass is not harvested, nitrogen removal is primarily through denitrification within the substrate. When the biomass is regularly harvested, a similar amount of nitrogen that has been assimilated by the macrophytes is removed. As the temperature falls and plant growth is reduced, nitrogen is stored in the sediment.

Most submerged macrophytes have roots and so need a substrate into which to be planted. The substrate plays an important role in the treatment process and should have a low nutrient content. While rooted macrophytes obtain phosphorus from the substrate and floating species exclusively from the water phase, submerged macrophytes obtain phosphorus from both sources. The overall removal of phosphorus is closely linked with plant uptake and, hence, biomass removal. In unharvested systems, sorption and precipitation are also important although this depends on substrate characteristics. These processes are summarized in Figs. 10.3 and 10.4. Gumbricht (1993a) ranked the importance of nitrogen removal mechanisms as sediment denitrification > assimilation > water column denitrification > sedimentation (Fig. 10.3). Assimilation is thought to be the most important removal mechanism in natural streams and can be more important than denitrification in submerged macrophyte ponds if regular harvesting is practiced.

Submerged macrophytes show increased uptake of nitrogen and phosphorus when the nutrients are abundant, with nitrogen uptake proportional to the concentration in the water. For example, the nitrogen uptake by *Myriophyllum aqua* growing in wastewater is between 0.05 and 1.26 g N $m^{-2}d^{-1}$, with mean uptake values of 0.78 g N $m^{-2}d^{-1}$ and 0.31 g P $m^{-2}d^{-1}$ (Nuttall, 1985). Gumbricht (1993b), while studying a full-scale plant, recorded its removal rates of total nitrogen (TN) and total phosphorus (TP) to be 0.3 and

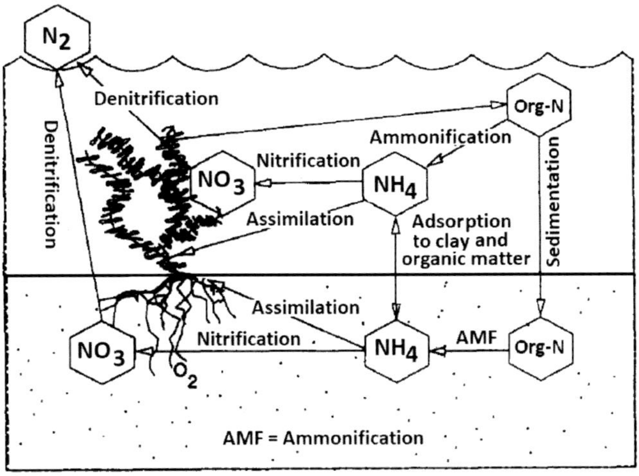

Fig. 10.3. Schematic of nitrogen cycling within a macrophyte pond (Gumbricht, 1993a). Reproduced with permission from Elsevier Publishing.

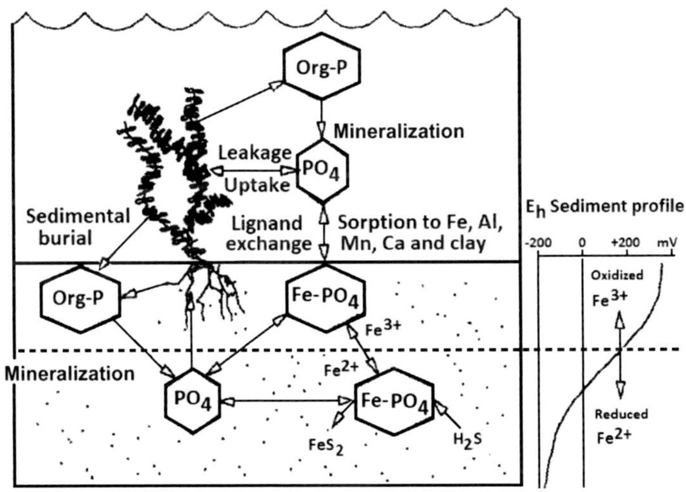

Fig. 10.4. Schematic of phosphorus cycling within a macrophyte pond (Gumbricht, 1993a). Reproduced with permission from Elsevier Publishing.

0.04 g $m^{-2}d^{-1}$, respectively. Once the biomass is detached or diesback in the winter, nutrients are quickly leached out, followed by a complete loss of nitrogen and phosphorus during decomposition.

The role of submerged macrophytes in constructed wetlands is considered in Sec. 13.2.

10.4 Floating Macrophytes

10.4.1 *Introduction*

This group of aquatic plants uses atmospheric oxygen and carbon dioxide, but obtain the remaining nutrients they require from water. Three groups of floating macrophytes are used in wastewater treatment: the water hyacinth (*Pontederia crassipes* syn. *Eichhornia crassipes*), duckweeds (*Spirodela* spp., *Lemna* spp., *Wolffia* spp. and *Wolffiella* spp.), and water ferns (*Salvernia* spp. and *Azolla* spp.) (Jackson and Gould, 1981; Mitchell, 1978; 1979; Finlayson, 1981). Treatment systems using floating plants are often confused with facultative ponds (Sec. 16.2). However, they are functionally very different as the floating plants release oxygen above the water surface, unlike the submerged planktonic algae in facultative ponds (Fig. 10.5). The dense layer of floating leaves reduces atmospheric oxygen diffusion, making such systems normally oxygen-deficient, with aerobic processes largely restricted to the plant root zone. The degree of oxygen depletion within the water phase of a pond containing floating plants is dependent on the organic loading rate, but is normally anoxic or anaerobic. Three primary treatment processes can be identified: (i) aerobic, anoxic and anaerobic metabolism by microorganisms on the plant roots and in the settled debris on the bottom of the pond, (ii) sedimentation of wastewater solids and autochthonous material (i.e., dead plant and animal debris generated within the system) on the bottom, and (iii) uptake of nutrients and some pollutants by the plants and their removal by harvesting.

Pollutants are removed from water by a number of different phytoremediation mechanisms:

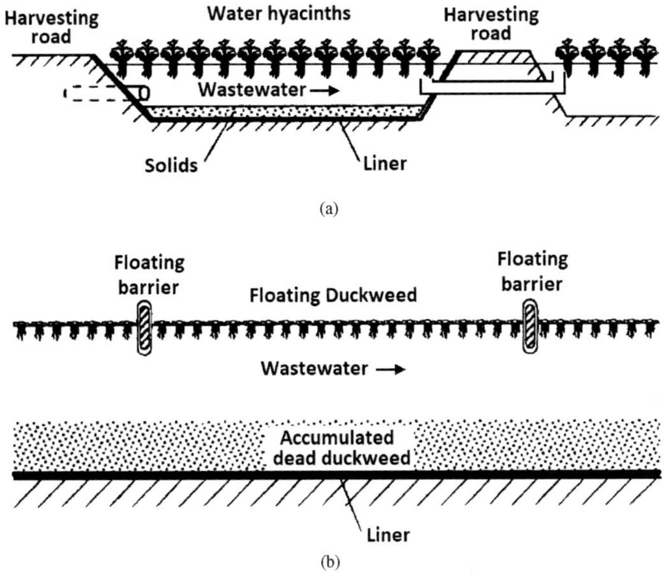

Fig. 10.5. Basic design of wastewater treatment systems employing floating macrophytes: (a) water hyacinth and (b) duckweed (Kadlec and Knight, 1996). Reproduced with permission from the Taylor & Francis Group.

Phytovolatilization, the uptake of organic compounds, their modification within the plant, and subsequent transpiration from the foliage into the atmosphere (Limmer and Burken, 2016).

Phytodegradation or phytotransformation is the breakdown of toxic organic compounds into non-toxic products by metabolic processes within the plant or by microorganisms found on the roots or within the rhizosphere (Newman and Reynolds, 2004).

Phytoextraction is the ability of plants to take up toxic compounds, especially from the environment, and to store them within the plant biomass. Some plants are natural hyperaccumulators; others have high biomass production while others are selectively bred for their ability to accumulate contaminants (Suman *et al.*, 2018).

Rhizofiltration is the sorption of contaminants onto or into plant roots, or via precipitation in the root zone.

Phytostabilization is the immobilization of contaminants in the environment rather than by adsorption or storage within the plant. This is done by chemicals released by the plant that bind or react with the contaminants, making them less toxic and less bioavailable.

The type of phytoremediation mechanisms employed depends primarily on the contaminants to be removed (Table 10.3) and is summarized in Fig. 10.6 (Ansari *et al.*, 2020).

Table 10.3. The main phytoremediation mechanisms involved with the removal of key contaminant groups (Ting *et al.*, 2018). Reproduced with permission from Elsevier Publishing.

Type of contaminants	Mechanisms
Organic compounds	• Phytoextraction: Direct uptake and accumulation of contaminants and metabolism in plant tissues
	• Phytovolatilization: Transpiration of volatile organic compounds (VOC) through the leaves
	• Rhizosphere bioremediation: Release of exudates that stimulate microbial activity and biochemical transformations in the soil
	• Phytotransformation: Enhancement of mineralization into relatively nontoxic constituents such as carbon dioxide, nitrate, chlorine and ammonia at the root-soil interface
Nitrogen	• Biological: Ammonification, nitrification, denitrification, plant uptake, biomass assimilation, dissimilatory nitrate reduction
	• Physico-chemical: Ammonia volatilization, adsorption
Metals	• Phytoextraction: Uptake and recovery of contaminants into above-ground biomass
	• Rhizofiltration: Filtering metals from water onto root systems
	• Phytovolatilization: Metal ions are converted into volatile state
Radionuclide	• Plant assimilation
	• Sedimentation of radionuclides-containing suspended materials on the plant's roots.

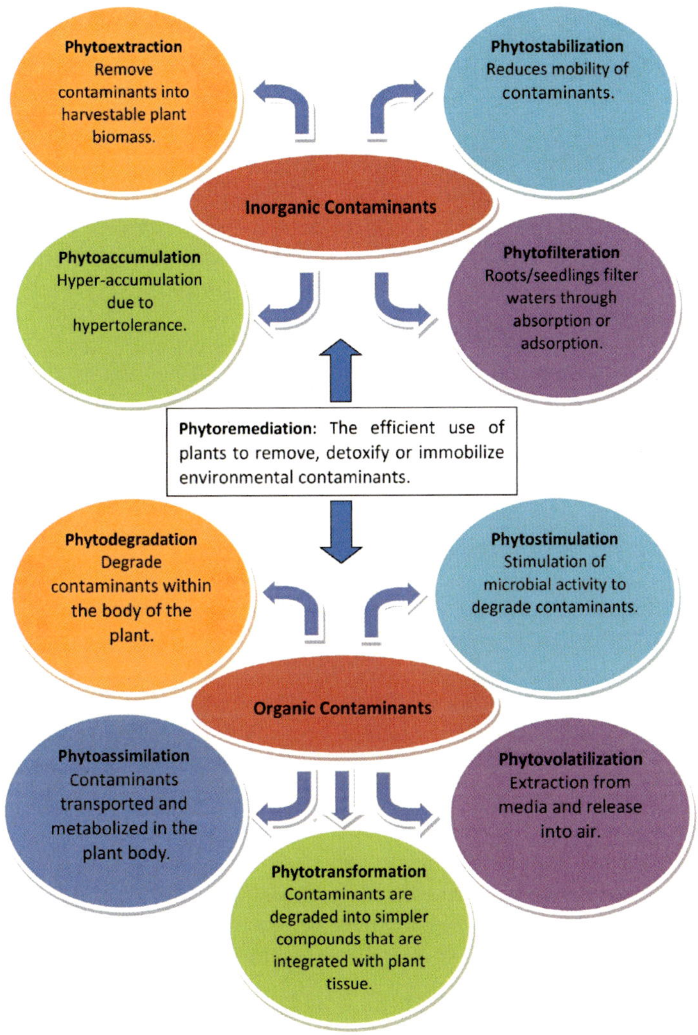

Fig. 10.6. Summary of phytoremediation processes for inorganic and organic contaminants (Ansari *et al.*, 2020). Reproduced with permission from the National Institute of Oceanography and Fisheries under a Creative Common License.

The application of floating plants for treatment has been limited because they rely on just one or two species to colonize the pond. Like all monoculture systems, this makes it vulnerable if the plant is either inhibited or eliminated. For example, water hyacinths are subject to attack by several serious pests and are also susceptible to cold weather. Although duckweeds are generally more tolerant to low temperatures, they can still be eliminated by extreme cold events. Once floating plants are lost, it may take months for plants to become re-established.

10.4.2 *Water hyacinth*

Although originally discovered growing in the Amazon River basin in Brazil in 1824, the water hyacinth (*Pontederia [=Eichhornia] crassipes*) is now found growing wild throughout the warmer parts of the world, including North America, Africa, Portugal, South and South East Africa, Australia, and New Zealand (Holm *et al.*, 1969) (Fig. 10.7). The water hyacinth is the largest of the floating plants, the most studied being, and the most widely used macrophyte employed in wastewater treatment. It is a rhizomatous plant with large glossy green leaves and a feathery unbranched root system. The pale lavender flowers with their beautiful splashes of yellow are very short-lived. The plant can reproduce asexually by the formation of stolons, a type of vegetative propagation, as well as sexually. New plants grow from the ends of each extended stolon, which can grow up to 30 cm in length depending on the available space. The plants grow rapidly in both vertical and horizontal directions, although initially, the hyacinths grow horizontally. When hyacinths are grown in ponds containing wastewater, they rapidly multiply and cover the available water surface. This crowding causes vertical growth with the plants growing more than 1 m in height. Interestingly, the more nutrient-rich the water is, the smaller the root system that develops, and in wastewater lagoons, the root system may be restricted to only 10 cm in length. The biology of the plant and its impact on aquatic ecosystems has been reviewed by Hailu and Degaga (2018).

Hyacinths are very much tropical plants requiring a minimum light intensity of 1,400 lm m^{-2} and preferring full sunlight for

Fig. 10.7. Different parts of water hyacinth (*Pontederia crassipes*): (a) Mature emergent leaves, (b) young plant spreading outwards, (c) whole plant showing rhizome, and (d) flowers (Rezania *et al.,* 2015b). Reproduced with permission from Elsevier Publishing.

maximum growth when their photosynthetic conversion efficiency approaches 4% (Penfound and Earle, 1958; Wolverston and McDonald, 1979; Eid and Shaltout, 2017). They have a growth range of between 10 and 35°C with optimum growth occurring between 25 and 27.5°C, but cannot tolerate cold temperatures with both the stem and leaves killed by frost. The central structure of the plant, from which the stem, roots and stolons arise, is the rhizome, which is similar in shape to a carrot and grows up to 20 cm in length. The rhizome is critical to the plant and if just 4 cm of the rhizome tip is removed, the plant dies. The tip floats just a few centimeters below the water surface, making it very vulnerable to the cold. The tip is killed and the plant subsequently decays if the water temperature approaches the freezing point. If the stem and leaves are killed by the frost, the plant rhizome will normally remain viable. However, as the dead vegetation dries, the bulk weight of the plant is reduced, so the rhizome is displaced less and rises closer to the water surface, making it more susceptible to possible damage from low temperatures. The northern limit of the plant in shallow waters is where the mean air temperature for the coldest month never falls below 1°C. Therefore, in practice, all the water hyacinth systems currently treating wastewater are located in tropical or semi-tropical climates (Middlebrooks *et al.*, 1982). When hyacinth systems have been attempted in cooler climates, they had to be housed in heated greenhouses with the temperature maintained within the optimum range. Although it has been suggested that methane produced from digesting the waste biomass in anaerobic digesters could be used to heat the greenhouses, it is doubtful if such a system could either be carbon-neutral or cost-effective. Also, where the system only operates for part of the year, there is the added expense of operating a culture unit to maintain sufficient hyacinths for introduction into the system each spring. Culturing hyacinths is straight-forward. For example, Sooknah and Wilkie (2004) were able to grow 1,610 new plants from an initial 10 plants within 10 months.

Hyacinths are freshwater plants and are inhibited by salinities of >1–6% (Haller *et al.*, 1974), but they can tolerate a wide pH range of 4 to 10 (Haller and Sutton, 1973). They are very disease resistant, with only three exotic insects being able to exert any significant effect on growth: the hyacinth weevils (*Neochetina eichhorniae* and *N. bruchi*) and the leaf-mining mite (*Orthogalumna terebrantis*). These insect species have been introduced in some areas of the world to control what has become an invasive plant, but other indigenous pests have had little effect on plant growth. However, the effect of insects feeding on hyacinths is generally thought to be beneficial as it stimulates the formation of new growth.

The hyacinth is one of the most productive macrophytes in the world, with its biomass able to double every 7 days, which is equivalent to a daily gain in biomass of 108 kg (DW) $ha^{-1}d^{-1}$. Standing crops of hyacinth can reach 2,500 g (DW) m^{-2}, although the exact biomass will depend on nutrient availability and the degree of crowding.

When grown in domestic wastewater, the standing crop reaches over 3,000 g (DW) m^{-2}, which is equivalent to 30 tons (DW) $ha^{-1}d^{-1}$ (Wooten and Dodd, 1976; Dinges, 1982). With a moisture content of 95%, this represents a vast quantity of biomass to be handled and eventually processed. The dense growth of hyacinths in a wastewater lagoon provides a unique and very stable environment. The extensive root system prevents horizontal movement, and particulate solids rapidly settle. Oxygen is released into the water via the roots, resulting in an active heterotrophic and autotrophic (nitrification) community and the lowest depths of the lagoon will be anoxic, allowing for denitrification. A complex micro and macrofauna develops on and within the root area of the lagoons with the most common groups being Branchiopoda, Oligochaeta, Talitridae, and Chironomidae (Diptera) (Barker *et al.*, 2014). However, *P. crassipes* is not considered an important habitat in freshwaters, rather a major invasive species that can severely damage freshwater ecosystems (Hailu and Degaga, 2018).

Hyacinth lagoons are primarily used for tertiary treatment and nutrient removal, and because they require warm temperatures, they would seem particularly suitable for use as a tertiary treatment phase after facultative stabilization ponds, which are widely used in warmer climates. They are also used for secondary treatment for small communities. In terms of treatment, most biochemical oxygen demand (BOD) and suspended solids reduction are due to settlement, whereas nitrogen and phosphorus removal are by the plants, with the former unaffected by the season but the latter severely reduced in the winter. The plants exhibit a luxury uptake of nutrients, far over normal requirements, but also accumulate other components of wastewater, such as heavy metals and synthetic organic compounds (Rezania *et al.*, 2015b). The uptake and accumulation of such compounds can be disadvantageous where the concentrations in the plant tissue prohibit the subsequent use of biomass for conversion into energy by digestion or as a feed supplement.

The design of hyacinth lagoons has been reviewed by the US Environmental Protection Agency (USEPA, 2011). Treatment objectives are important to ensure sufficient retention time, but the limiting factors in design appear to be the periodic removal of settled solids and the harvesting of hyacinths. Lagoons are normally large, varying in size from 0.4 to 4 ha. Long narrow channels with plug flow appear most efficient as it prevents short-circuiting and allows easy harvesting of the macrophytes, although the settlement is less efficient. Various modifications to this basic design have been made (Dinges, 1978; 1982; Hermosillo and Sarquis, 1990), for example, a zig-zag channel configuration has been used to increase settlement as well as ensuring adequate access for harvesting equipment (Fig. 10.8).

Adequate root depth must be provided for the hyacinths to enable the roots to come into contact with the majority of the liquid passing through the lagoon. For example, a depth of 0.4 m ensures complete wastewater contact with the root system, ensuring maximum nutrient removal. However, the depth can vary from 0.4 to 1.2 m, with the deeper lagoons allowing space for settlement and anoxic activity. Although deeper ponds reduce the total land area

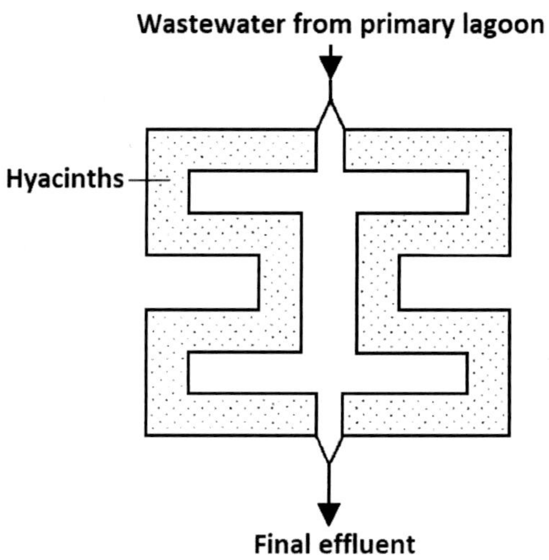

Fig. 10.8. Zig-zag design of the culture basin for water hyacinths.

required, they tend to become completely anaerobic at night if they are completely covered with plants. Hydraulic loading is more critical than organic loading for wastewaters that have already received some treatment, with hydraulic loadings varying from 240 to 3,570 $m^3ha^{-1}d^{-1}$ for domestic effluents. However, in practice to produce a final effluent with a BOD <10 mg L^{-1}, suspended solids <10 mg L^{-1}, total Kjeldahl nitrogen <5 mg L^{-1}, and total phosphorus <5 mg L^{-1} from a secondary effluent, then the hydraulic loading rate should not exceed 1,870 $m^3ha^{-1}d^{-1}$. Stronger effluents require lower hydraulic loading rates. The retention time should not be less than 5 days. For secondary treatment, organic loading should not exceed 100 kg BOD $ha^{-1}d^{-1}$ or 35 kg BOD $ha^{-1}d^{-1}$ for tertiary treatment. There have been a number of kinetic models that describe water hyacinth ponds, although the one by Polprasert and Khatiwada (1998) also integrates the role of heterotrophic bacteria, both suspended and attached to the roots and submerged section of leaves as a biofilm.

To achieve maximum nutrient removal, the hyacinth biomass necessary to absorb the nutrients are maintained by harvesting at a rate that balances net plant productivity. Frequent harvesting is necessary to ensure that the hyacinth population is kept in an active growing phase. The initial uptake of nutrients is rapid, with the amount absorbed related to the concentration of nutrients in the water. The removal rate of nutrients can be directly related to the lagoon area. For example, a 2.1-ha hyacinth lagoon in Florida receiving 3,800 m^3 of secondary-treated domestic wastewater each day removed 80% of the nitrogen and 44% of the available phosphorus (Cornwell *et al.*, 1977). The productivity of the plant fluctuated seasonally, according to the temperature and other climatic factors such as light intensity and day length; therefore, removal efficiency will also fluctuate and harvesting has to be adjusted accordingly.

Numerous pilot studies on hyacinth lagoons have been conducted and there is a considerable amount of performance data available. The comparative pilot study by Hauser (1984) at Roseville, California, is typical. Three separate ponds were operated using secondary effluent from the Roseville Sewage Treatment Works. Pond 1 was unmanaged, with the hyacinths allowed to control their own density on the pond, whereas Pond 2 was aerated, using a submerged perforated tubing connected to a compressor, to ensure an aerobic aquatic environment, but the hyacinths were not harvested. Only the first two-thirds of the pond were aerated with the pipes spread 3 feet apart, whereas the remaining third was left to allow settlement and filtration. In Pond 3, the plants were harvested but not aerated. Harvesting in Pond 3 was carried out every 2 weeks during the growing season, with 20% of the surface area cleared each time. By alternating the harvesting location and not redistributing the plants after harvesting, the period between the disturbance of the hyacinths was maximized. Initially, all three ponds were operated in an unmanaged mode, thus any difference in performance during maturation was due solely to unequal plant cover. All three ponds produced good final effluents with significant reductions in BOD and suspended solids of between 30 and 50% reduction in the winter and 70% for the warmer growing season (May to October). The final effluent concentration was independent of the influent concentration, suggesting that the potential BOD reduction in these ponds was only partially utilized. Efficiency in suspended solids removal varied between the ponds more than the BOD removal. The unmanaged pond (Pond 1) produced a consistently good effluent with low suspended solids concentration throughout the year, whereas in the other two ponds, high suspended solids concentrations were occasionally recorded. As the BOD was not high on these occasions, it suggested that the high solids in Pond 2 were inert particles disturbed by the aeration system as they accumulated, and in Pond 3, as the high suspended solids concentration was associated with harvesting, they were probably fragments of water hyacinths (Table 10.4). To control this "harvest induced" rise in suspended solids, Hauser (1984) suggests that a portion of the pond near the effluent end should be left unharvested or a separate non-harvested polishing pond could be used. Dinges and Doersam (1987) have described a full-scale hyacinth treatment plant in Austin, Texas. The Analândia Water Treatment Plant in São Paulo, Brazil, also uses hyacinths to remove turbidity from impounded river water, and improve overall quality, before supply. Hyacinths are grown in a canal with a total area of 1,590 m^2, a depth of 0.7 m, and a hydraulic retention

Table 10.4. Performance summary of three experimental hyacinth ponds. Pond 1 was unmanaged, Pond 2 was aerated, and Pond 3 was harvested (Hauser, 1984). Reproduced with permission from the Wastewater Treatment Federation.

Parameter	Influent concentration, mg L^{-1} (except pH)		Mean pond loading (all ponds) kg ha^{-1})	Effluent concentration, mg L^{-1} (except pH)					
				Pond 1 (unmanaged)		Pond 2 (aerated)		Pond 3 (harvested)	
	Mean	S.d.		Mean	S.d.	Mean	S.d.	Mean	S.d.
Nov 1981–Apr 1982									
BOD_5	11.6	6.5	18	4.9	3.6	3.6	2.5	3.5	1.9
SS	10.7	6.5	16	2.8	2.4	4.6	6.9	3.0	2.1
pH	7.8	0.2	—	6.9	0.3	6.9	0.3	7.1	0.3
NH_4 as N	14.1	2.7	21	6.1	4.6	1.8	2.4	9.5	3.3
Organic N	2.3	1.0	3.5	1.4	0.7	0.9	0.5	1.3	0.6
$NO_3 + NO_2$ as N	0.3	0.3	0.5	3.6	3.5	7.1	2.9	2.7	2.2
Total N	16.7	2.7	25	11.1	2.9	9.9	4.1	13.5	3.0
Alkalinity	132	14	200	97	35	64	17	109	22
May 1982–Oct 1982									
BOD_5	8.6	4.2	13	2.3	1.3	1.9	0.3	2.5	0.9
SS	7.1	4.2	11	2.0	2.0	2.6	1.0	5.4	7.0
pH	7.8	0.2	—	6.5	0.2	6.6	0.2	6.6	0.2
NH_4 as N	15.8	3.1	24	4.6	2.4	0.2	0.1	3.7	1.2
Organic N	2.3	1.9	3.5	1.1	0.5	1.2	1.3	1.2	0.6
$NO_3 + NO_2$ as N	0.2	0.1	0.3	2.7	2.7	3.8	2.4	3.2	2.4
Total N	18.2	2.5	28	8.4	3.0	5.2	3.0	8.0	3.1
Alkalinity	117	15	178	78	17	51	11	67	9

time (HRT) of 21 hours. The water subsequently flows through soil filters planted with rice (*Oriza sativa*) (Salati *et al.*, 1999).

In a comparison of nitrogen removal technologies, Ting *et al.* (2018) found phytoremediation and the use of water hyacinth, in particular, to be the most cost-effective and environmentally sustainable option. The plant is the most efficient macrophyte for nutrient removal and is considered to be a hyperaccumulator of nitrogen (Fox *et al.*, 2008). The two major mechanisms for ammonia reduction in hyacinth ponds are bacterial nitrification and plant uptake. The primary function of water hyacinths in the bacterial nitrification process is to provide sufficient support for the development of nitrifying bacteria. The nitrogen is converted from ammonia to nitrate with the possibility of denitrification in the lower anoxic layers of the ponds. However, nitrogen is also removed by plant uptake and is incorporated into the plant biomass. Harvesting is thought to encourage plant uptake by eliminating the inhibitory effects of crowding on plant growth, although it may limit nitrification by reducing the availability of support structures as well as removing a portion of the nitrifying bacteria with the harvested plants (Hauser, 1984). Extreme rainfall events increase the dissolved oxygen concentration in hyacinth ponds due to an influx of oxygenated surface runoff, which affects the relative rates of nitrification and denitrification. In pilot plant studies, Qin *et al.* (2020) recorded high removal rates of nitrogen and phosphorus regardless of dissolved oxygen concentrations caused by precipitation events or high ammonia concentrations (>7 mg NH_4^+–N L^{-1}). However, during periods of extreme rainfall, the denitrification was suppressed by elevated dissolved oxygen (DO) caused by aerated stormwater entering the pond, resulting in increased nitrate in the final effluent, whereas outside these events, >99% of the total nitrogen was removed as nitrogen gas by denitrification.

Water hyacinths are easily harvested as they float on the surface of the water. Where there is open water, the leaves of the plant act as sails with the wind blowing the plants to the leeward side of the pond. Hyacinths are generally collected and chopped manually on experimental ponds, although there is a variety of mechanical harvesters commercially available for larger installations. These normally operate from special platforms and combine a feed or collection system such as a rotary head, a chopper, and a conveyor to transport the material to trucks. There are numerous potential uses for harvested hyacinths, although these depend to some extent on the composition of the wastewater used to grow the plants. Hyacinths grown on domestic wastewater can be used for livestock and fish feed (Gunnarsson and Petersen, 2007), a protein supplement, compost or soil amendment and fertilizer (Vidya and Girish, 2014), paper manufacture (De Groote *et al.*, 2003), bioethanol production (Eshtiaghi *et al.*, 2012; Krishnan *et al.*, 2020), pyrolysis (Rahman, 2018), biochar production (Tripathi *et al.*, 2016), livestock bedding, and bioenergy (Rezania *et al.*, 2015a;

Gaurav *et al.*, 2020). Rich in hemicellulose (48.7%), cellulose (18.2%), lignin (3.5%), and crude protein (13.3%), water hyacinths are ideal for the production of biogas, especially when mixed with other wastes, as well as other biofuels (Mitan, 2019). Hyacinths can be readily digested in conventional mesophilic stirred digesters at low loadings and high retention times (Klass and Ghosh, 1980; Ghosh *et al.*, 1980; Chin and Goh, 1978) (Table 10.1). Better yields have been obtained using hyacinths mixed with primary sewage sludge and by using digesters in conjunction with upflow anaerobic filters (Chynoweth *et al.*, 1981; 1982). They can also be dried to form briquettes for burning in stoves with a calorific value of 14.55 MJ kg^{-1} or 8.3 GJ m^{-3}, which is similar to charcoal (Gunnarsson and Petersen, 2007; Munjeri *et al.*, 2016). Bote *et al.* (2020) give a detailed example of small-scale biogas production in India using the macrophyte, with the waste digester sludge used to produce briquettes. This is a sustainable alternative to using other solid fuels and is best suited to those areas where the plants can be naturally sun-dried. They are then chopped into 6-mm pieces and compressed into either pellets or briquettes.

Chopped, pelleted or ensiled hyacinths can be fed to cattle and sheep. Ensiling requires the addition of about 4% of citrus pulp or corn to act as a readily available carbohydrate source and can be fed directly to dairy cattle. The main value of hyacinth for livestock is as a source of roughage, minerals and energy (Bagnall *et al.*, 1974; Zerinque *et al.*, 1979). A valuable compost can be produced from whole or chopped hyacinths by storing wet plants aerobically for 1 to 6 months, followed by drying and grinding. Growth is poor in pure compost, but when mixed with sand at a ratio of 3:1, excellent growth is obtained, with the composted hyacinths able to retain water unusually well. Water hyacinths have also been successfully vermicomposted (Balasubramanian and Bai, 1995; Gajalakshmi *et al.*, 2001).

Hyacinths grown on industrial wastewater accumulate high concentrations of heavy metals, making them unsuitable for consumption or disposal to land as compost or amendment. Much research has been done on the ability of water hyacinth to accumulate heavy metals, with copper most efficiently taken up (Cu > Zn > Ni > Pb > Cd), and most metals taken up by the roots and subsequently translocated into the shoots and leaves where they concentrate (Liao and Chang, 2004; Lu *et al.*, 2004; Jadia and Fulekar, 2009). The uptake of metals, organic and inorganic contaminants are reviewed by Rezania *et al.* (2015b). As long as the heavy metal concentration does not inhibit digestion, methane can be produced with the residual sludge used for metal recovery of particularly valuable or toxic metals such as silver, gold, cadmium, mercury, lead, and all the base metals. Alternatively, the plants can be dried, then burnt, and the metals recovered from the ash.

10.4.3 *Duckweeds and water ferns*

Two other groups of floating plants are also widely used for wastewater treatment: duckweed and the water fern.

Duckweeds (Lemnaceae)

Small-scale duckweed production has been common in many Asian countries for at least a century, primarily to feed pigs, ducks and poultry, based on waste being allowed to settle in a pond that overflows into a second pond approximately 0.5 m in depth, where a duckweed mat develops which can be harvested daily. In Northern Thailand and Myanmar (former Burma), duckweed was also used as a vegetable, being a rich source of vitamin A (Leng, 1999). While duckweeds are usually indigenous, reducing the threats of the spread of invasive species to nearby surface waters make them an ideal macrophyte for wastewater treatment, though their small size and structure make them difficult to manage and harvest. The development of larger duckweed treatment systems for wastewater has been hampered by technical difficulties and harvesting problems and has been superseded at the commercial level by the development of high-rate algal ponds (Chap. 11). However, they are still used for small to medium on-site wastewater treatment, especially tertiary treatment combined with biomass reuse.

There are over 40 species of *Lemnaceae* in five genera: *Spirodela*, *Landoltia*, *Lemna*, *Wolffia*, and *Wolffiella*, which are widely distributed and found in all parts of the world except deserts and the polar regions. Duckweeds comprise a flattened leaf-like frond without a stem, with some species even lacking roots. Reproduction is primarily by vegetative multiplication, with biomass doubling times of between 1.0 and 5.3 days under optimum conditions. They are resistant to diseases or pests and will grow in slightly saline waters. Duckweed can grow over a wide temperature range of 6 to 30°C but can become stressed under direct sunlight in open ponds due to overheating. While they are tolerant of a wide pH range (5–9), in practice, the optimum pH is between 6.5 and 7.0 in order to maintain ammonia as the ammonium ion to optimize uptake, reduce toxicity, and any ammonia volatilization losses. They float on the surface of the water, forming dense mats, but unlike hyacinths, they provide much less support for underwater microfauna. The fronds form a single layer on the water surface until crowding occurs, then the thickness of the surface layer may increase to form an extensive mat on the water surface several centimeters thick, thus making harvesting easier. The water beneath a dense mat will have a lowered dissolved oxygen concentration, and if the organic loading is high, anaerobiosis may occur. Duckweed will grow well even on raw wastewater, improving the water quality by direct plant uptake of organic and mineral material, suppression of algal growth, and by creating a suitable habitat for the growth of zooplankton, especially *Daphniidae* (Dinges, 1982).

Duckweed ponds support diverse meso and macrofauna as well as a range of microalgal (Phytoplanktonic) species (e.g. species of the genera *Euglenophyceae* and *Chlorophyceae).* Zooplankton includes rotifers, which are particularly common, as are *Branchiopoda* (*Daphniidae*) and a range of ciliate protozoans, both free-swimming (e.g. *Paramecium* spp.) and attached (e.g. *Vorticella* spp.) species. Macroinvertebrates including Hemiptera (e.g. *Notonectidae*), and *Insecta* (*Dytiscidae Chironomidae, Culicidae*, and *Heteroptera*) are recorded, but other groups such as *Gastropoda* can become problematic (Ennabili *et al.,* 2019). The limited free water area significantly reduces mosquito breeding success (Tariq *et al.,* 2009).

Duckweed ponds operate in a similar way to facultative ponds with duckweed, phytoplankton, and both aerobic and anaerobic

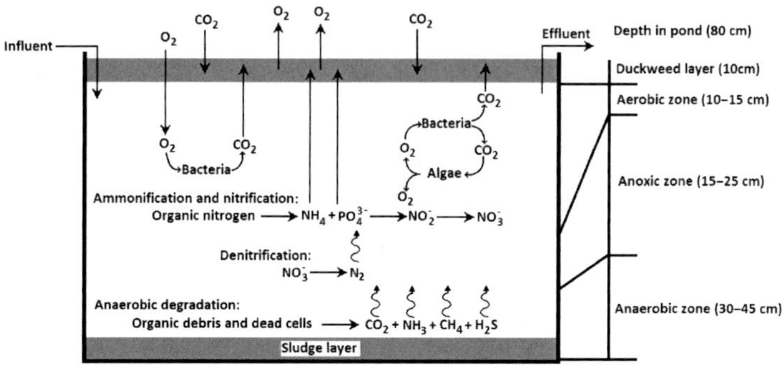

Fig. 10.9. Biological processes in a duckweed treatment pond.

bacteria involved in the treatment. There are four distinct zones in ponds: a surface layer or floating mat comprising duckweed, an aerobic zone layer immediately below, followed by an oxygen or redox gradient creating an anoxic, and possibly an anaerobic zone depending on depth, sludge accumulation, and its organic content (Fig. 10.9). The total nitrogen removal is by a combination of plant uptake and bacterial nitrification-denitrification. The phosphorus removal is principally by plant uptake, adsorption and settlement.

In early experiments, Harvey and Fox (1973) grew *Lemna minor* in laboratory-scale units to evaluate the nutrient removal efficiency by using 80 aquaria filled with secondary wastewater to a depth of 450 mm, providing a surface area of 500 cm^2. The temperature was maintained at 24°C with the light intensity held at 11,022 lm m^{-2} using 12-hour photoperiods. Over a 10-day test period, frond doubling occurred every 4 days, with the harvested plants containing 4.6% of nitrogen and 0.8% of phosphorus (DW). In a more recent study, *Lemna gibba* productivity was recorded at 28.4 tons (DW) $ha^{-1}yr^{-1}$ in tertiary treatment ponds and successfully reduced the level of nutrients in the final effluent through plant uptake and bacterial nitrification-denitrification in the root zone, which extended 10 to 15 cm into the pond (Ennabili *et al.*, 2019).

There is growing interest in free water surface (FWS) systems using *Lemna minor* in combination with subsurface flow-constructed wetlands for enhanced nutrient removal when final effluent discharges are to nutrient-sensitive rivers and lakes (Masi *et al.*, 1999) (Sec. 13.2). High nutrient content is essential in order to maintain optimal uptake by the duckweed as well as maintaining biomass production. The removal rates for phosphorus remain fairly constant at between 90 and 95%.

Nutrient removal by *Lemna japonica* 0234 was directly compared to the water hyacinth using dual-channel pilot-scale reactors for over a year. The TN removal was the same for both macrophytes at 0.4 g m^2d^{-1}, although the duckweed removed less TP at 0.1 g m^2d^{-1}. A higher proportion of the nitrogen was fixed by the duckweed (60%) compared to hyacinths (47%). The total biomass production of duckweed was less than half of that of water hyacinth, although the former was rich in protein (33.3%), amino acids (25.8%), starch (40.2%), and had a flavonoid content nearly seven times higher, making it more valuable and versatile for reuse (Zhao *et al.*, 2014). The breakdown of amino acids in both the duckweed and water hyacinth is compared in Fig. 10.10.

Two pilot-scale ponds were operated in series to treat weak wastewater (average: 157 mg COD L^{-1}, 71 mg BOD L^{-1}, 73 mg TN L^{-1}, 8 mg TP L^{-1}). With a low organic loading rate of 39 g COD $ha^{-1}d^{-1}$,

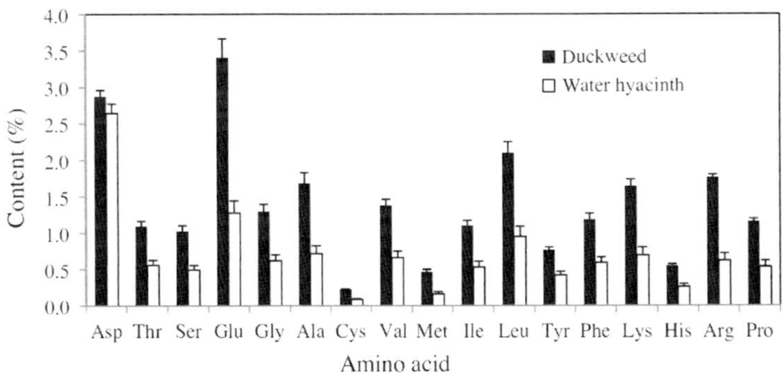

Fig. 10.10. Comparison of the amino-acids found in duckweed and water hyacinth during the summer months in pilot-scale ponds (Zhao *et al.*, 2014). Reproduced with permission from Elsevier Publishing.

18.1 kg TN $ha^{-1}d^{-1}$, and 2 kg TP $ha^{-1}d^{-1}$, the duckweed ponds achieved 79, 93 and 84% removals, respectively (Mohedano *et al.*, 2019) (Fig. 10.11). This study also found that CO_2 was fixed within the plant biomass (*Landoltia punctata*, formally *Spirodela punctata*) at a rate three times greater than what was emitted. The actual emissions over the 15-month study period were 3,048 to 6,017 mg CO_2 m^2d^{-1} compared to the 19,592 to 40,052 mg CO_2 m^2d^{-1} fixed by the duckweed. Methane was not produced from the pond sediment,

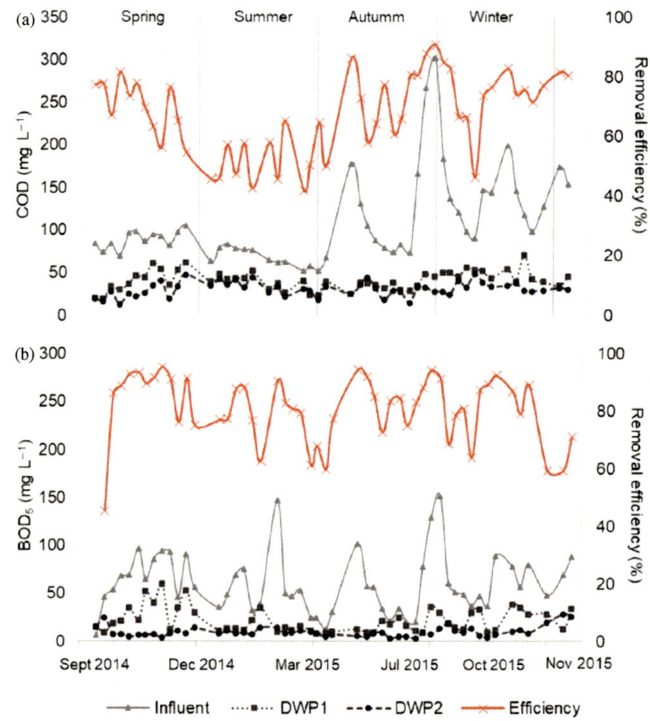

Fig. 10.11. (a) Influent COD and (b) BOD (mg L^{-1}) and the final effluent quality after the first (DWP 1) and second stage (DWP 2) treatment in pilot-scale duckweed ponds. The percentage removal is also given (Mohedano *et al.*, 2019). Reproduced with permission from Elsevier Publishing.

making them extremely climate-friendly treatment units. In a separate study, settled municipal wastewater was successfully treated by using two macrophyte ponds in series; the first containing the microalgae *Chlorella sorokiniana*, which removed the COD and partially removing nitrogen and phosphorus, while the second pond containing *Lemna minor* removed the remaining nitrogen. Overall, the removals were COD, 99%; TN, 88%; and TP, 91%, with 10 times more microalgal biomass produced than duckweed (Kotoula *et al.*, 2020).

There is little research on the optimal design for duckweed ponds, but as the macrophyte readily grows on all types of natural water bodies, the design may not be critical as long as the water surface is sheltered and relatively still (Al-Nozaily *et al.*, 2000a; 2000b). Generally, the actual shape and depth of ponds are determined primarily by the harvesting method employed. Duckweed is normally harvested using a skimming device similar to those used for removing oil from the surface of the water. Microstrainers, as used in water treatment, can also be used for larger operations. At small on-site ponds, the duckweed can be manually skimmed off the surface using a fine net as required.

An HRT of between 10 and 20 days is required to reduce the COD or BOD from 30 to 20 mg L^{-1}. Longer retention times tend to reduce the protein content of the biomass. The depth varies between 0.5 to 1.5 m, with shallow ponds used for rootless species. An optimum water velocity of 0.1 m s^{-1} is recommended to prevent the duckweed mat from being disturbed, and must not exceed 0.3 m s^{-1} to prevent solids from being discharged with the final effluent. Narrow plug flow reactors that are 5 to 15 m wide are normally used with an ideal total length:width ratio of 38:1 (Alaerts *et al.*, 1996). To prevent wind-creating areas of clear water on the pond surface, barriers or booms are placed every 10 to 20 m, creating cells. These can be simple floating barriers made from bamboo or actual physical barriers, each with a different positioned gap to allow mixing as the wastewater moves from one cell to another (Fig. 10.12). In practice, the strength of raw wastewater should not exceed 350 mg L^{-1} with the organic loading estimated using the required percentage BOD

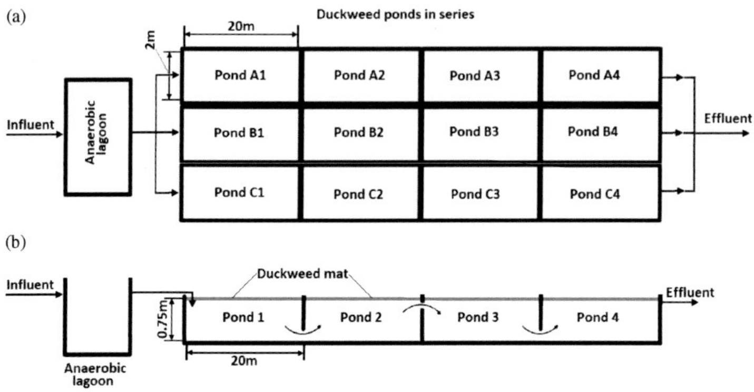

Fig. 10.12. Schematic layout of a duckweed wastewater treatment pond: (a) horizontal, and (b) vertical section.

removal and temperature (Eqs. [10.2] and [10.3]) (Smith and Moelyowati, 2001):

90% of BOD removal:

$$\lambda = 0.2995\ T + 3.3308. \tag{10.2}$$

95% of BOD removal:

$$\lambda = 0.2302\ T + 2.5601, \tag{10.3}$$

where λ is the BOD loading rate (g $m^3 d^{-1}$) and T is the temperature (°C).

Duckweed ponds can be integrated with standard anaerobic lagoons, facultative and maturation ponds. Alternatively, complete treatment multiple duckweed ponds are used in series, including a primary anaerobic stage and several secondary stages followed by a tertiary stage to help with pathogen reduction and ensure that the final effluent is aerated. However, Smith and Moelyowati (2001) recommend incorporating anaerobic lagoon and maturation pond stages for pre and post-treatment, respectively, with duckweed ponds providing a secondary treatment (Fig. 10.13).

Much interest has been shown in the use of plants for poultry and livestock feeds as well as for the removal of nutrients from

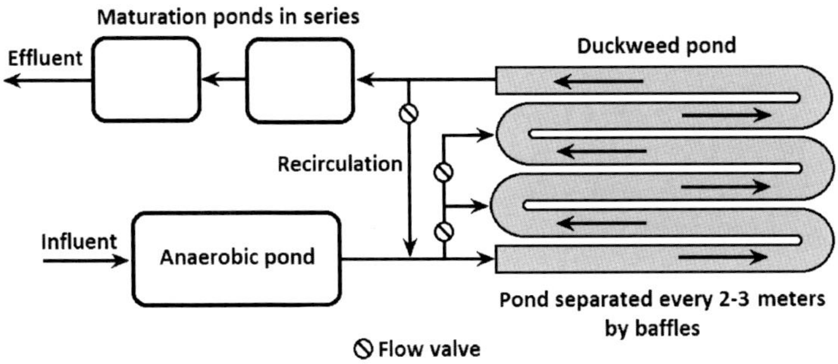

Fig. 10.13. Recommended design of an integrated anaerobic lagoon, duckweed and maturation ponds.

secondary-treated wastewaters (Culley and Epps, 1973; Harvey and Fox, 1973; Hillman and Culley, 1978). Duckweed can concentrate a wide range of minerals and trace elements often missing from animal feeds and basic fodder crops, such as straw, and so makes a valuable animal feed supplement. The crude protein content of duckweed increases with an increasing concentration of ammonium in the water, rising from 15% (DW) at 1–4 mg N L^{-1} to 37% at 10–15 mg N L^{-1}. No increase in crude protein is seen at ammonium concentrations in excess of 15 mg N L^{-1}. Above >60 mg N L^{-1}, toxicity becomes increasingly apparent due to higher concentrations of free ammonia. Fat content is between 4 and 5% (DW). However, a high water content (the average water content for *Lemna* and *Spirodela* varies between 92 and 97%) and problems in removing the water have restricted their full development as an agricultural crop. The harvesting rate is generally between 80 and 160 kg $ha^{-1}d^{-1}$ wet weight (WW), although when nutrients become limiting, this should be increased to prevent the dieback of duckweed. Dry matter yields of duckweed vary between 20 and >100 t $ha^{-1}yr^{-1}$, averaging out at about 40 t $ha^{-1}yr^{-1}$ under ideal operating conditions, and falling to 5 to 20 t $ha^{-1}yr^{-1}$ under sub-optimum conditions (Leng, 1999).

Like other macrophytes, duckweeds can accumulate heavy metals. Arsenic, for example, is a particularly toxic metal that can easily transfer into the food chain via rice and is known to contaminate the drinking water of tens of millions of people. *Lemna gibba*, *Lemna minor* and *Spirodela polyrhiza* have been found to be able to efficiently remove arsenic from contaminated water, with *Wolffia globose* having an uptake efficiency up to 10 times greater than other duckweeds or the *Azolla* species and is able to accumulate >1,000 mg As kg^{-1} (DW), making it the ideal candidate for arsenic phytoremediation (Rahman and Hasegawa, 2011).

A full-scale duckweed pond system at Mirzapur, Bangladesh, is described by Alaerts *et al.* (1996) and the application of duckweed ponds for sustainable farming has been extensively reviewed by Leng (1999) for the Food and Agricultural Organization. However, not all duckweed ponds are successful, and this is explored by Nhapi *et al.* (2002), who operated two full-scale ponds treating domestic wastewater in Zimbabwe that served populations of 5,000 and 10,000. They found that poor design, especially excessive depth and short-circuiting, infrequent harvesting of biomass, and high levels of ammonia leading to duckweed toxicity, were examples of the problems encountered. They recommended recycling effluent, where necessary, to dilute the influent within design loadings.

An innovative project in Ireland has seen the development of a recirculating closed-loop aquaculture system for perch and trout using duckweed ponds for treating the effluent. The system comprises four 1,000 m^3 fishponds with the water recirculated every four hours through a duckweed treatment pond with a surface area of 1.2 ha (Fig. 10.14). The system has been designed and constructed using sustainable and circular economy principles. There is a minimum of concrete used in the construction with ponds lined with natural glacial till, with the small amount of power required coming from an adjacent wind farm. The pond is a joint project primarily between Bord Iascaigh Mhara (Irish Sea Fisheries Board) and Bord na Móna (Irish Peat Board), having been constructed on abandoned land once used for peat production in Muntlucas, County Offaly. As a closed-loop system, it is self-sufficient in water which is topped up by rainwater stored in a small reservoir. Fish are kept to a maximum stocking density of <20 kg m^{-3} with a projected output of 25 tonnes of perch per annum (equivalent to 140,000 fish) and 12 tonnes of trout. Fish are fed on organic feed supplemented by the natural biota in the

Fig. 10.14. Aerial view of the Oasis Fish Farm in Ireland. The water storage reservoir is located bottom center of the site with the four fishponds spaced around it, each separated into four sections. Water from the fishponds flow through channels on the outer sides of the site to enter either end of the duckweed treatment pond. The treatment pond is divided into two sections, both discharging treated water at the center, which flows back to the fishponds via the centrally located channels. Water is moved around the site by numerous surface paddle wheels that also oxygenate the water, using electricity powered by the adjacent wind farm. Image by Gary McCoy and reproduced with permission of Bord Iascaigh Mhara.

ponds. There are plans to use the harvested duckweed for protein recovery. Water quality was found to be excellent and supported the fishery without any reported problems (O'Neill *et al.*, 2020). This demonstrates that it is possible to have low-energy, sustainable aquaculture systems with zero pollutant discharge.

Water ferns (Salveniaceae)

Salveniaceae consist of some 16 species in two genera, *Salvenia* and *Azolla*. Little work has been done on this group, although several workers feel they are less suitable for wastewater treatment than other macrophytes as they produce large amounts of detritus within the pond. The majority of studies have employed *Azolla* spp. especially *A. microphyla* (Noorjahan and Jamuna, 2015), *A. pinnata* (Akinbile *et al.*, 2019), and *A. filiculoides* (Zhao and Duncan, 1997), formerly *A. caroliniana* (Reddy and Debusk, 1982). The biology and taxonomy of *Azolla* are reviewed by Dunham and Fowler (1987).

Azolla spp. have scale-like leaves (Fig. 10.15) and are unusual as they have a symbiotic relationship with at least one cyanobacterium *Anabaena azollae*, which lives within cavities in the leaves and is dependent on its host (Fig. 10.16) (Costa *et al.*, 2009). The cyanobacterium fixes nitrogen from the atmosphere as ammonium, so that fern growth is only limited by sufficient phosphorus availability. It does not produce seeds but grows by vegetative multiplication only. With maximum growth rates at 25°C, the *Azolla* spp. suffer less pest or disease damage at lower temperatures, although it must always

Fig. 10.15. Key features of *Azolla filiculoides* (Booy *et al.*, 2020). Reproduced with permission from GB Non-native Species Secretariat, London.

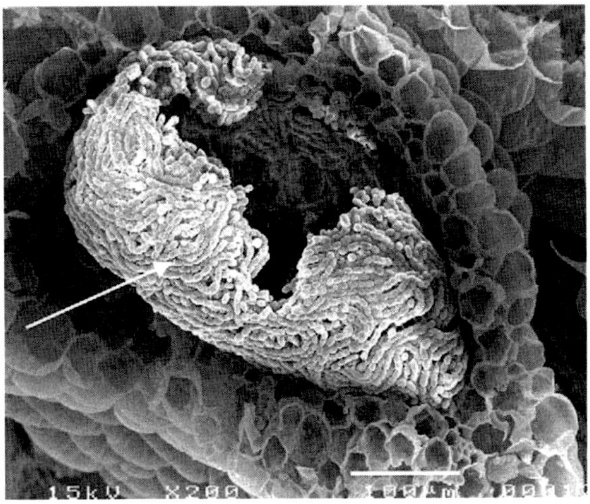

Fig. 10.16. Scanning electron microscope image of cyanobacterial filaments of *Anabaena azollae* (arrow) growing within mature cavities of *Azolla filiculoides*. The bar represents 100 μm (Costa *et al.*, 2009). Reproduced with permission from Elsevier Publishing.

remain in the water as it rapidly dies from desiccation. Growth rates are influenced by a number of key factors, including plant density, temperature, phosphorus concentration in the water, and solar radiation (Reddy and Debusk, 1982). The *Azolla* gains all its nutrients, except for carbon and nitrogen, from the water via its roots. They are very robust and tolerant plants but are inhibited by salinities of <1.5%.

Azolla has traditionally been used as a biofertilizer in paddy fields throughout China and North Vietnam (Watanabe *et al.*, 1977), containing 2 to 5% of nitrogen and 0.3 to 6.0% of phosphorus (DW). It produces about 40 tons (fresh weight) of biomass per hectare per rice crop, which is equivalent to 80 kg N ha^{-1}. More recently, *Azolla* has been used as a feed for livestock, especially poultry, being rich in protein, essential amino acids, vitamins and minerals. Water ferns can be grown on secondary and tertiary sewages for biomass production, mainly for use as a fertilizer (Arora and Saxena, 2005).

While the *Azolla* spp. are not readily digested on their own, they compost well (Harvey and Fox, 1973). The anaerobic digestion of the *Azolla* biomass is best achieved by mixing it with animal slurry at a ratio of 1:2.5. Kumar *et al.* (2020a) compared several different biomass ratios using *A. pinnata* and cattle slurry, using novel modeling techniques to explore performance. They identified that optimum biogas yields were obtained using 50% of the *Azolla* biomass at 35°C, although the best organic carbon and COD removals were obtained at 40% of *Azolla* biomass loadings (Eqs. [10.4] to [10.6]):

Biogas production (Y_1) (mL)

$$Y_1 = 3525.62 + 113.832X_1 + 424.313X_2 - 51.023X_1X_2 - 719.672X_1^2 - 1036.176X_2^2. \quad (10.4)$$

Methane production (Y_2) (%)

$$Y_2 = 55.103 + 1.330X_1 + 4.666X_2 - 1.001X_1X_2 - 8.856X_1^2 - 10.862X_2^2. \quad (10.5)$$

COD reduction (Y_3) (%)

$$Y_3 = 51.021 + 1.785X_1 + 6.061X_2 - 1.377X_1X_2 - 8.131X_1^2 - 8.896X_2^2, \quad (10.6)$$

where X_1 is the biomass dose (%), and X_2 is the temperature (°C) (Fig. 10.17).

Azolla is known to be very effective at removing nutrients from water (Forni *et al.*, 2001) as well as metals and other contaminants (Cohen-Shoel *et al.*, 2002). *A. filiculoides* has been widely used to recover a range of toxic and useful metals from aqueous solution, including gold (Zhao and Duncan, 1997; 1998; Fogarty *et al.*, 1999; Antunes *et al.*, 2001). Noorjahan and Jamuna (2015) carried out laboratory-scale wastewater treatment studies that showed good removal efficiencies by *A. microphylla* after 96 hours contact of total dissolved solids (TDS), 63.4 and 71.5%; BOD, 80.5 and 83.6%; and COD, 73.6 and 20%, respectively, with metal removals of Cu, 83.5 and 98.7%; and Zn, 99.2 and 100% from primary and secondary wastewater, respectively. A similar study in Thailand by Jangwattana

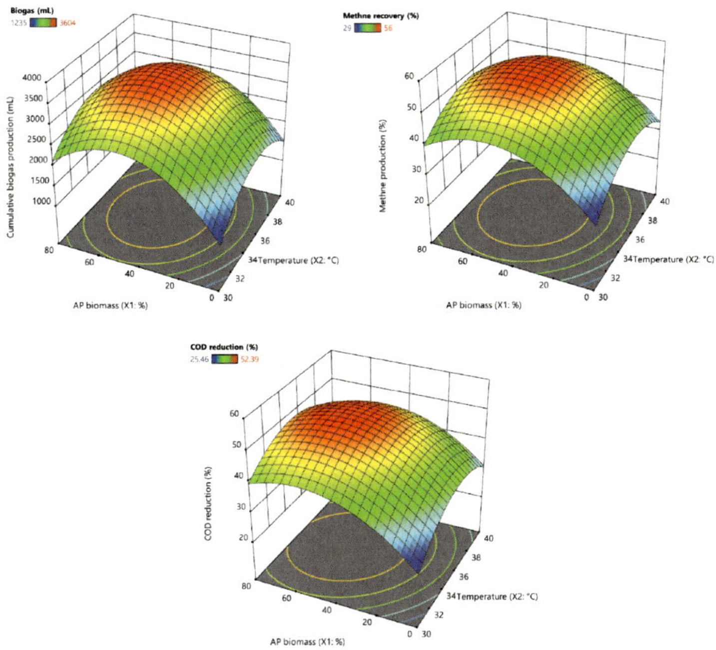

Fig. 10.17. Novel 3D surface plots showing the effect of Azolla biomass loading in cattle slurry and temperature on biogas production (mL), methane content (%), and COD reduction (%) during anaerobic digestion (Kumar *et al.*, 2020a). Reproduced with permission from Elsevier Publishing.

and Boonthaliwai (2010) looked at the treatment of poultry wastewater by *A. pinnata* with the biomass produced used as a soil conditioner. Akinbile *et al.* (2016) also used *A. pinnata* to successfully treat domestic wastewater in constructed wetlands, while Kumer *et al.* (2020b) successfully utilized this species to treat industrial wastewater. *Azolla filiculoides* have a similar removal efficiency of nitrogen from wastewater to *Lemna*, even when grown in relatively high ammonium concentrations (34 mg N L^{-1}) with the *Azolla* biomass doubling every 5 to 6 days (Costa *et al.*, 2009).

Water ferns are very invasive and difficult to eradicate once established, which is why they are not generally employed within Europe for wastewater treatment (Booy *et al.*, 2020). They can be easily transferred from one water body to another on the feet and plumage of gulls and waterfowl, which is why ponds should be under cover or covered with fine netting in areas where the species used is not indigenous.

There have been a number of studies directly comparing duckweeds to water ferns for wastewater treatment, including toxicity and treatment studies that show little difference in their abilities to remove the BOD, COD, nitrogen and phosphorus (Amare *et al.*, 2018). Bianchi *et al.* (2020) found little difference in their ability to remove metals or pharmaceuticals, although *Lemna minuta* showed greater inhibition for certain compounds than *A. filiculoides*. The dual use of duckweed (*Landoltia punctata*) and water ferns (*A. filiculoides*) in the treatment of wastewater had been proposed by Muradov *et al.* (2014), who found that the differences in their removal capabilities over time complemented each other, thus optimizing overall performance. Ponds with either duckweed and ferns suppress odor release and other algal growth. Importantly, they also severely reduce mosquito breeding success (Smith and Moelyowati, 2001; Amare *et al.*, 2018), with some duckweeds reported as having insecticidal properties (Eid *et al.*, 1992a; 1992b).

After secondary treatment, many effluents will have comparatively low nitrogen concentrations with most of the phosphorus remaining, which will result in inhibition of other potential macrophytes that will require molar ratios of N:P of between 10 and 16:1. The key advantage of the *Azolla* to other macrophytes is their ability not to become nitrogen limited, their scavenging ability for phosphorus preventing eutrophication, and their high propensity to remove metals. The use of water ferns in wastewater treatment and agriculture has been reviewed by Dinges (1982) and Smith and Cervantes (1987).

References

Akinbile, C.O., Ikuomola, B.T. and Ralphael, O.D. (2019) Development and performance evaluation of low-cost wastewater treatment plant.

Sustainable Water Resources Management, **5,** 1217–1226. https://doi.org/10.1007/s40899-018-0295-8

Akinbile, C.O., Ogunrinde, T.A., Man, H.C. and Aziz, H.A. (2016) Phytoremediation of domestic wastewaters in free water surface constructed wetlands using *Azolla pinnata. International Journal of Phytoremediation,* **18**(1), 54–61. doi:10.1080/15226514.2015.1058330

Al-Nozaily, F., Alaerts, G. and Veenstra, S. (2000a) Performance of duck-weed-covered sewage lagoons-I. Oxygen balance and COD removal. *Water Research,* **34**(10), 2727–2733.

Al-Nozaily, F., Alaerts, G. and Veenstra, S. (2000b) Performance of duck-weed-covered sewage lagoons-II. Nitrogen and phosphorus balance and plant productivity. *Water Research,* **34**(10), 2734–2731.

Alaerts, G.J., Mahbubar, R. and Kelderman, P. (1996) Performance analysis of a full-scale duckweed covered sewage lagoon. *Water Res*earch, **30**(4), 843–852.

Ali, H. (2010) Biodegradation of synthetic dyes — A review. *Water Air and Soil Pollution,* **213**, 251–273. http://dx.doi.org/10.1007/s11270-010-0382-4.

Amare, E., Kebede, F. and Mulat, W. (2018) Wastewater treatment using *Lemna minor* and *Azolla filiculoides* in tropical semi-arid regions of Ethiopia. *Ecological Engineering,* **120**, 464–473.

Amenorfenyo, D.K., Huang, X., Zhang, Y., Zeng, Q., Zhang, N., Ren, J. and Huang, Q. (2019) Microalgae brewery wastewater treatment: Potentials, benefits and the challenges. *International Journal of Environmental Research and Public Health,* **16**, 1910. https://doi.org/10.3390/ijerph16111910

Ansari, A.A., Naeem, M., Gill, S.S. and Al Zuaibr, F.M. (2020) Phytoremediation of contaminated waters: An eco-friendly technology based on aquatic macrophytes application. *Egyptian Journal of Aquatic Research,* **46**(4), 371–376. https://doi.org/10.1016/j.ejar.2020.03.002

Antunes, A.P.M., Watkins G.M. and Duncan, J.R. (2001) Batch studies on the removal of gold (III) from aqueous solution by *Azolla filiculoides. Biotechnology Letters,* **23**, 249–251.

Arora, A. and Saxena, S. (2005) Cultivation of *Azolla microphylla* biomass on secondary-treated Delhi municipal effluents. *Biomass and Bioenergy,* **29**, 60–64.

Badiou, P., Page, B. and Ross, L. (2019) A comparison of water quality and greenhouse gas emissions in constructed wetlands and conventional retention basins with and without submerged macrophyte management. *Ecological Engineering*, **127**, 292–301. https://doi.org/10.1016/j.ecoleng.2018.11.028

Bagnall, L.L., Furman, T.S., Hentges, J.F., Nolan, W.J. and Shirley, R.L. (1974) Feed and fiber from effluent-grown water hyacinth. Report 660/2-74-041. US Environmental Protection Agency, Cincinnati, OH.

Bahr, T.G., King, D.L., Johnson, H.E. and Kerns, C.L. (1977) Municipal wastewater recycling: Production of algae and macrophytes for animal food and other uses. *Developments in Industrial Microbiology*. **18**, 121–140.

Balasubramanian, P.R. and Bai, R.K. (1995) Recycling of cattle dung, biogas plant-effluent and water hyacinth in vermiculture. *Bioresource Technology*, **52**(1), 85–87.

Barker, J.E., Hutchens, J.J. and Luken, J.O. (2014) Macroinvertebrates associated with water hyacinth roots and a root analog. *Freshwater Science*, **33**(1), 159–167.

Becker, E.W. (1983) Limitations of heavy metal removal from wastewater by means of algae. *Water Research*, **17**, 459–466.

Bianchi, E., Biancalani, A., Berardi, C., Antal, A., Fibbi, D., Coppi, A., Lastrucci, L., Bussotti, N., Colzi, I., Renai, L., *et al.* (2020) Improving the efficiency of wastewater treatment plants: Bio-removal of heavy-metals and pharmaceuticals by *Azolla filiculoides* and *Lemna minuta*. *Science for the Total Environment*, **746**, 141219. https://doi.org/10.1016/j.scitotenv.2020.141219

Bishop, P.L. and Eighmy, T.T. (1989) Aquatic wastewater treatment using *Elodea nuttalli*. *Journal of the Water Pollution Control Federation*, **61**, 641–648.

Booy, O., Wade, M. and White, V. (2020) Water fern. GB Non-native Species Secretariat, London. https://www.biodiversityireland.ie/wordpress/wp-content/uploads/Water-Fern-v3b.pdf

Bote, M.A., Naik, V.R. and Jagdeeshgouda, K.B. (2020) Production of biogas with aquatic weed water hyacinth and development of briquette making machine. *Materials Science for Energy Technologies*, **3**, 64–71. https://doi.org/10.1016/j.mset.2019.09.001

Braun, R. (1982) *Biogas-Methangarung organischer Abfallstoffe*. Springer, Wien.

Brennan, L. and Owende, P. (2010) Biofuels from microalgae — A review of technologies for production, processing, and extractions of biofuels and co-products. *Renewable and Sustainability Energy Reviews*, **14**, 557–577. http://dx.doi.org/10.1016/j.rser.2009.10.009.

Cai, T., Park, S.Y. and Li, Y. (2013) Nutrient recovery from wastewater streams by microalgae: Status and prospects. *Renewable and Sustainable Energy Reviews*, **19**, 360–369.

Chin, K.K. and Goh, T.N. (1978) Bioconversion of solar energy: Methane production through water hyacinth. In *Clean Fuels From Biomass and Wastes*. Institute of Gas Technology, Washington, DC, pp. 215–228.

Chisti, Y. (2007) Biodiesel from microalgae. *Biotechnology Advances*, **25**, 294–306. http://dx.doi.org/10.1016/j.biotechadv.2007.02.001.

Chynoweth, D.P., Ghosh, S. and Henry, M.P. (1981) Biogasification of blends of water hyacinths and domestic sludge. In *Proceedings of the 1981 International Gas Research Conference*. Los Angeles, pp. 742–755.

Chynoweth, D.P., Dolenc, D.A., Ghosh, S., Henry, M.P. Jerger, D.E. and Srivastava, V.J. (1982) Kinetics of advanced digester design for anaerobic digestion of water hyacinth and primary sludge. *Biotechnology and Bioengineering Symposium* **12**, 381–398.

Cohen-Shoel, N., Barkay, Z., Ilzycer, D., Gilath, L. and Tel-Or, E. (2002) Biofiltration of toxic elements by Azolla biomass. *Water, Air and Soil Pollution*, **135**, 93–104.

Cornwell, D.A., Zoltek, J., Patrinely, C.D., Furman, T. de S. and Kim, J.I. (1977) Nutrient removal by water hyacinths. *Journal of the Water Pollution Control Federation*, **49**(1), 57–65.

Costa, M.L., Santos, M.C.R., Carrapica, F. and Pereira, A.L. (2009) Azolla-Anabaena's behaviour in urban wastewater and artificial media-influence of combined nitrogen. *Water Research*, **43**, 3743–3750.

Cuellar-Bermudez, S.P., Aguilar-Hernandez, I., Cardenas-Chavez, D.L., Ornelas-Soto, N., Romero-Ogawa, M.A. and Parra-Saldivar, R. (2015) Extraction and purification of high-value metabolites from microalgae: Essential lipids, astaxanthin and phycobiliproteins. *Microbial Biotechnology*, **8**, 190–209. http://dx.doi.org/10.1111/1751-7915.12167.

Cuellar-Bermudez, S.P., Aleman-Nava, G.S., Chandra, R., Garcia-Perez, J.S., Contreras-Angulo, J.R., Markou, G., Muylaert, K., Rittmann, B.E. and Parra-Saldivar, R. (2017) Nutrients utilization and contaminants removal. A review of two approaches of algae and cyanobacteria in wastewater. *Algal Research*, **24**, 438–449.

Culley, D.D. and Epps, E.A. (1973) The use of duckweed for waste treatment and animal feed. *Journal of the Water Pollution Control Federation*, **45**, 337–347.

De Groote, H., Ajuonu, O., Attignon, S., Djessou, R. and Neuenschwander, P. (2003) Economic impact of biological control of water hyacinth in Southern Benin. *Ecological Economics*, **45**, 105–117.

Dinges, R. (1978) Upgrading stabilization pond effluent by water hyacinth culture. *Journal of the Water Pollution Control Federation*, **50**(5), 833–845.

Dinges, R. (1982) *Natural Systems for Water Pollution Control.* Van Nostrand Reinhold, New York.

Dinges, R. and Doersam, J. (1987) The Hornsby bend hyacinth facility in Austin, Texas. *Water Science and Technology*, **19**(10), 41–49.

Dunham, D.G. and Fowler, K. (1987) Taxonomy and species recognition in Azolla Lam. In *Azolla Utilization* (eds. W.H. Smith and E.P. Cervantes). International Rice Research Institute, Laguna, Philippines, pp. 7–16.

Eid, E.M. and Shaltout, K.H. (2017) Growth dynamics of water hyacinth (*Eichhornia crassipes*): A modelling approach. *Rendiconti Lineci*, **28**, 169–181.

Eid, M.A.A. Kandil, M.A.E., Moursy, E.B. and Sayed, G.E.M. (1992a) Effect of duckweed, *Lemna minor* vegetations on the mosquito *Calex bibiens pipiens. Insect Science and its Application*, **13**, 357–361.

Eid, M.A.A., Kandil, M.A.E., Moursy, E.B. and Sayed, G.E.M. (1992b) Bioassay of duckweed vegetation extracts. *Insect Science and its Application*, **13**, 741–748.

El-Sheekh, M.M., Gharieb, M.M. and Abou-El-Souod, G.W. (2009) Biodegradation of dyes by some green algae and cyanobacteria. *International Journal of Biodeterioration and Biodegradation*, **63**, 699–704. http://dx.doi.org/10.1016/j.ibiod.2009.04.010.

Ennabili, A., Ezzahri, J. and Radoux, M. (2019) Performance of *Lemna gibba* bioreactor for nitrogen and phosphorus retention, and biomass production in Mediterranean climate. *Journal of Environmental Management*, **252**, 109627. https://doi.org/10.1016/j.jenvman.2019.109627

Eshtiaghi, M.N., Yoswathana, N., Kuldiloke, J. and Ebadi, A.G. (2012) Preliminary study for bioconversion of water hyacinth (*Eichhornia crassipes)* to bioethanol. *African Journal of Biotechnology*, **11**, 4921–4928.

Filip, S.D., Peters, T., Adams, V.D. and Middlebrooks, E.J. (1979) Residual heavy metal removal by an algae-intermittent sand filtration system. *Water Research*, **13**, 305–313.

Finlayson, C.M. (1981) Aspects of the hydrobiology of the Lake Moondarra-Leichhardt River water supply system, Mount Isa. Ph.D. thesis. James Cook University, North Queensland.

Fogarty, R.V., Dostalek, P., Patzak, M., Votruba, J., Tel-Or, E. and Tobin, J.M. (1999) Metal removal by immobilised and non-immobilised *Azolla filiculoides*. *Biotechnology Techniques*, **13**, 53–58.

Forni, C., Chen, J., Tancioni, L. and Grilli., C.M. (2001) Evaluation of the fern *Azolla* for growth, nitrogen of the phosphorus removal from wastewater. *Water Research*, **35**(6), 1592–1598.

Fox, L.J., Struik, P.C., Appleton, B.L. and Rule, J.H. (2008) Nitrogen phytoremediation by water hyacinth (*Eichhornia crassipes* [Mart.] Solms). *Water Air and Soil Pollution*, **194**, 199–207.

Gajalaskshmi, S., Ramasamy, E.V. and Abbasi, S.A. (2001) Potential of two epigeic and two anecic earthworm species in vermicomposting of water hyacinth. *Bioresource Technology*, **76**(3), 177–181.

Gaurav, G.K., Mehmood, T., Cheng, L., Klemeš, J.J. and Shrivastava, D.K. (2020) Water hyacinth as a biomass: A review. *Journal of Cleaner Production*, **277**, 122214. https://doi.org/10.1016/j.jclepro.2020.122214

Ghosh, S., Henry, M.P. and Klass, D.L. (1980) Bioconversion of water hyacinth-coastal bermuda grass-MSW-sludge blends to methane. *Biotechnology and Bioengineering Symposium* **10**, 163–187.

Gumbricht, T. (1993a) Nutrient removal processes in freshwater submersed macrophyte systems. *Ecological Engineering*, **2**, 1–30.

Gumbricht, T. (1993b) Nutrient removal capacity in submersed macrophyte pond systems in a temperate climate. *Ecological Engineering*, **2**, 49–61.

Gunnarsson, C.C. and Petersen, C.M. (2007) Water hyacinths as a resource in agriculture and energy production: A literature review. *Waste Management*, **27**, 117–129.

Hailu, A. and Degaga, E.G. (2018) Water hyacinth (*Eichhornia crassipes*) biology and its impacts on ecosystem, biodiversity, economy and human well-being. *Journal of Life Science and Biomedicine*, **8**(6), 94–100.

Haller, W.T. and Sutton, D.L. (1973) Effect of pH and high phosphorus concentrations on growth of water hyacinth. *Hyacinth Control Journal*, **11**, 59–69.

Haller, W.T., Sutton, D.L., and Barlowe, W.C. (1974) Effects of salinity on growth of several aquatic macrophytes. *Ecology*, **55**, 891–901.

Hammer, D.A. (ed.) (1989) *Constructed Wetlands for Wastewater Treatment: Municipal, Industrial and Agricultural*. Lewis Publishers, Chelsea, MI, USA.

Harding, J.P.C. and Whitton, B.A. (1981) Accumulation of zinc, cadmium and lead by field populations of *Lemanea*. *Water Research*, **15**, 301–319.

Harvey, R.M. and Fox, J.L. (1973) Nutrient removal using *Lemma minor*. *Journal of the Water Pollution Control Federation*, **45**, 1928–1938.

Hauser, J.R. (1984) Use of water hyacinth aquatic treatment systems for ammonia control and effluent polishing. *Journal of the Water Pollution Control Federation*, **56**(3), 219–225.

Hawkes, H.A. (1983) Activated sludge. In *Ecological Aspects of Used Water Treatment* (eds. C.R. Curds and H.A. Hawkes). Academic Press, London, pp. 77–162.

Hermosillo, O.M. and Sarquis, S. (1990) Design considerations for wastewater treatment with water hyacinth *E. crassipes*. *Environmental Technology*, **11**(7), 669–674.

Hillman, W.S. and Culley, D.D. (1978) The uses of duckweed. *American Science*, **66**, 442–451.

Ho, S.H., Chen, C.Y., Lee, D.J. and Chang, J.S. (2011) Perspectives on microalgal CO_2-emission mitigation systems — A review. *Biotechnology Advances*, **29**, 189–198. http://www. sciencedirect.com/science/article/pii/S0734975010001412

Hoh, D., Watson, S. and Kan, E. (2016) Algal biofilm reactors for integrated wastewater treatment and biofuel production: A review. *Chemical Engineering Journal*, **287**, 466–473. http://dx.doi.org/10.1016/j.cej.2015.11.062.

Holm, L.G., Weldon, L.W. and Blackburn, R.D. (1969) Aquatic weeds. *Science*, **166**, 699–709.

Jackson, J.D. and Gould, B.W. (1981) Sewage treatment with water weeds. Proceedings of the Australian Water and Wastewater Association. Perth, Australia, pp. 7–15.

Jadia, C.D. and Fulekar, M.H. (2009) Phytoremediation of heavy metals: Recent techniques. *African Journal of Biotechnology*, **8**, 921–928.

Jangwattana, R. and Boonthaliwai, C. (2010) Using *Azolla pinnata* for wastewater treatment from poultry farm. *International Journal of Environmental and Rural Development*, **1**(2), 23–27.

Jewell, W.J. (1971) Aquatic weed decay: Dissolved oxygen utilization and nitrogen and phosphorus regeneration. *Journal of the Water Pollution Control Federation*, **43**, 1457–1467.

Johnson, I., Sithik, M. and Kumar, M. (2019) Cyanobacteria/microalgae for distillery wastewater treatment — Past, present and the future. In *Microbial Wastewater Treatment* (eds. M.P. Shah and S. Rodriguez-Couto). Elsevier, London, pp. 195–236.

Kadlec, R.H. and Knight, R.L. (1996) *Treatment Wetlands*. Lewis Publishers, Boca Raton.

Kadlec, R.H., Knight, R.L., Vymazal, J., Brix, H., Cooper, P. and Haberl, R. (2000) Constructed wetlands for pollution control: Processes, performance, design and operation. Scientific and Technical Report No 8. International Water Association, IWA Publishing, London.

Kaushik, P. and Malik, A. (2009) Fungal dye decolourization: Recent advances and future potential. *Environment International*, **35**, 127–141. http://dx.doi.org/10.1016/j.envint.2008.05.010.

Klass, D.L. and Ghosh, S. (1980) Methane production by anaerobic digestion of water hyacinth (*Eichornia crssipes*). Paper presented at Fuels from Biomass and Wastes. American Chemical Society, San Francisco.

Kotoula, D., Iliopoulou, A., Irakleous-Palaiologou, E., Gatidou, G., Aloupi, M., Antonopoulou,P., Fountoulakis, M.S. and Stasinakis, A.S. (2020) Municipal wastewater treatment by combining in series microalgae *Chlorella sorokiniana* and macrophye *Lemna minor*: Preliminary results. *Journal of Cleaner Production*, **271**, 122704. https://doi.org/10.1016/j.jclepro.2020.122704

Krishnan, S., Ahmad, M.F., Zainuddin, N.A., Din, M.F.M., Rezania, S., Chelliapan, S., Taib, S.M., Nasrullah, M. and Wahid, Z.A. (2020) Bioethanol production from lignocellulosic biomass (water hyacinth): A biofuel alternative. In *Bioreactors: Sustainable Design and Industrial Applications in Mitigation of GHG Emissions* (eds. L. Singh, A. Yousuf and D.M. Mahapatra). Elsevier, London, pp. 123–143. https://doi.org/10.1016/B978-0-12-821264-6.00009-7

Kumar, V., Kumar, P., Kumar, P. and Singh, J. (2020a) Anaerobic digestion of *Azolla pinnata* grown in integrated industrial effluent for enhanced biogas production and COD reduction: Optimization and kinetics studies. *Environmental Technology and Innovation*, **17**, 100627.

Kumar, V., Kumar, P., Singh, J. and Kumar, P. (2020b) Potential of water fern (Azolla pinnata R.Br.) in phytoremediation of integrated industrial effluent of SIIDCUL, Haridwar, India: Removal of physicochemical and heavy metal pollutants. *International Journal of Phytoremediation*, **22**(4), 392–403. doi:10.1080/15226514.2019.1667950

Larsdotter, K. (2006) Wastewater treatment with microalgae — A literature review. *Vatten*, **62**, 31–38.

Leng, R.A. (1999) Duckweed: A tiny aquatic plant with enormous potential for agriculture and environment. Food and Agriculture Organization, Rome. http://www.fao.org/ag/AGAInfo/resources/documents/DW/Dw2.htm

Liao, S. and Chang, W. (2004) Heavy metal phytoremediation by water hyacinth at constructed wetlands in Taiwan. *Photogrammetric Engineering & Remote Sensing*, **54**, 177–185.

Limmer, M. and Burken, J. (2016) Phytovolatilization of organic contaminants. *Environmental Science and Technology*, **50**(13), 6632–6643

Lu, X., Kruatrachue, M., Pokethitiyook, P. and Homyok, K. (2004) Removal of cadmium and zinc by water hyacinth *Eichhornia Crassipes. ScienceAsia*, **30**, 93–103.

Masi, F., Martinuzzi, N., Loiselle, S., Peruzzi, P. and Bacci, M. (1999) The tertiary treatment pilot plant of publisher SPA (Florence, Tuscany): A multistage experience. *Water Science and Technology*, **40**(3), 195–202.

McNabb, C.D. (1976) The potential of submersed vascular plants for reclamation of wastewater in temperate zone ponds. In *Biological Control of Water Pollution* (eds. J. Tourbier and R.W. Pierson). University of Pennsylvania Press, Pennsylvania, pp. 123–132.

Middlebrooks, E.J., Middlebrooks, C.H., Reynolds, J.H., Walters, G.Z., Reed, S.C. and George, D.B. (1982) *Wastewater Stabilization Lagoon Design, Performance and Upgrading*. Macmillan Publishing, New York.

Mitan, N.M.M. (2019) Water hyacinth: Potential and threat. *Materials Today: Proceedings*, **19**, 1408–1412.

Mitchell, D.S. (1978) The potential for wastewater treatment by aquatic plants in Australia. *Water*, **5**(4), 15–17.

Mitchell, D.S. (1979) The potential for using aquatic plants to treat effluent. Research Conference on Piggery Effluent Utilization and Disposal. Australian Pig Industry Research Committee, Melbourne.

Mohedano, R.A., Tonon, G., Costa, R.H.A., Pelissari, C. and Fiho, P.B. (2019) Does duckweed ponds used for wastewater treatment emit or sequester greenhouse gases? *Science of the Total Environment*, **691**, 10443–10500. https://doi.org/10.1016/j.scitotenv.2019.07.169

Mulamoottil, G., McBean, E.A. and Rovers, F. (eds.) (1998) *Constructed Wetlands for the Treatment of Landfill Leachates*. Lewis Publishers, Boca Raton, FL, USA.

Munjeri, K., Ziuku, S., Maganga, H., Siachingoma, B. and Ndlovu, S. (2016) On the potential of water hyacinth as a biomass briquette. *International Journal of Energy and Environmental Engineering*, **7**, 37–43.

Muradov, N., Taha, M., Miranda, A.F., Kadali, K., Gujar, A., Rochfort, S., Stevenson, T., Ball, A.S. and Mouradov, A. (2014) Dual application of duckweed and azolla plants for wastewater treatment and renewable fuels and petrochemicals production. *Biotechnology for Biofuels*, 7, 30.

Nakajima, A., Horikoshi, T. and Saskaguchi, T. (1981) Studies on the accumulation of heavy metal elements in biological systems. XVII. Selective accumulation of heavy metal ions by *Chlorella regularis*. *European Journal of Applied Microbiology and Biotechnology*, **12**, 76–83.

Newman, L.A. and Reynolds, C.M. (2004) Phytodegradation of organic compounds. *Current Opinions in Biotechnology*, **15**(3), 225–230.

Nhapi, I., Dalu, J., Ndamba, J., Siebel, M.A. and Gijzen, H.J. (2002) An evaluation of duckweed-based pond systems as an alternative option for decentralized treatment and reuse of wastewater in Zimbabwe. *Water Science and Technology*, **48**(2), 323–330.

Noorjahan, C.M. and Jamuna, S. (2015) Biodegradation of sewage wastewater using *Azolla microphylla* and its reuse for aquaculture of fish *Tilapia mossambica*. *IOSR Journal of Environmental Science, Toxicology and Food Technology*, **9**(3), 75–80.

Nuttall, P.M. (1985) Uptake of phosphorus and nitrogen by *Myriophyllum aquaticum (Velloza)* Verd. growing in wastewater treatment system. *Australian Journal of Marine Research*, **36**, 493–507.

O'Neill, E.A., Stejskal, V., Clifford, E. and Rowan, N.J. (2020) Novel use of peatlands as future locations for sustainable intensification of freshwater aquaculture production — A case study from the Republic of Ireland. *Science of the Total Environment*, **706**, 136044.

Oswald, W.J. (1972) Complete waste treatment in ponds. Proceedings of the Sixth International Water Pollution Research Conference. Pergamon, London.

Penfound, W.T. and Earle, T.T. (1958) The biology of the water hyacinth. *Ecology Monographs*, **18**, 447–472.

Polprasert, C. and Khatiwada, N.R. (1998) An integrated kinetic model for water hyacinth ponds used for wastewater treatment. *Water Research*, **32**(1), 179–185.

Priya, B., Uma, L., Ahamed, A.K., Subramanian, G. and Prabaharan, D. (2011) Ability to use the diazo dye, C.I. Acid Black 1 as a nitrogen source by the marine cyanobacterium *Oscillatoria curviceps* BDU92191. *Bioresources Technology*, **102**, 7218–7223. http://dx.doi.org/10.1016/j.biortech.2011.02.117.

Qin, H., Diao, M., Zhang, Z., Visser, P.M., Zhang, Y., Wang, Y. and Yan, S. (2020) Responses of phytoremediation in urban wastewater with water hyacinths to extreme precipitation. *Journal of Environmental Management*, **271**, 110948. https://doi.org/10.1016/j.jenvman.2020.110948

Rahman, M.A. (2018) Pyrolysis of water hyacinth in a fixed bed reactor: Parametric effects on product distribution, characterization and syngas evolutionary behavior. *Waste Management*, **80**, 310–318. https://doi.org/10.1016/j.wasman.2018.09.028

Rahman, M.A. and Hasegawa, H. (2011) Aquatic arsenic: Phytoremediation using floating macrophytes. *Chemosphere*, **83**, 633–646.

Reddy, K.R. and Debusk, W.F. (1982) Phosphorus removal by *Azolla caroliniana* cultured in nutrient enriched waters. In *Practical Application of* Azolla *for Rice Production*. Proceedings of the International Workshop: Developments in Plant and Soil Sciences. International Rice Research Institute, Laguna, Phillipines, pp. 151–162.

Rezania, S., Ponraj, M., Din, M.F.M., Songip, A.R., Sairan, F.M. and Chelliapan, S. (2015a) The diverse applications of water hyacinth with main focus on sustainable energy and production for new era: An overview. *Renewable and Sustainable Energy Reviews*, **41**, 943–954.

Rezania, S., Ponraj, M., Talaiekhozani, A., Mohamad, S.E., Din, M.F.M., Taib, S.M., Sabbagh, F. and Sairan, F.M. (2015b) Perspectives of phytoremediation using water hyacinth for removal of heavy metals, organic and inorganic pollutants in wastewater. *Journal of Environmental Management*, **163**, 125–133.

Ringuelet, R. (1983) Un bel example d'épuration des eaux domestiques poussé jusqu'à la décontamination fécale avec des méthodes naturelles douces et rustiques: le bassin versant de l'étang de Salces-Leucate. *Techniques et Sciences Municipales: Eau*, **4**, 187–201.

Rose, A.H. (ed.) (1979) Microbial biomass. In *Economic Microbiology, Vol. 4*. Academic Press, London.

Salati, E., Salati, E., Salati, E., Tauk-Tornisielo, S.M. Brega, D.F. and Bombonato, C.J. (1999) Use of the HDS (hydric depuration with soils) associated to other constructed wetland techniques for public water supply. *Water Science and Technology*, **40**(3), 203–210.

Sarkar, P. and Dey, A. (2021) Phycoremediation, an emerging technique for dye abatement: An overview. *Process Safety and Environmental Protection*, **147**, 214–225.

Shelef, G., Moraine, R. and Oron, G. (1978) Photosynthetic biomass production from sewage. *Ergebnisse der Limnologie*, **11**, 3–14.

Smith, M. and Moelyowati, I. (2001) Duckweed based wastewater treatment (DWWT): Design guidelines for hot climates. *Water Science and Technology*, **43**(11), 291–299.

Smith, W.H. and Cervantes, E.P. (eds.) (1987) Azolla utilization. Proceedings of the Workshop on Azolla Use. International Rice Research Institute, Laguna, Philippines.

Solovchenko, A., Verschoor, A.M., Jablonowski, N.D. and Nedbal, L. (2016) Phosphorus from wastewater to crops: An alternative path involving microalgae. *Biotechnology Advances*, **34**(5), 550–564. doi:10.1016/j.biotechadv.2016.01.002

Sooknah, R.D. and Wilkie, A.C. (2004) Nutrient removal by floating aquatic macrophytes cultured in anaerobically digested flushed dairy manure wastewater. *Ecological Engineering*, **22**, 27–42.

Suman, J., Uhlik, O., Vikorova, J. and Macek, T. (2018) Phytoextraction of heavy metals: A promising tool for clean-up of polluted environment. *Frontiers in Plant Science*, **9**, 1476.

Sutton, D.L. (1974). Utilization of *Hydrilla* by white amur. *Hyacinth Control Journal*, **12**, 66–70.

Taiganides, E.P., Chou, K.C. and Lee, B.Y. (1979) Animal waste management and utilization in Singapore. *Agricultural Wastes*, **1**, 129–141.

Tariq, R.M., Naqvi, S.N.H. and Zafar, S.M.N. (2009) Two indigenous aquatic weeds *L. minor* and *Spirodella* spp., gave promising biological control of mosquito larvae with rainbow fish on field level in Karachi, Sindh, Pakistan. *Pakistan Journal of Botany*, **1**, 269–276.

Ting, W.H.T., Tan I.A.W., Salleh, S.F. and Waha, N.A. (2018) Application of water hyacinth (*Eichhornia crassipes)* for phytoremediation of ammoniacal nitrogen: A review. *Journal of Water Process Engineering*, **22**, 239–249. https://doi.org/10.1016/j.jwpe.2018.02.011

Tripathi, M., Sahu, J.N. and Ganesan, P. (2016) Effect of process parameters on production of biochar from biomass waste through pyrolysis: A review. *Renewable and Sustainable Energy Reviews*, **55**, 467–481. https://doi.org/10.1016/j.rser.2015.10.122.

USEPA. (2011) Principles of design and operations of wastewater treatment pond systems for plant operators, engineers, and managers. EPA/600/R-111/088. National Risk Management Research Laboratory, US Environmental Protection Agency, Cincinnati, OH.

Vidya, S. and Girish, L. (2014) Water Hyacinth as a green manure for organic farming. *International Journal of Research in Applied, Natural and Social Sciences*, **2**, 65–72.

Wang, L., Min, M., Li, Y., Chen, P. and Chen, Y. (2010) Cultivation of green algae Chlorella sp. in different wastewaters from municipal wastewater treatment plant. *Applied Biochemistry and Biotechnology*, **162**, 1174–1186. https://doi.org/10.1007/s12010-009-8866-7

Watanabe, I., Espinas, C.R., Berja, N.S. and Alimango, B.V. (1977) Utilisation of *Azolla-Anabaena* complex as a nitrogen fertilizer for rice. IRRI Research Paper Series. International Rice Research Institute, Laguna, Phillipines.

Wolverton, B.C. and McDonald, R.C. (1979) Upgrading facultative wastewater lagoons with vascular aquatic plants. *Journal of the Water Pollution Control Federation*, **51**, 305–313.

Wooten, J.W. and Dodd, J.D. (1976) Growth of water hyacinths in treated sewage effluent. *Economic Botany*, **30**, 29–37.

Zeraatkar, A.K., Ahmadzadeh, H., Talebi, A.F., Moheimani, N.R. and McHenry, M.P. (2016) Potential use of algae for heavy metal bioremediation, a critical review. *Journal of Environmental Management*, **181**, 817–831.

Zerinque, S.P., Rusoff, L.L. and Wolverton, B.C. (1979) Water-hyacinth — A source of roughage for lactating cows. February Meeting of the American Dairy Science Association (Southern Division). New Orleans, Louisiana.

Zhao, M. and Duncan, J.R. (1997) Batch removal of sexivalent chromium by *Azolla filiculoides*. *Biotechnology and Applied Biochemistry*, **26**, 172–179.

Zhao, M. and Duncan, J.R. (1998) Removal and recovery of nickel from aqueous solution and electroplating rinse effluent using *Azolla filiculoides*. *Process Biochemistry*, **33**, 249–255.

Zhao, Y., Fang, Y., Jin, Y., Huang, J., Bao, S., Fu, T., He, Z., Wang, F. and Zhao, H. (2014) Potential of duckweed in the conversion of wastewater nutrients to valuable biomass: A pilot-scale comparison with water hyacinth. *Bioresource Technology*, **163**, 82–91. http://dx.doi.org/10.1016/j.biortech.2014.04.018

11

High-rate Algal Ponds and Photobioreactors

11.1 Design and Operational Basis of HRAPs and PBRs

Microalgae are grown in either open ponds or high-rate algal ponds (HRAPs) or transparent walled closed-systems known as photobioreactors (PBRs). High-rate algal ponds are long shallow channels that are 20 to 50 cm deep, with a retention time of 1 to 3 days. The whole pond is kept aerobic by maintaining a high algal concentration and using some form of mechanical mixing (Figs. 11.1 and 11.2). The flow in ponds is generated by some form of pump or paddle wheel, which often provides sufficient mixing. Extra mixing is sometimes required, which is normally carried out for short periods at night, to prevent the formation of a sludge layer. Mixing may be required for short periods during the day to prevent a rise in pH in the surface water due to high photosynthetic activity. The hydrodynamics and oxygen balance in HRAPs is discussed by El Ouarghi *et al.* (2000). The pond is commissioned in the same way as a facultative pond except that continuous loading should not be permitted until an algal bloom has developed. Loading depends on insolation, and in California, the average loading throughout the year is 134 kg BOD $ha^{-1}d^{-1}$, reaching an optimum summer loading of 366 kg BOD $ha^{-1}d^{-1}$ (Oswald *et al.*, 1985). Strong organic sewages inhibit the photosynthetic action due to high ammonia concentrations, which results

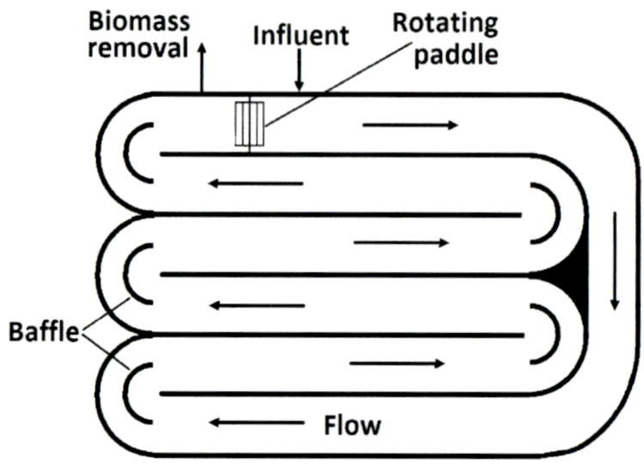

Fig. 11.1. Typical layout of a high-rate algal pond.

Fig. 11.2. Common designs for open and closed microalgal wastewater treatment systems (de Vree *et al.*, 2015). Reproduced with permission from the authors and Springer Nature under a Creative Commons License.

in the pond becoming anaerobic. Although currently used in California and Israel, few large-scale HRAPs have been developed and they remain largely experimental or for small-scale applications.

PBRs come in many different layouts from simple tubular systems to more complex, bubble columns, flat-panel or air-lift designs, each with their own advantages and disadvantages that have been reviewed by Duan and Shi (2014) (Fig. 11.3). Tubular PBRs are a horizontal or vertical array of transparent tubing, either glass or plastic, which is kept constantly moving and mixed by a circulation pump. Immediately after the pump, carbon dioxide (CO_2) gas is added to ensure

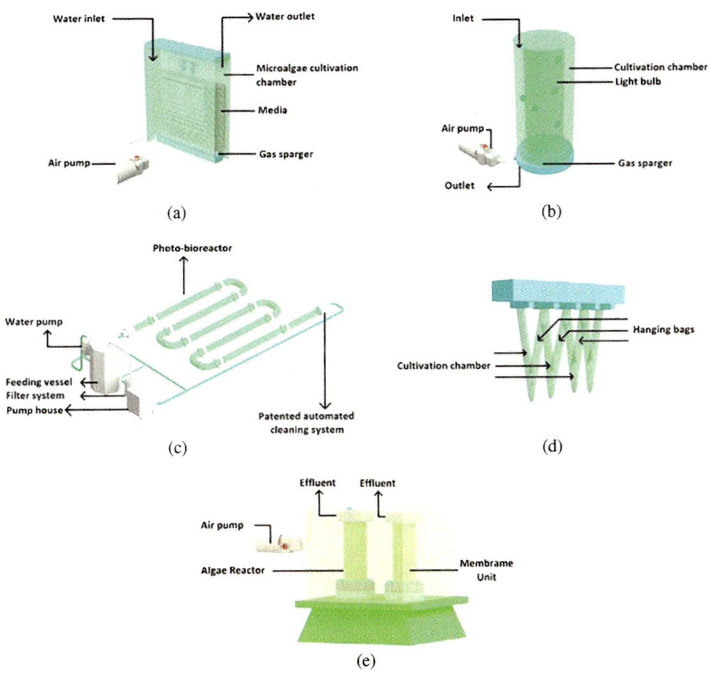

Fig. 11.3. Common designs of PBRs: (a) flat plate, (b) column, (c) tubular, (d) soft-frame, and (e) hybrid system (Vo *et al.*, 2018). Reproduced with permission from Elsevier Publishing.

maximum time for the algal to utilize the carbon source. Bubble column PBRs are vertical transparent cylindrical columns that are 20 to 35 cm in diameter into which CO_2 gas is pumped through the base. This causes both mixing and maximizes gas exchange. Flat plate PBRs are flattened transparent boxes that are orientated either vertically or inclined to make optimal use of the light source. The boxes are joined in series and each box may contain sections to encourage mixing. Air-lift column PBRs are similar to bubble column systems except that they have a central column up into which the CO_2 or air is released, causing an air-lift effect creating three-dimensional mixing (Fig. 11.4). Theoretically, PBRs can produce 426 m^3 of algae biomass $ha^{-1}yr^{-1}$, although current designs and operational practice is producing a tenth of this amount of biomass at between 40 and 53 $m^3ha^{-1}yr^{-1}$. Key operational factors include the selection of algal species, light intensity and periodicity, temperature, CO_2 requirements, wastewater characteristics, and nutrient availability.

In both open and closed systems, light limitation is the main problem. Due to the density of microalgae in ponds, unless some sort

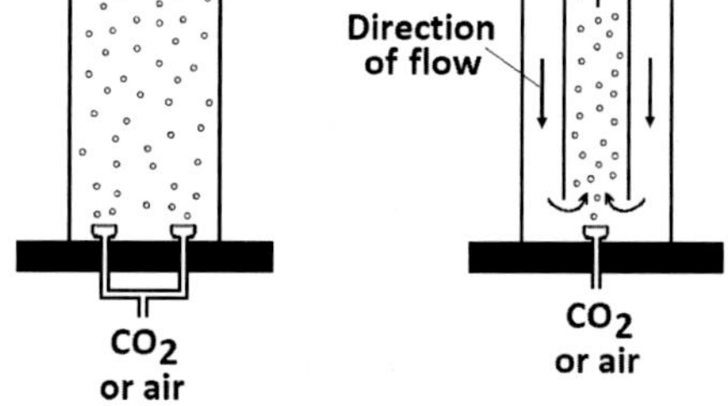

Fig. 11.4. Comparison of (a) a simple bubble column PBR with (b) an air-lift column PBR.

of mixing is used, then light penetration is limited to 5 cm or less with light intensity falling rapidly over this small distance. Agal development can also be reduced in closed PBRs due to a lack of light, although some designs comprising thin flat plates try to overcome the problem by increasing the surface area:volume ratio. The downside is that the algal biomass per unit volume of this PBR design is significantly reduced while the pumping costs are increased, making the operation far more expensive than standard open HRAPs. Higher transparent surface areas also increase the risk of biofilm development on the transparent surface, reducing light penetration and possible contamination. So occasional cleaning is necessary, adding to the operational costs.

Open ponds are cheap to construct and operate, although large areas are required as ponds are only 40 to 50 cm in depth. Unless there is some form of mixing, only a thin layer of surface algae receives sufficient sunlight, although aeration is costly and can cause unwanted bacterial development. Passive mixing by creating turbulent flow is a simple but effective alternative. Outdoor operation is problematic as well as dependent on climate conditions. For example, in some marginal climatic zones (e.g. temperate), HRAPs cannot operate during winter months with the additional problem of start-up each year. On the other hand, countries with the most sunlight also tend to have the highest evaporation rates requiring careful water management of larger ponds. Open ponds are always susceptible to contamination problems, with the more aggressive algae dominating, often eliminating more desirable species. However, they are preferred for general biomass, animal feed, and general protein production.

Closed PBRs are far more expensive, both to construct and operate, then open systems and so need to produce three times more algae biomass to be competitive. They are preferred for algae-based biofuel production and pharmaceutical use to optimize product content (e.g. oil droplets), as contamination is less and competition from other algal species can be more easily controlled. Although there is no evaporation loss, they cannot be used in very sunny locations due to overheating, so they tend to be located in indoor-controlled environments. Currently, most commercial systems are operated in greenhouses often employing artificial heat and lighting as well as supplementing the CO_2 to help maximize productivity. The small amounts of CO_2 in the atmosphere are insufficient for maximum microalgal growth due to the limitations of mass transfer of the gas. This problem is solved by using waste CO_2 from flue gases optimizing biomass development while at the same time capturing and sequestering CO_2, which otherwise would escape to the atmosphere, adding to global warming. Small bubble infusers are used to ensure maximum gas transfer as well as stripping out other gases including oxygen (O_2) and nitrogen (N_2). Optimum CO_2 concentrations for maximum microalgal growth are between 5 and 10%, above which inhibition gradually occurs (Razzak *et al.*, 2017). Ideally, systems should be integrated with existing factory waste streams, usually in the food processing sector, using waste heat and flue CO_2 to optimize algal growth and reduce both carbon emissions and nutrient loads from the factory.

A comparison between open and closed systems is summarized in Table 11.1.

Table 11.1. Comparison between open and closed algal reactors (Xu *et al.*, 2009). Reproduced with permission from Wiley-VCH Verlag GmbH & Co.

	Open systems	Closed systems
Contamination risk	High	Low
CO_2 losses	High	Low
Evaporative losses	High	Low
Light use efficiency	Poor	Excellent
Area/volume ratio	Low	High
Area required	High	Low
Process control	Difficult	Easy
Biomass productivities	Low	High
Investment costs	Low	High
Operation costs	Low	High
Harvesting costs	High	Relatively low
Scale-up	Easy	Difficult

11.2 The Role of Microalgae in Wastewater Treatment

Microalgae provide a unique opportunity for wastewater treatment, being able to utilize nutrients without the need for costly aeration as well as removing greenhouse gases from the atmosphere (Amenorfenyo *et al.*, 2019) (Fig. 11.5). Microalgae are known to produce thousands of useful byproducts and being able to bioaccumulate pollutants from water. However, HRAPs and PBRs are not designed for optimum purification efficiency but maximum algal production (Mara and Pearson, 1998), with the biomass primarily used for high-quality algal protein or biofuel (Oswald, 1995) (Sec. 11.3). Both green and blue-green (cyanobacteria) algae are used for wastewater treatment and biomass production. In wastewater treatment, autotrophic growth is the main treatment process converting sunlight, CO_2 and nutrients into biomass via photosynthesis. Other algae are capable of heterotrophic growth using organic carbon or combining both heterotrophic and autotrophic growth (i.e., mixotrophic growth) (Subashchasndrabose *et al.*, 2013). In mixed culture HRAPs, all these

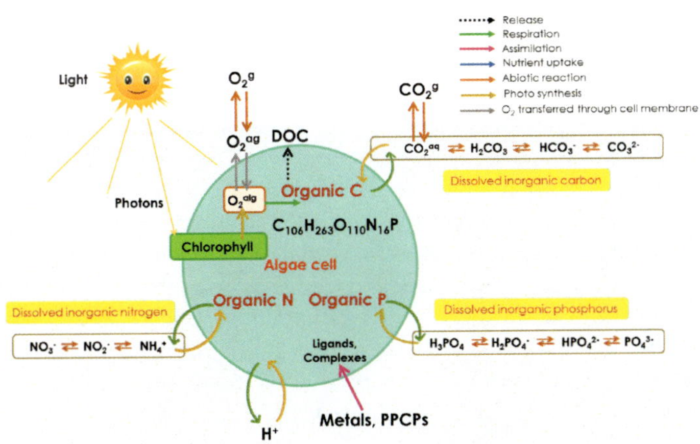

Fig. 11.5. Summary of the key biochemical pathways in microalgal HRAP and PBR systems (Vo *et al.*, 2018). Reproduced with permission from Elsevier Publishing.

metabolic groups may be found with their relative importance dependant on wastewater type and operating conditions. Globally, algae are very important in sequestering CO_2 and producing the oxygen we need.

11.2.1 *Chlorophytes (green algae)*

Green algae (e.g. *Chlorella* spp. and *Scenedesmus* spp.) are very effective in the removal of nitrogen (N) and phosphorus (P) from wastewaters, with *Chlorella* spp. having removal rates ranging from 23 to 100%, depending on initial concentrations, and being very tolerant to ammonia (Cai *et al.*, 2013). The optimum N:P ratio for the healthy development of green algae is 7.2:1. Some *Chlorella* species are heterotrophic or mixotrophic so that they can consume organic carbon from the waste streams, especially those treated anaerobically and which contain acetate (Bhatnagar *et al.*, 2010; Su *et al.*, 2011).

Much research on microalgae has involved *Scenedesmus* spp., which form large colonies consisting of 2, 4, 8, 16, or 32 cells, making them easier to harvest. Removal efficiencies of nitrogen are similar in practice to *Chlorella* spp. ranging from 30 to 100%, although they have a higher optimum N:P ratio of 30:1, resulting in the algae often being nitrogen-limited, causing a reduced phosphorus removal rate (20–50%). Phosphorus removal by microalgae is largely controlled by factors such as the N:P ratio and pH value. Enhanced phosphorus uptake from secondary wastewaters has been observed after subjecting microalgae to P-starvation for 7 days, subsequently achieving 88.7, 89.5 and 27.5% removals of dissolved inorganic phosphorus after 24 hours by *Desmodesmus communis*, *Tetradesmus obliquus* and *Chlorella protothecoides*, respectively, rising to >99% after 10 days, and achieving very low effluent concentrations (<0.1 mg L^{-1}) and enhanced growth rates. The excess phosphorus is stored in the cells as polyphosphate reaching concentrations of 250 ug mg^{-1} in *Chlorella protothecoides* (Lavrinovičs *et al.*, 2020) (Fig. 11.6). The removal efficiencies of nutrients by a range of chlorophyte and cyanobacterial species are compared by Cai *et al.* (2013). Both the *Scenedesmus* and *Chlorella* spp. have similar cellular compositions of carbohydrates

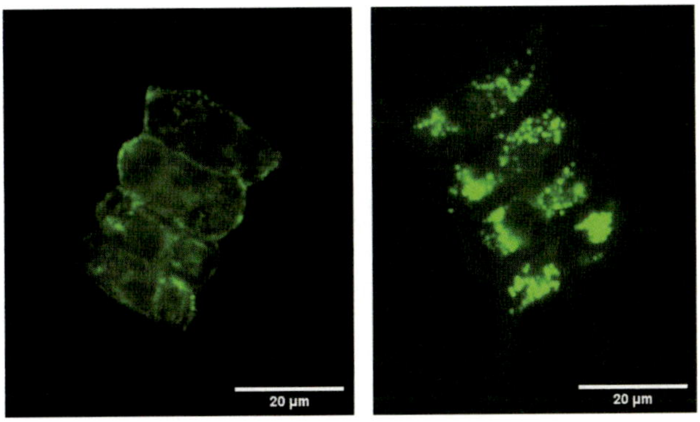

Fig. 11.6. Luxury uptake of phosphorus after 7 days of starvation by *Desmodesmus communis*. Before (*left*) and after (*right*) starvation, with polyphosphate granules clearly shown stored within the cell (Lavrinovičs *et al.*, 2020). Reproduced with permission from Elsevier Publishing.

(10–17% and 12–17% w/w, respectively), protein (50–56 and 51–58% w/w), and lipids (12–14 and 14–22% w/w) (Bruton *et al.*, 2009) (Table 11.2). HRAPs often use a consortium of green-algal species. For example, Tripathi and Shukla (1991) used *C. vulgaris, S. falcatus, Chlamydomonas miriabilis*, and *Microcystis aeruginosa* as a treatment step for municipal wastewater, achieving a 58% reduction in ammonium and 34% reduction in phosphates.

The successful removal of nutrients and heavy metals from wastewaters by microalgae (Zeraatkar *et al.*, 2016) has led to their extensive use in the removal of pharmaceuticals and organic contaminants, primarily using the green algae *Chlamydomonas* spp. (e.g. *C. mexicana*), *Chorella* spp. (e.g. *C. pyrenoidosa, C. sorokiniana, Chlorella vulgaris*) and *Scenedesmus* spp. (e.g. *S. obliquus*). The removal mechanisms are similar to that of bacteria and fungi, principally, bioadsorption, bioaccumulation, biodegradation, photo-degradation, and volatilization (Fig. 11.7). This is explored in detail with tables for the removal efficiency for a wide range of pharmaceuticals by microalgae in Hena *et al.* (2021), with good removal efficiencies reported for many common pharmaceuticals compounds. In practice, this appears

Table 11.2. Summary of the reported carbohydrate, protein and lipid content of microalgae used in wastewater treatment.

	Carbohydrate	Protein	Lipid
		% dry w/w	
Chlorophytes (Green algae)			
Ankistrodesmus sp.	10.8	31.1	24–31
Botryococcus braunii	14.1	22	14.5/25–75
Chlamydomonas reinhardtii	17	48	16.6–25.3
Chlorella pyrenoidas	26	57–60.4	2–37
Chlorella vulgaris	9–17	51–58	5–58
Dunaliella salina	16.3–32	25.7–57	6–25.
Dunaliella tertiolecta	12.2	20.0	15.0/16.7–71
Haematococcus pluvialis	27	48	15/25
Spirogyra sp.	33–64	6–20	11–21
Scenedesmus quadricauda	NM	47	1.9–18.4
Scenedesmus obliquus	10–17	11.8–56	11–55
Scenedesmus dimorphus	16–52	8–43	16–40
Cyanobacteria			
Anabaena cylindrica	25–30	43–56	4–7
Aphanizomenon flos-aquae	23	62	3
Arthrospira maxima	13–16/20	60–71	6–7
Chlorogloeopsis fritschii	44	50	7
Synechococcus sp.	15	63	11 5
Spirulina (Arthrospira) platensis	8–24.9	31.4–68.2	4–16.6
Spirulina maxima	13–16	60–71	4–9
Other Phylum			
Nannochloropsis sp.	35.9	28.8	12–53
Nitzschia sp.	9.2	16.8	12.1/16–47
Prymnesium parvum	25–33	28–45	22–38
Porphyridium cruentum	32.1/40–57	28–39	6.5/9–18.8/60.7

far more complex than laboratory-scale trials have suggested, with Maryjospeh and Ketheesan (2020) questioning the sustainability of such processes as the algal biomass often becomes heavily contaminated while toxic byproducts are often produced during treatment

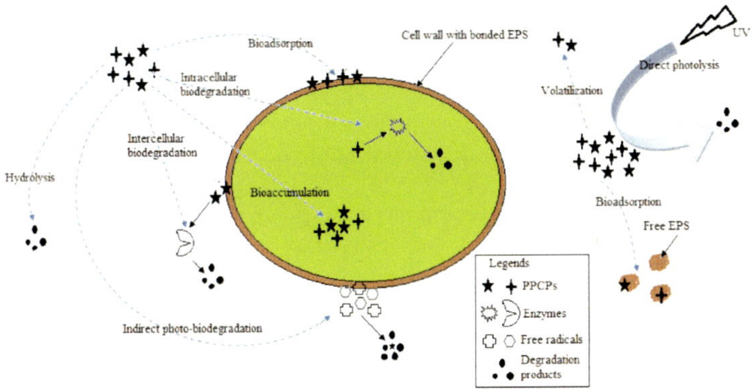

Fig. 11.7. The removal mechanisms used by microalgae to remove pharmaceutical compounds from wastewater (Hena *et al.*, 2021). Reproduced with permission from Elsevier Publishing.

and released with the wastewater. Hena *et al.* (2021) in a detailed review suggest that hybrid systems may offer the best treatment options for full-scale microalgal treatment of pharmaceuticals, including microalgal-UV, microalgal-activated sludge, microalgae-constructed wetland systems, and microalgae-membrane PBRs.

11.2.2 *Cyanobacteria (blue-green algae)*

Cyanobacteria are widely used in HRAPs due to being highly adaptive, able to fix nitrogen and survive a wide range of environmental conditions, especially light levels. They can grow on a wide variety of nutrient sources, and industrial, domestic and agricultural wastewaters, including digester leachate (Markou and Georgakakis, 2011). The key species used in wastewater treatment include *Anabaena* spp., *Arthrospira* spp. (formally *Spriulina*), *Oscillatoria* spp., *Phormidium* spp., and *Nostoc* spp. (El-Bestawy, 2008). Apart from their ability to use the nitrogen and phosphorus in wastewater, they also can accumulate high levels of lipids within their cells, making them ideal as a substrate for biofuel production (Arias *et al.*, 2020; Gorain *et al.*, 2020) (Table 11.2). In a pilot study using a mixed soil culture of cyanobacteria to treat municipal wastewater, Arias *et al.* (2020) were able to successfully treat the wastewater and produce carbohydrate-rich biomass. At a hydraulic retention time of 10 days and restricting the influent nutrient concentrations to <7.07 mg TN L^{-1} and <0.86 mg TP L^{-1} by dilution, overall removal efficiencies were >95% for total nitrogen (TN), 35–78% for total phosphorus (TP), >93% for total organice carbon (TOC), and >82% for total inorganic carbon (TIC). The high removal rates created nitrogen limitation, which caused the filamentous cyanobacteria to accumulate carbohydrates up to 48% of the cell weight. With a biomass production rate of 0.05 to 0.07 mg $L^{-1}d^{-1}$, high nutrient loads at a shorter hydraulic retention time (HRT) resulted in poor nutrient removal (TN, 66–78%; TP, 27–58%) due to carbon limitation and favored the development of green algae. The cellular carbohydrate concentration fell to <14%.

Lake Mariut in Alexandria, Egypt, suffers from pollution from both industrial and domestic effluents, resulting in excess growth of pollution tolerant cyanobacteria. El-Bestawy (2008) compared isolated cyanobacterial strains from the lake for use in phytoremediation studies of the wastewater. The primary-treated wastewater was treated in batch reactors with a hydraulic retention time of 7 days using axenic cyanobacterial strains of *Anabaena oryzae*, *A. variabilis*, and *Tolypothrix ceytonica*. Each species had different removal capabilities and contaminant tolerances, but all proved to be efficient in removing biochemical oxygen demand (BOD) and chemical oxygen demand (COD), with a maximum removal rate of 89 and 74%, respectively, by the *Anabaena* spp. *Tolypothrix ceytonica* had the highest removals of total suspended solids (TSS), fats, oils and grease (FOG), and the metals Cu and Zn at 64, 94, 86, and 95%, respectively.

Most wastewaters are treated using activated sludge, which is a high-energy input treatment system. Microalgae offer a cheaper and more sustainable option where oxygen is supplied entirely by photosynthesis rather than aeration devices (Ferreira *et al.*, 2017). One disadvantage is high turbidity levels, which reduces light penetration, so some form of pretreatment is often necessary. Although widely used in water treatment, electrocoagulation can be used to clarify secondary wastewaters prior to phytoremediation (Liu and Liu, 2016). The process is ideal for industrial wastewaters where apart from suspended solids, high removal rates of color and true turbidity can also be achieved. Electrocoagulation works by employing a sacrificial anode, which is oxidized, causing metallic ions into the solution. These cations adsorb pollutants and fine solids forming particles dense enough to freely settle (Garcia-Segura *et al.*, 2017; Moussa *et al.*, 2017). Several successful pilot operations have shown that electrocoagulation can be successfully used in conjunction with microalgae for the treatment of industrial wastewaters. For example, in order to treat brewery wastewater, Papadopoulos *et al.* (2020b) compared Fe and Al electrodes and recorded 100% TSS removal, with between 80 and 88% decolorization after 30 minutes of contact time. This was followed by a treatment using cyanobacteria with a 15-day HRT, which achieved 90% removal of sCOD as well as any residual metals from the electrocoagulation stage. Overall, the removal rates for the two-stage process was 89.1% for NO_3, 100% for NH_4, 89.4% for total Kjeldahl nitrogen (TKN), 98.5% for TP, 91.6% for COD, and 100% for TSS. Biomass yields were 525 and 740 mg L^{-1} using the Al and Fe electrodes, respectively, with a carbohydrate concentration of 50% making the biomass ideal for bioethanol production.

The role of cyanobacteria in wastewater treatment and their effect on bacteria and protozoa, including the effects of cyanobacterial toxins is reviewed by Martins *et al.* (2011).

11.2.3 *Microalgal aggregates*

Microalgal-only systems can have a number of operational problems, such as inhibition of growth due to certain wastewater contaminants and biomass recovery, which are largely overcome by using microalgal-bacterial-based systems. The relationship, as seen in facultative ponds, is highly beneficial in terms of nutrient release from oxidation of the organic matter and CO_2 by the bacteria and oxygen generation by the microalgae, making such systems very sustainable. However, both the bacteria and some microalgae release extracellular polymeric substances (EPS), which allow bacteria-microalgal aggregates to form by adsorption (Gray, 2021). These aggregates allow for much higher organic and nutrient removal rates, are easy to separate and recover, and provide a more robust biological treatment capability (Quijan *et al.*, 2017). Papadopoulos *et al.* (2020b) successfully employed cyanobacterial-bacteria aggregates using a filamentous cyanobacterium species, *Leptolyngbya* sp., to treat brewery wastewater. The initial concentration and removal rates achieved were: 2,272 mg L^{-1} (95%) sCOD, 100 mg L^{-1} (90%) TN, 31 mg L^{-1} (80%) NO_3-N, 4 mg L^{-1} (92%) NH_4-N, 4.9 mg L^{-1} (73%) TP, and 4.8 mg L^{-1} (74%) PO_4^{3}-P at 28°C. Heterotrophic bacteria are primarily responsible for sCOD removal and their activity dominates initially but as organic strength falls and more nutrients are released, then the concentration of cyanobacteria increases, eventually comprising 55% of the total biomass (Fig. 11.8a). The concentration of carbohydrates and lipid of the aggregates also increased over time, making the biomass ideal for bioethanol production (Fig. 11.8b).

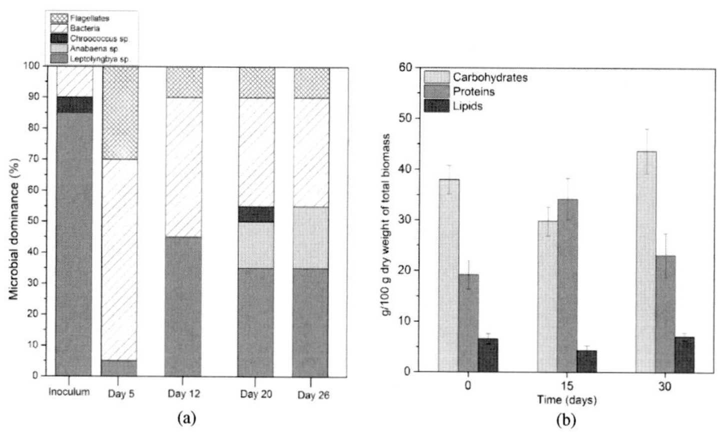

Fig. 11.8. (a) Variation in carbohydrate, protein and lipid concentration of cyanobacteria-bacterial aggregate biomass, and (b) changes in the microbial diversity of cyanobacteria-bacterial aggregate biomass over time in reactors treating raw brewery wastewater (Papadopoulos *et al.*, 2020a). Reproduced with permission from Elsevier Publishing.

11.3 Algal Productivity, Harvesting Methods and Uses of the Algal Crop

11.3.1 *Algae as a source of protein and other nutrients*

The proposition that the growth of photosynthetic algae could be used for single-cell protein (SCP) synthesis, with energy derived from sunlight and carbon from carbon dioxide in the atmosphere, has been considered for a long time. In fact, algae once formed part of the staple diet of native tribes that live near Lake Chad in Africa, who still collect matted algae from pools around the shore of the lake and dry it in the sun. As Norris (1981) points out, the equation of "*free sunlight + free carbon dioxide + cheap inorganic nitrogen = valuable protein*" is an attractive one, and it is bound to continue to attract attention. Algae have a high nutritive value, with green algae, such as *Chlorella* and *Scenedesmus*, containing 50% protein (dry weight [DW]), and the large blue-green alga *Spirulina*, 60 to 70% protein. These values are only approximate and depend largely on the nature of the substrate. For example, Milner (1953) found that by altering the growth conditions, he could control the cellular constituents of *Chlorella* over a range of 7 to 88% protein (DW), 6 to 36% carbohydrates, and 1 to 75% true fat. The protein content in algal cells decreases and the lipid content generally increases under conditions of nitrogen limitation, while other nutritional limitations also have strong effects on the composition of microalgae (Waslien, 1975; Benemann *et al.*, 1979). Table 11.3 summarizes the chemical components of the most important algae in SCP production: *Chlorella*, *Scenedesmus* and *Spirulina*. However, SCP production using algae is fraught with difficulties. Cultivation must take place in diluted aqueous solution and as cell yields are generally low, recovery is difficult. This makes the resultant protein rather expensive in comparison with other protein sources. Light availability is the critical factor controlling algal growth and although artificial light could be used, it is limited to high-value byproducts grown in PBRs. Sunshine is the only practical light source, and this restricts the location of outdoor algal ponds to tropical areas. Furthermore, the algae must be grown in thin layers to ensure maximum penetration and utilization of sunlight,

Table 11.3. Chemical composition of selected microalgae (Rose, 1979). Reproduced with permission from Elsevier Publishing.

Component (% DW)	*Chorella pyrenoidosa* 7–11–05	*Scenedesmus* sp.	*Chlorella–Scenedesmus* (10:1, sewage-grown)	*Spirulina maxima*
Protein*	55.5	53.0	41.8	55.5
Fat	7.5	13.0	7.2	12.7
Carbohydrate	17.8	13.5	27.4	17.4
Ash	8.25	6.5	19.1	7.4
Moisture	7.0	6.0	4.5	7.0
Crude fibre	3.1	8.0	—	—
Ascorbic acid	1.46	—	3.96	1.03
β-Carotene	5.02	—	6.02	2.25
Pantothenic acid	1.12	1.2	0.46	—
Pyridoxin	0.3	—	0.11	0.043
Thiamin	0.77	0.17	0.115	0.138
Riboflavin	—	0.42	0.269	0.285

* Protein contents were calculated by multiplying the value for the total nitrogen content by 6.25. True protein contents are about two-thirds of the values quoted.

unless constantly mixed. Most economic systems employ very shallow lagoons but in hot climates, this results in a high evaporation rate of the water, which is normally a scarce and expensive medium in those areas best suited to algal ponds. Finally, as the algal cultures are not grown under strict aseptic conditions, they often become contaminated by other microorganisms. Few species of algae are suitable for SCP production because of their indigestibility, toxicity, presence of silicate or carbonate shells, and the lack of economic cultivation technology (Benemann *et al.*, 1979).

A pilot HRAP has been constructed in Mexico to produce algal SCP using sunlight and atmospheric CO_2 for energy and substrate, respectively. The pilot plant uses the alga *Spirulina platensis*, which has helically coiled cells that readily intermesh to form large aggregates that can be easily and cheaply separated and is able to produce about one ton of algal biomass daily. However, the limitations of high capital costs, moderate yields, poor product quality control, and a lack of markets have restricted the practical applications of algal cultures for food or feed production (Benemann *et al.*, 1977a; Prentis, 1984).

When algal culture is combined with wastewater treatment, the production of SCP is no longer the primary objective but merely a part of the treatment process that produces a useful by-product that can contribute to the economics of the process. Growing algae in sewage have received much attention in recent years, and several plants have been built to take advantage of this "cash crop". In hot and arid areas, the culture of algae may be an economically competitive use of land, and certainly, ponds of this type are the most cost-effective method of treating wastewaters under such climatic conditions. Generally, algal ponds are constructed on unproductive land so long as sufficient sunlight is available. These systems offer a degree of flexibility so that either algal production or treatment efficiency can be optimized. What is particularly important in some arid coastal regions is that the system operates equally well with wastewater or a mixture of wastewater and sea water.

Algal ponds are very efficient in terms of BOD and nutrient removal. In arid regions, the final effluent, after algal separation, may be more valuable than the algae because of the scarcity of water. In Haifa, Israel, a pilot algal pond system was found to be as effective as any other treatment process, more so if demand existed for irrigation water and SCP simultaneously with wastewater treatment (Table 11.4) (Shelef *et al.*, 1978). Dubinsky *et al.* (1979) summarized the problems of exploiting algal mass culture that must be overcome before it becomes an economic reality (Fig. 11.9). Although the physical configuration of algal ponds is not critical, with circular, square and rectangular tanks all successfully employed, the important design criteria include water depth, mixing intensity, retention time, and organic loading. However, there is a limited number of interrelated operational variables that can be adjusted during the pond operation (Table 11.5).

The major restriction of algal ponds for sewage treatment is the relatively large area of land required, about 1 ha per 1,000 people, and the dependence on continuous sunlight throughout the year. Their use to date, therefore, has been restricted to smaller, isolated

Table 11.4. Performance of a high-rate oxidation pond at Technion, Haifa, Israel (Shelef *et al.*, 1978). Reproduced with permission from John Wiley & Sons.

	Raw sewage (mg L^{-1})	Pond effluent (mg L^{-1})	Final effluent (mg L^{-1})*
Suspended solids	240	268	515
BOD_5			
total	330	106	10
dissolved	—	12	5
COD			
total	750	670	148
dissolved	—	64	46
Nitrogen			
total	86	71	20
dissolved	—	18	12
Phosphorus			
total	16	10	1
Coliforms per 100 ml	6×10^7	3.5×10^5	8×10^3

* Following flocculation and flotation.

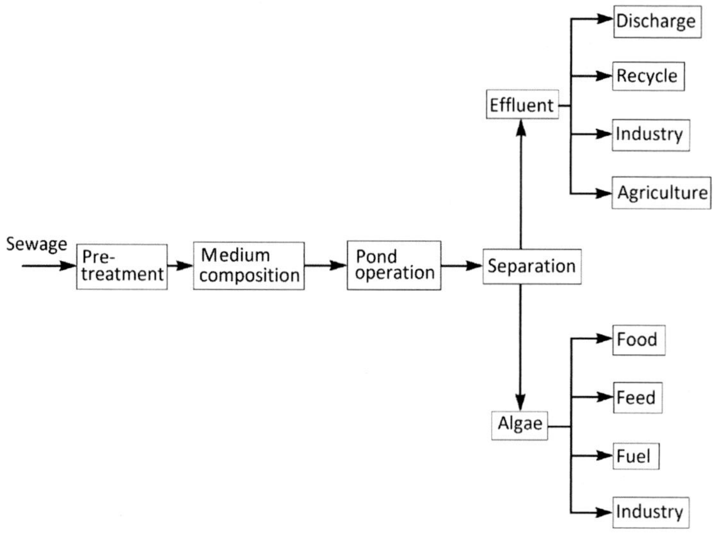

Fig. 11.9. Exploitation of algal mass culture.

Table 11.5. Major operational variables in lagoons for the production of algal biomass (Rose, 1979). Reproduced with permission from Elsevier Publishing.

Parameter	Control method(s)	Normal limits
Algal concentration	Harvesting, dilution, recycle, inoculation	150–700 mg L^{-1}
Depth	Dilution, harvesting	20–50 cm
Hydraulic retention time	Dilution	1.5–6.0 days
Phytoplankton retention time	Biomass recycle, dilution	1.5–6.0 days
Zooplankton retention time	Harvesting-recycling with DSM screen	0.5–6.0 days
Hydraulic loading	Dilution (applicable to oxidation ponds)	2–20 cm d^{-1}
pH value	Carbon dioxide, dilution	6.0–10.5
Nutrient additions	Add with dilution water or independently	Should not be limiting
Oxygen tension	Mixing, carbonation	0–25 mg L^{-1}
Light absorption	First four parameters, mixing	Absorption of 99–99.9% of incident light

communities, mainly in Australia and the southern United States. Between 40 and 100 tons (DW) of algae can be produced per hectare per year. Shelef *et al.* (1978) harvested over 110 tons per hectare per annum from their high-rate algal pond. Over 70% of the dry matter was algae with a protein content of between 42 and 48%. In Bezonbagh, Nagpur, India, a yield of 123.5 tons per hectare per annum was obtained from an experimental pond, which had a very high yield on an area basis, compared with other agricultural crops grown in the region (Arceivala *et al.*, 1970).

Although algae have only normally been exploited in facultative and high-rate ponds, much attention has also been paid to the exploitation of algae growing in maturation ponds in the United Kingdom (Sec. 16.4). Although algae in maturation ponds are beneficial in terms of phosphorus and nitrogen removal, their presence in the final effluent can be negative, increasing its suspended solids and turbidity, reducing the mortality rate of fecal bacteria, and because of the increase in pH caused by their photosynthetic activity, there is also an increased risk of ammonia toxicity to fish (Hawkes, 1983). Suggestions have been made that cladocerans could be used as grazers to remove the algae from the effluent and that the cladocerans could then be harvested as a commercial product as food for aquarium fish. The harvesting would prevent overcrowding, encourage maximum growth rates, and minimize fluctuations in the cladoceran population (Green and Watts, 1973). In practice, this has proved very difficult to operate and a rather unreliable means of removing algae, although the presence of a large cladoceran population plays a significant role in reducing the density of algal cells in ponds and final effluent (Hawkes, 1983).

Algae can be used to supplement the diet of intensively reared animals, including fish (Shelef *et al.,* 1978). For example, Carp (*Cyprinus carpio*) and St. Peter's fish (*Tilapia galila*) in which 30% of the algae in the fish's diet replaced 85% of the fish-meal portion, which is usually 15% of the fish commercial diet (Hepher and Schoeder, 1974). In controlled experiments, the weight gain of these fish was consistently higher when fed on the algae-supplemented diet. Even when algae have been chemically separated and contain 4% aluminum, they can be successfully used to replace 25 to 40% of the soya bean and fish-meal protein fed to chickens (Mukadi and Berk, 1975) and fish (Hepher and Schoeder, 1974) (Sec. 16.4).

11.3.2 *Algae as a biofuel*

The production of algae on wastewater can be linked to biogas production (Weissman *et al.,* 1979). Nitrogen and phosphorus can be removed from the effluent emanating from another secondary treatment unit by using a tertiary lagoon containing algae. Blue-green algae are especially suitable as they are filamentous and comparatively easy to harvest, and are also a suitable substrate for methane production by anaerobic digestion, with the final digested sludge being rich in nitrogen and phosphorus (Benemann *et al.*, 1977a; 1977b) (Fig. 11.10). In algal fermentation for biogas, 60 to 65% of the

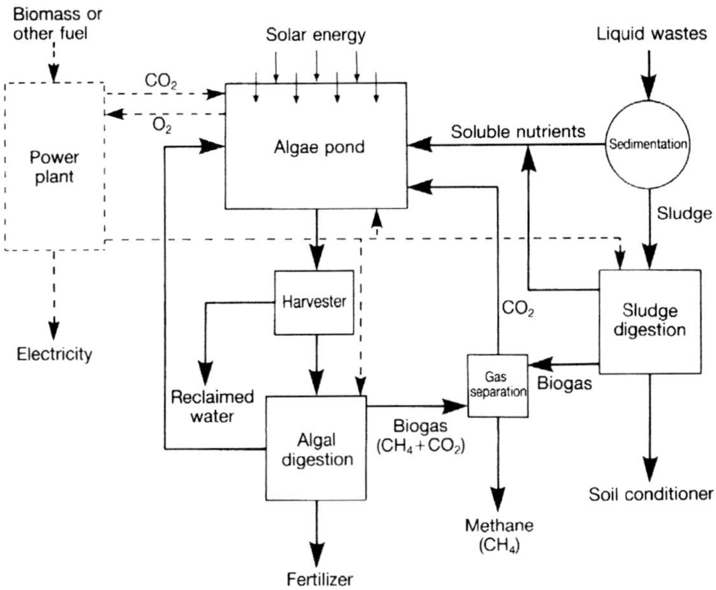

Fig. 11.10. Integrated algal biomass and methane digestion process.

potential energy content of algae can be converted into methane, with the blue-green algae *Osciallatoria* and *Spirulina* being slightly more digestible than single-cell algae, such as *Euglena* and *Scenedesmus* (Benemann *et al.*, 1977b). Anderson *et al.* (1980) have proposed a scheme that is almost totally integrated, from the recovery of biogas from fermenter effluents to almost complete recycling of N and other nutrients to the utilization of solar energy for thermophilic digestion to produce ethanol (Fig. 11.11). Although the scheme is purely hypothetical, progress has been achieved on different elements, and this may well be the type of integrated fermentation system used for the treatment of wastewater in the future. Algae can produce up to 40 times more energy per hectare than maize.

There is increasing interest in the use of algae for oil and biodiesel production. Those microalgae that have a lipid content >15% are all

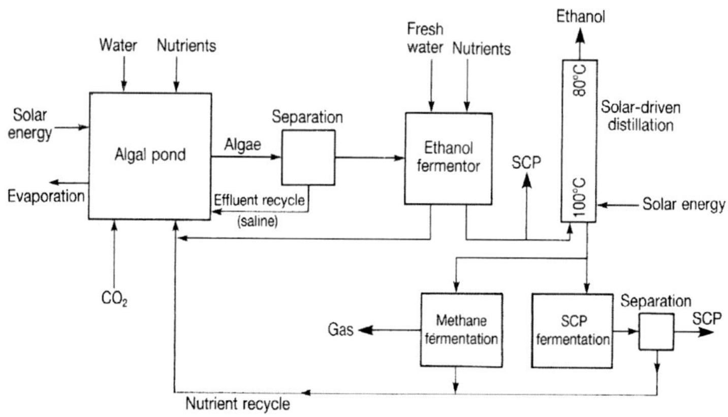

Fig. 11.11. Integrated process for the recovery of biogas, single-cell protein (SCP) and ethanol from algal fermentation.

potentially suitable for biodiesel production, with some having lipid contents of up to 85%. The lipids produced by microalgae are triglycerides more commonly known as triacylglycerols (TAG) and are separated after harvesting and turned into fatty acid methyl esters (i.e., biodiesel, often referred to as FAME) by the transesterification process (Mata *et al.,* 2010). This involves the use of alcohol (usually methanol) and a catalyst (usually NaOH or H_2SO_4) that converts TAGs to diglycerides and then to monoglycerides, which, in turn, are converted to esters (biodiesel) and the byproduct, glycerol, see Eq. (11.1).

$$\begin{array}{ccccccc} CH_2\text{–}OCOR_1 & & & & R_1\text{–}COOCH_3 & & CH_2\text{–}OH \\ | & & & & & & | \\ CH\text{–}OCOR_2 & + & 3\ HOCH_3 & \xrightleftharpoons{\text{Catalyst}} & R_2\text{–}COOCH_3 & + & CH\text{–}OH \\ | & & & & & & | \\ CH_2\text{–}OCOR_3 & & & & R_3\text{–}COOCH_3 & & CH_2\text{–}OH \\ \text{Triglyceride (Lipids)} & & \text{Methanol} & & \text{Methyl esters (Biodiesel)} & & \text{Glycerol} \end{array} \quad (11.1)$$

Compared to terrestrial plants, microalgal have very high growth rates and phenomenal biodiesel production potential exceeding

Table 11.6. The oil (TAG) content, yields and land requirement of microalgae compared with terrestrial biodiesel feedstock (Mata *et al.*, 2010). Reproduced with permission from Elsevier Publishing.

Plant source	Oil content (% dry weight)	Oil yield (L oil/ ha/year)	Land use (m^2/ year/kg biodiesel)	Biodiesel productivity (kg biodiesel/ ha/year)
Corn/Maize (*Zea mays* L.)	44	172	66	152
Hemp (*Cannabis sativa* L.)	33	363	31	321
Soybean (*Glycine* max L.)	18	636	18	562
Jatropha (*Jatropha curcas* L.)	28	741	15	656
Camelina (*Camelina sativa* L.)	42	915	12	809
Canola/rapeseed (*Brassica napus* L.)	41	974	12	862
Sunflower (*Helianthus annuus* L.)	40	1,070	11	946
Castor (*Ricinus communis*)	48	1,307	9	1,156
Palm oil (*Elaeis guineensis*)	36	6,366	2	4,747
Microalgae (low oil content)	30	58,700	0.2	51,927
Microalgae (medium oil content)	50	97,800	0.1	86,515
Microalgae (high oil content)	70	136,900	0.1	121,104

100,000 kg $ha^{-1}yr^{-1}$ (Kadir *et al.*, 2018) (Table 11.6). The lipid content of microalgae varies according to species and growth conditions, reaching maxima in excess of 75% DW under stressed conditions in some species (Table 11.7). The species currently considered the best options for commercial biodiesel production are *Chlorella protothecoides*, *C. vulgaris*, *Nannochloropsis* sp., and *Neochloris oleabundans*. The performance of microalgal biodiesel is reviewed by Deshmukh *et al.* (2019).

In theory, microalgal-derived biodiesel could completely replace its fossil-based equivalent, significantly reducing greenhouse gas emissions globally, although the high production costs associated with algal biodiesel make this currently unfeasible. Pilot-scale

Table 11.7. The potential production rates of lipids by microalgae for biodiesel production (Mata *et al.*, 2010). Reproduced with permission from Elsevier Publishing.

Microalgae species	Lipid content (% dry weight biomass)	Lipid productivity (mg L^{-1} d^{-1})	Volumetric productivity of biomass (g L^{-1} d^{-1})	Areal productivity of biomass (g m^{-3} d^{-1})
Ankistrodesmus sp.	24.0–31.0	—	—	11.5–17.4
Botryococcus braunii	25.0–75.0	—	0.02	3.0
Chaetoceros muelleri	33.6	21.8	0.007	—
Chaetoceras calcitrans	14.6–16.4/39.8	17.6	0.04	—
Chlorella emersonii	25.0–63.0	10.3–50.0	0.036–0.041	0.91–0.97
Chlorella protothecoides	14.6–57.8	12–14	2.00–7.70	—
Chlorella sorokiniana	19.0–22.0	44.7	0.23–1.47	—
Chlorella vulgaris	5.0–58.0	11.2–40.0	0.02–0.20	0.57–0.95
Chlorella sp.	10.0–48.0	42.1	0.02–25	1.61–16.47/25
Chlorella pyrenoidosa	2.0	—	2.90–3.64	72.5–130
Chlorella	18.0–57.0	18.7	—	3.50–13.90
Chlorococcum sp.	19.3	53.7	0.28	—
Crypthecodinium cohnii	20.0–51.1	—	10	—
Dunaliella salina	6.0–25.0	116.0	0.22–0.34	1.6–3.5/20–38
Dunaliella primolecta	23.1	—	0.09	14
Dunaliella tertiolecta	16.7–71.0	—	0.12	—
Dunaliella sp.	17.5–67.0	33.5	—	—
Ellipsoidion sp.	27.4	47.3	0.17	—
Euglena gracilis	14.0–20.0	—	7.70	—
Haematococcus pluvialis	25.0	—	0.05–0.06	10.2–36.4
Isochrysis galbana	7.0–40.0	—	0.32–1.60	—
Isochrysis sp.	7.1–33	37.8	0.08–0.17	—
Monodus subterraneus	16.0	30.4	0.19	—
Monallanthus salina	20.0–22.0	—	0.08	12
Nannochloris sp.	20.0–56.0	60.9–76.5	0.17–0.51	—
Nannochloropsis oculata.	22.7–29.7	84.0–142.0	0.37–0.48	—
Nannochloropsis sp.	12.0–53.0	37.6–90.0	0.17–1.43	1.9–5.3
Neochloris oleoabundans	29.0–65.0	90.0–134.0	—	—
Nitzschia sp.	16.0–47.0			8.8–21.6

(*Continued*)

Table 11.7. (*Continued*)

Microalgae species	Lipid content (% dry weight biomass)	Lipid productivity (mg L^{-1} d^{-1})	Volumetric productivity of biomass (g L^{-1} d^{-1})	Areal productivity of biomass (g m^{-3} d^{-1})
Oocystis pusilla	10.5	—	—	40.6–45.8
Pavlova salina	30.9	49.4	0.16	—
Pavlova lutheri	35.5	40.2	0.14	—
Phaeodactylum tricomutum	18.0–57.0	44.8	0.003–1.9	2.4–21
Porphyridium cruentum	9.0–18.8/60.7	34.8	0.36–1.50	25
Scenedesmus obliquus	11.0–55.0	—	0.004–0.74	—
Scenedesmus quadricauda	1.9–18.4	35.1	0.19	—
Scenedesmus sp.	19.6–21.1	40.8–53.9	0.03–0.26	2.43–13.52
Skeletonema sp.	13.3–31.8	27.3	0.09	—
Skeletonema costatum	13.5–51.3	17.4	0.08	—
Spirulina platensis	4.0–16.6	—	0.06–4.3	1.5–14.5/24–51
Spirulina maxima	4.0–9.0	—	0.21–0.25	25
Thalassiosira pseudonana	20.6	17.4	0.08	—
Tetraselmis suecica	8.5–23.0	27.0–36.4	0.12–0.32	19
Tetraselmis sp.	12.6–14.7	43.4	0.30	—

demonstration plants have been constructed next to power stations where the waste CO_2 and heat can be used to accelerate algal growth, enabling frequent harvesting and reducing costs. As microalgae are approximately 50% carbon, every kilogram (DW) of microalgae biomass produced results, in 1.83 kg CO_2 being sequestrated (Laamanen and Scott, 2020), see Eq. (11.2):

$$\left(0.5\frac{\text{g C}}{\text{g algae}}\right)\left(\frac{44\text{ g CO}_2/\text{mol}}{12\text{ g C/mol}}\right)=\frac{11\text{ g CO}_2}{6\text{ g algae}}=1.83\frac{\text{g CO}_2}{\text{g algae}}. \quad (11.2)$$

11.3.3 *Harvesting algae*

The major problems of successful algal culture are harvesting the algae and maintaining the desired balance of species in the pond, which is normally achieved by separation and recirculation of algae. Harvesting the algae has proved the most difficult problem to resolve, with the size of algal cells an important consideration in species selection as small unicellular species are difficult to separate from the water phase.

Biomass recovery from HRAPs and PBRs is normally a two-step process: actual harvesting of the biomass and subsequent dewatering. There is a wide range of filtration and gravitational separation and dewatering unit processes, both in water and wastewater treatment, although the former is more applicable to algal harvesting/dewatering. Among those evaluated for microalgae are: centrifugation, chemical flocculation, cross-flow filtration, hollow fiber filtration, belt filter press, micro-straining, discrete sedimentation, bioflocculation, autoflocculation, gravity settlement, dissolved air flotation, suspended air flotation, and froth flotation (Sim *et al.*, 1988; Japar *et al.*, 2017; Razzak *et al.*, 2017). These are summarized in Fig. 11.12, with harvesting methods compared in Table 11.8 (Kadir *et al.*, 2018). Coagulation — flocculation using non-toxic polyelectrolyte polymers is the most effective separation process for small unicellular algae, creating chemically induced flocs that will settle or can be removed by microstraining (Vandamme *et al.*, 2013). Centrifugation is the most reliable and effective method but is not generally economically viable. Filtration and screening are usually only partially effective, especially for the single-celled green algae, although they may be satisfactory for the filamentous and colonial species. Algal settling and flotation using chemicals, such as lime, alum or polyelectrolytes, is effective but expensive both in capital and operating costs. Also, the algal chemical mixtures cannot be used for SCP and have limited value for agricultural use. Benemann and Weissman (1977) have obtained 90% removal efficiencies of algae, resulting in a ten-fold increase in concentration, by using micro-strainers. The device is a rotating drum

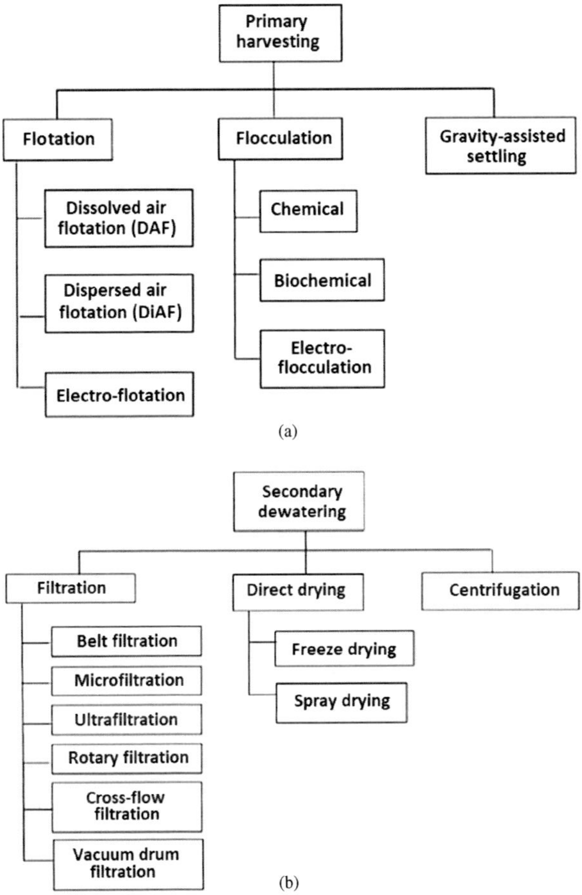

Fig. 11.12. Key harvesting methods for (a) microalgae and (b) dewatering options (Kadir *et al.*, 2018). Reproduced with permission from Elsevier Publishing.

Table 11.8. The advantages and disadvantages of the most commonly employed harvesting methods for microalgae (Kadir *et al.*, 2018). Reproduced with permission from Elsevier Publishing.

Harvesting method	Advantages	Disadvantages
Auto and bio-flocculation	• Inexpensive method, no cell damage, non-toxic to microalgae biomass, high efficiency	• Changes in cellular composition, possibility of microbiological contamination
Chemical coagulation/ flocculation	• Simple and fast method, no energy requirements	• Chemical flocculants may be expensive and toxic to microalgae biomass, recycling of culture medium is limited
Ultrafiltration	• Can handle delicate cells	• High capital and operational costs
Electrical-based processes	• Applicable to a wide variety of microalgae species, do not require the addition of chemical flocculants	• Poorly disseminated, high energetic and equipment costs
Filtration	• High recovery efficiencies, allows the separation of shear sensitive species	• The possibility of clogging increases operational costs, membranes should be regularly cleaned
Gravity sedimentation	• Simple and the inexpensive method	• Time-consuming, possibility of biomass deterioration, low concentration of the algae cake
Flotation	• Feasible for large scale applications, low cost method and space requirements, short operation times	• Generally, requires the use of chemical flocculants • Unfeasible for marine microalgae harvesting

covered with a fine mesh screen (25–100 μm). The pond effluent enters the bottom of the drum and as it slowly rotates, the excess water drains through the mesh, leaving the algae held to the inside of the drum. A pressurized jet at the top of the drum dislodges the algae into a collecting trough. This system costs 10% of the chemical treatment and leaves the algae free from contamination so that it can be used for SCP.

11.4 Microalgae and Sustainability

Growing algal biomass offers a huge opportunity in dealing with residual nutrients and trace contaminants in wastewaters that are difficult and costly to remove by conventional treatment, by treating waste streams as a resource rather than an environmental problem. The ability to create useful byproducts as well as sequester greenhouse gases creates a truly sustainable wastewater treatment system and is at the core of the circular economy principles (Chap. 2). Improved techniques for biofuel production and the development of new byproducts such as bioplastics from wastewater using are important tools for achieving sustainable treatment (López Rocha *et al.*, 2020).

However, although systems are already in operation, more research is urgently needed to develop more efficient photoreactors that can produce large volumes of algal biomass cheaply. Recent developments include illuminating the biomass internally so that the algae are illuminated three-dimensionally; growing algae in sealed plastic bags aerated with waste CO_2 via tubes that float in illuminated, heated water tanks, or by placing low-energy lights fitted in a spiral pattern to a rotating vertical shaft that is placed within a large PBR, which creates mixing and provides lights, allowing for a compact, large-volume PBR to be created.

Currently, the majority of microalgal systems use suspended cultures of cells using HRAP or PBR designs; however, immobilization techniques are being pioneered especially for the treatment of difficult waste streams and contaminants such as pharmaceuticals, including antibiotics (He and Xue, 2010). Passive, natural and artificial immobilization techniques have all been successfully employed with microalgae (Eroglu *et al.*, 2015). The matrix in which the microalgae are embedded protects them from direct contact with contaminants or

Table 11.9. Comparison of advantages and disadvantages of the various algal reactors using both suspended and immobilized microalgae in wastewater treatment (Hena *et al.*, 2021). Reproduced with permission from Elsevier Publishing.

	Suspended culture technique			Immobilized culture technique	
	Advantages	**Limitations**		**Advantages**	**Limitations**
High rate algal pond	• It costs less in installation and scaling-up, as well as maintenance with simple operational steps. • Volatilization mechanism of PPCPs removal is high.	• Light can only be utilized from the top of the surface; to increase the light utilization efficiency, the pond should be shallow and large in area, which uses large land area and a paddlewheel to circulate the culture. • Since it is an open system, it has a high risk of contamination and high evaporation, and culture parameter cannot be controlled. • Harvesting is the most energy-consuming step.	Algal turf scrubber	• It costs less in installation and scaling-up, as well as maintenance with simple operational steps. • High evaporation favors the volatilization mechanism of PPCPs removal. • Harvesting is easy.	• Only surface cells can utilize light efficiently, and those that are embedded inside will start losing the capability to take part in indirect light photodegradation. • To increase the efficiency of the technique, the turf is supposed to be thin and large in area, which consequently covers a large area of land.
Stirred-tank algal photo-bioreactor	• Optimization and maintenance of culture parameter is easy. • Since contamination is low so biomass productivity is high. • It requires less space than a high-rate algal pond.	• It costs more in installation, scaling-up, and maintenance. • Operation is complicated. • Poor light utilization efficiency. • Harvesting is not cost-effective.	Rotating algae biofilm photoreactor	• It requires a small space and can work with a deep tank since every rotational disk gets a chance to be exposed to light periodically. • Harvesting is easy. • It has high gas transfer efficiency. • Scaling-up and maintenance are easy.	• It consumes high energy in the rotating disk. • Poor in controlling species; since the system is open, hence contamination is high. • Gradual increase in biomass causes complexity in the rotational pattern, which needs to be supervised constantly.
Plane algal photo-bioreactor	• The closed system of a plane algal photobioreactor facilitates high light utilization efficiency. • Better control on culture conditions. • Low risk of contamination. • High biomass production.	• Expensive installation. • Required optimization of parameters during scaling-up causes complications during operation and makes it unsuitable for wastewater treatment. • Harvesting is not cost-effective.	Plane algal-biofilm photoreactor	• It possesses high light utilization efficiency. • It shows better control of culture conditions. • Low risk of contamination. • High biomass production. • It is more suitable for the production of valuable biomass products. • Harvesting of biomass is easy.	• Expensive installation. • Required optimization of parameters during scaling-up causes complications during operation and makes it unsuitable for wastewater treatment. • Requires a large space.

(*Continued*)

Table 11.9. (*Continued*)

	Suspended culture technique			Immobilized culture technique	
	Advantages	**Limitations**		**Advantages**	**Limitations**
Tubular algal photo-bioreactor	• The closed system of the tubular algal photobioreactor makes it possible for better control on culture parameters. • High biomass production is due to the low risk of contamination.	• Installation is expensive. • Aeration drecreases with the length of tubes, which requires multiple aeration ports. • Cleaning and scaling ups are challenging. • Harvesting is costly.	Tubular algal biofilm photoreactor	• Developed biofilm on tubular wall causes light to get utilized efficiently by microalgae. • Closed system prevents it from getting contaminants.	• It causes a high cost for installation. • This system inhibits the direct photodegradation. • Cleaning of the tube is difficult.
Algal membrane photo-bioreactor	• Harvesting and scaling-up of the system are easy. • Suitable for wastewater treatment. • Membrane provides a second line of PPCPs removal from wastewater. • High biomass production is due to the membrane.	• Installation is costly and complicated to operate. • Membrane fouling is a serious problem.			

operational extremes. Natural immobilization tends to be more difficult to achieve as it relies on a consortium of microorganisms and the algae to form a biofilm on a substrate by releasing EPS. Within the biofilm, extracellular enzymes are released that degrade contaminants that are adsorbed onto adsorption sites. Artificial immobilization entraps the microalgae into a matrix of a natural polymer such as alginate, carrageenan, chitosan, or agar; alternatively, they are adsorbed onto the surface of the polymer (Hameed and Ebrahim, 2007). The advantages and disadvantages of current suspended and immobilized microalgal systems are compared in Table 11.9.

References

Amenorfenyo, D.K., Huang, X., Zhang, Y., Zeng, Q., Zhang, N., Ren, J. and Huang, Q. (2019) Microalgae brewery wastewater treatment: Potentials, benefits and the challenges. *International Journal of Environmental Research and Public Health*, **16**, 1910. https://doi.org/10.3390/ijerph16111910

Anderson, R., Heden, C.-G. and Williams, L. (1980) The potential of algae in decentralized bioconversion systems. In *Bioresources for Development* (eds. A. King, H. Cleveland and G. Streatfield). Pergamon, New York, pp. 177–183.

Arceivala, S.J., Lakshminarayana, J.S.S., Alagarsamy, S.R. and Sastry, C.A. (1970) *Waste Stabilization Ponds: Design, Construction and Operation in India*. Central Public Health Engineering Research Institute, Nagpur.

Arias, D.M., Uggetti, E. and García, J. (2020) Assessing the potential of soil cyanobacteria for simultaneous wastewater treatment and carbohydrate-enriched biomass production. *Algal Research*, **51**, 102042. https://doi.org/10.1016/j.algal.2020.102042

Benemann, J.R. and Weissman, J.C. (1977) Biophotolysis: Problems and prospects. In *Microbial Energy Conversion* (eds. H.G. Schlegel and I. Barnea). Pergamon, Oxford, pp. 413–426.

Benemann, J.R., Weissman, J.C. and Oswald, W.J. (1979) Algal biomass. In *Microbial Biomass* (ed. A.H. Rose). Academic Press, London, pp. 177–206.

Benemann, J.R., Weissmann, J., Koopman, B. and Oswald, W.J. (1977a) Energy production by microbial photosynthesis. *Nature*, **268**, 19–23.

Benemann, J.R., Koopman, B.L., Weissman, J.C., Eisenberg, D.M. and Oswald, W.J. (1977b) Cultivation on sewage of microalgae harvestable by microstrainers. Final Report. Sanitary Engineering Research Laboratory, University of California, Berkeley, CA.

Bhatnagar, A., Bhatnagar, M., Chinnasamy, S. and Das, K.C. (2010) *Chlorella minutissima* — A promising fuel alga for cultivation in municipal wastewaters. *Applied Biochemistry and Biotechnology*, **161**, 523–536.

Bruton, T., Lyons, H., Lerat, T., Stanley, M. and Rasmussen, M.B. (2009) A review of the potential of marine algae as a source of biofuel in Ireland. Sustainable Energy, Dublin.

Cai, T., Park, S.Y. and Li, Y. (2013) Nutrient recovery from wastewater streams by microalgae: Status and prospects. *Renewable and Sustainable Energy Reviews*, **19**, 360–369.

de Vree, J.H., Bosma, R., Janssen, M., Barbosa, M.J. and Wijffels, R.H. (2015) Comparison of four outdoor pilot-scale photobioreactors. *Biotechnology for Biofuels*, **8**, 215. https://doi.org/10.1186/s13068-015-0400-2

Deshmukh, S., Kumar, R. and Bala, K. (2019) Microalgae biodiesel: A review on oil extraction, fatty acid composition, properties and effect on engine performance and emissions. *Fuel Processing Technology*, **191**, 232–247.

Duan, Y. and Shi, F. (2014) Bioreactor design for algal growth as a sustainable energy resource. In *Reactor and Process Design in Sustainable Energy Production* (ed. F. Shi). Elsevier, New York, pp. 27–60.

Dubinsky, Z., Berner, J. and Aavonson, S. (1979). Potential of large-scale algal culture for biomass and lipid production in arid lands. In *Biotechnology in Energy Production and Conservation* (ed. C.D. Scott). John Wiley, London, pp. 57–68.

El Ouarghi, H., Boumansour, B.E., Dufayt, O., El Hamouri, B. and Vasel, J.L. (2000) Hydrodynamics and oxygen balance in a high-rate algal pond. *Water Science and Technology*, **42**(10–11), 349–356.

El-Bestawy, E. (2008) Treatment of mixed domestic–industrial wastewater using cyanobacteria. *Journal of Industrial Microbiology and Biotechnology*, **35**, 1503–1516. https://doi.org/10.1007/s10295-008-0452-4

Eroglu, E., Smith, S.M. and Raston, C.L. (2015) Application of various immobilization techniques for algal bioprocesses. In *Biomass and Biofuels from Microalgae* (eds. N.R. Moheimani, M.P. McHenry, K. de Boer and P.A. Bahri). Springer International Publishing Chem, pp. 19–44.

Ferreira, A., Ribeiro, B., Marques, P.A.S.S., Ferreira, A.F., Dias, A.P., Pinheiro, H.M., Reis, A. and Gouveia, L. (2017) *Scenedesmus obliquus* mediated brewery wastewater remediation and CO_2 biofixation for green energy purposes. *Journal of Cleaner Production*, **165**, 1316–1327.

Garcia-Segura, S., Eiband, M.M.S.G., de Melo, J.V. and Martínez-Huitle, C.A. (2017) Electrocoagulation and advanced electrocoagulation processes: A general review about the fundamentals, emerging applications and its association with other technologies. *Journal of Electroanalytical Chemistry,* **801**, 267–299.

Gorain, P.C., Paul, I., Bhadoria, P.S. and Pal, R. (2020) An integrated approach towards agricultural wastewater remediation with fatty acid production by two cyanobacteria in bubble column phytoreactors. *Algal Research*, **42**, 101594.

Gray, N.F. (2021) *Fixed-Film Reactors in Wastewater Treatment.* World Scientific Publishing, Singapore.

Green, J. and Watts, E. (1973). Sewage effluent lagoons: Zooplankton in theory and practice. In *Symposium on Sewage Effluent as a Water Resource.* Institution of Public Health Engineers, London, pp. 36–44.

Hameed, M. and Ebrahim, O. (2007) Biotechnological potential uses of immobilized algae. *Journal of Agricultural Biology*, **9**, 183–192.

Hawkes, H.A. (1983) Stabilization ponds. In *Ecological Aspects of Used Water Treatment, Vol. 2. Biological Activities and Treatment Processes* (eds. C.R. Curds and H.A. Hawkes). Academic Press, London, pp. 163–217.

He, S. and Xue, G. (2010) Algal-based immobilization process to treat the effluent from a secondary wastewater treatment plant (WWTP). *Journal of Hazardous Materials*, **178**, 895–899.

Hena, S., Gutierrez, L. and Croupe, J.-P. (2021) Removal of pharmaceutical and personal care products (PPCPs) from wastewater using micro-algae: A review. *Journal of Hazardous Materials*, **403**, 124041.

Hepher, B. and Schroeder, G. (1974) Wastewater utilization in integrated aquaculture and agriculture systems. In *Wastewater Use in the Production of Food and Fibre.* Research and Development, US Environmental Protection Agency, Washington, DC, pp. 13–20.

Japar, A.S., Takriff, M.S. and Yasin, N.H.M. (2017) Harvesting microalgal biomass and lipid extraction for potential biofuel production: A review. *Journal of Environmental Chemical Engineering*, **5**, 555–563.

Kadir, W.N.A., Lam, M.K., Uemura, Y., Lim, J.W. and Lee, K.T. (2018) Harvesting and pre-treatment of microalgae cultivated in wastewater for biodiesel production: A review. *Energy Conservation and Management*, **171**, 1416–1429.

Laamanen, C.A. and Scott, J.A. (2020) Microalgae biofuel bioreactors for mitigation of industrial CO_2 emissions. In *Bioreactors* (eds. L. Singh, A. Yousuf and D.M. Mahapatra). Elsevier, London, pp. 1–16. https://doi.org/10.1016/B978-0-12-821264-6.00001-2

Lavrinovičs, A., Mežule, L. and Juhna, T. (2020) Microalgae starvation for enhanced phosphorus uptake from municipal wastewater. *Algal Research*, **52**, 102090. https://doi.org/10.1016/j.algal.2020.102090

Liu, Z. and Liu, Y. (2016) Synergistic integration of electrocoagulation and algal cultivation to treat liquid anaerobic digestion effluent and accumulate algal biomass. *Process Biochemistry*, **51**(1), 89–94.

López Rocha, C.J., Alvarez-Castillo, E., Yáñez, M.R.E., Bengoechea, C., Guerrero, A. and Ledesma, M.T.O. (2020) Development of bioplastics from a microalgae consortium from wastewater. *Environmental Management*, **263**, 110353.

Mara, D.D. and Pearson, H.W. (1998) Design manual for waste stabilization ponds in Mediterranean countries. Lagoon Technology International, Leeds.

Markou, G. and Georgakakis, D. (2011) Cultivation of filamentous cyanobacteria (blue-green algae) in agro-industrial wastes and wastewaters: A review. *Applied Energy*, **88**, 3389–3401.

Martins, J., Peixe, L. and Vasconcelos, V.M. (2011) Unravelling cyanobacteria ecology in wastewater treatment plants (WWTP). *Microbial Ecology*, **62**, 241–256.

Maryjospeh, S. and Ketheesan, B. (2020) Microalgae based wastewater treatment for the removal of emerging contaminants: A review of challenges and opportunities. *Case Studies in Chemical and Environmental Engineering*, **2**, 100046. https://doi.org/10.1016/j.cscee.2020.100046

Mata, T.M., Martins, A.A. and Caetano, N.S. (2010) Microalgae for biodiesel production and other applications: A review. *Renewable and Sustainable Energy Reviews*, **14**, 217–232.

Milner, H.W. (1953) Algae as food. *Scientific American*, **189**(4), 31–35.

Moussa, D.T., El-Naas, M.H., Nasser, M. and Al-Marri, M.J. (2017) A comprehensive review of electrocoagulation for water treatment: Potentials and challenges. *Journal of Environmental Management*, **186**(1), 24–41.

Mukadi, S. and Berk, Z. (1975). Feeding experiments of chicken with sewage grown algae. Second Progress Report, Techmon, Haifa.

Norris, J.R. (1981). Single-cell protein production. In *Essays in Applied Microbiology* (eds. J.R. Norris and M.H. Richmond). John Wiley, London.

Oswald, W.J. (1995) Ponds in the twenty-first century. *Water Science and Technology*, **31**(12), 1–8.

Papadopoulos, K.P., Econoomou, C.N. and Tekerlekopoulou, A.G. (2020a) Two-step treatment of brewery wastewater using electrocoagulation and cyanobacteria-based cultivation. *Journal of Environmental Management*, **265**, 110543. https://doi.org/10.1016/j.jenvman.2020.110543

Papadopoulos, K.P., Economoua, C.N., Dailianisb, S., Charalampousb, N., Stefanidouc, N., Moustaka-Gounic, M., Tekerlekopouloud, A.G. and Vayenasa, D.V. (2020b) Brewery wastewater treatment using cyanobacterial-bacterial settleable aggregates. *Algal Research*, **49**, 101957. https://doi.org/10.1016/j.algal.2020.101957

Prentis, S. (1984). *Biotechnology: A New Industrial Revolution.* Orbis Publishing, London.

Quijano, G., Arcila, J.S. and Buitron, G. (2017) Microalgal-bacterial aggregates: Applications and perspectives for wastewater treatment. *Biotechnology Advances*, **35**, 772–781.

Razzak, S.A., Ali, S.A.M., Hossain, M.M. and deLasa, H. (2017) Biological CO_2 fixation with production of microalgae in wastewater. A review. *Renewable and Sustainable Energy Reviews*, **76**, 379–390.

Rose, A.H. (ed.) (1979) *Microbial Biomass.* Academic Press, London.

Shelef, G., Moraine, R. and Oron, G. (1978) Photosynthetic biomass production from sewage. *Ergebnisse der Limnologie*, **11**, 3–14.

Sim, T., Goh, A. and Becker, E. (1988) Comparison of centrifugation, dissolved air flotation and drum filtration techniques for harvesting sewage-grown algae. *Biomass*, **16**, 51–62.

Su, H., Zhang, Y., Chunmin, W., Zhou, X., and Li, J. (2011) Cultivation of *Chlorella pyrenoidosa* in soybean processing wastewater. *Bioresource Technology*, **102**, 9884–9890.

Subashchasndrabose, S.R., Ramakrishnan, B., Megharaj, M., Venkateswarlu, K. and Naidu, R. (2013) Mixotrophic cyanobacteria and microalgae as distinctive biological agents for organic pollutant degradation. *Environment International*, **51**, 59–72. https://doi.org/10.1016/j.envint.2012.10.007

Tripathi, B.D. and Shukla, S.C. (1991) Biological treatment of wastewater by selected aquatic plants. *Environmental Pollution*, **69**, 69–78.

Vandamme, D., Foubert, I. and Muylaert, K. (2013) Flocculation as a low-cost method for harvesting microalgae for bulk biomass production. *Trends in Biotechnology,* **31**, 233–239.

Vo, H.N.P., Ngo, H.H., Guo, W., Nguyen, T.M.H., Liu, Y., Liu, Y., Nguyen, D.D. and Chang, S.W. (2018) A critical review on designs and applications of microalgae-based photobioreactors for pollutants treatment. *Science of the Total Environment*, **651**, 1549–1568. https://doi.org/10.1016/j.scitotenv.2018.09.282

Waslien, C.I. (1975) Unusual sources of proteins for man. *Critical Review of Food Science and Nutrition*, **6**, 77–151.

Weissman, J.R., Eisenberg, D.M. and Benemann, J.R. (1979) Cultivation of nitrogen-fixing blue-green algae on ammonia depleted effluents from sewage oxidation ponds. In *Biotechnology in Energy Production And Conservation* (ed. C.D. Scott). John Wiley, London, pp. 57–68.

Xu, L., Weathers, P.J., Xiong, X.-R. and Liu, C.-Z. (2009) Microalgal bioreactors: Challenges and opportunities. *Engineering in Life Science,* **9**(3), 178–189. https://doi.org/10.1002/elsc.200800111

Zeraatkar, A.K., Ahmadzadeh, H., Talebi, A.F., Moheimani, N.R. and McHenry, M.P. (2016) Potential use of algae for heavy metal bioremediation, a critical review. *Journal of Environmental Management*, **181**, 817–831.

12

Wetlands

12.1 Introduction

The position of natural wetlands has resulted in many of these habitats receiving pollution from a wide variety of diffuse and point sources, including wastewater, for centuries. Their natural ability to assimilate and break down pollutants entering the water cycle has been a vital part of catchment self-purification capacity. Since the early 1970s, however, considerable interest has been shown in the use of wetlands for complete, secondary or tertiary treatment of wastewaters as a point source. Natural wetlands include freshwater and saltwater marshes, swamps, fens, peat bogs, and cypress domes, all of which have been used for wastewater treatment (Tilton *et al.,* 1976; Sprangler *et al.*, 1976; Tourbier *et al.*, 1976; Fetter *et al.,* 1978; Good *et al.,* 1978; Dolan *et al.*, 1981; Dierberg and Brezonik, 1983; Mitsch and Gosselink, 2015). Apart from self-purification of the main water body, wetlands perform other vital functions within catchments, including groundwater recharge, flood mitigation, self-purification of the main water body, providing important and diverse habitats for a wide range of flora and fauna, and, increasingly carbon sequestration. In recent years, natural wetlands have come under considerable pressure from humans for a variety of purposes, especially agriculture, when they have been drained or filled (Cui *et al.*, 2020).

Where natural wetlands have been adequately managed, the damage from wastewater discharge has been minimal and in some particular wetland sites, the addition of extra nutrients has increased the productivity of the habitat, resulting in improved wildlife habitat and enhancing its overall aesthetic value (Dinges, 1982). Unfortunately, it has been strongly and incorrectly argued that the use of wetlands for this purpose may be a significant factor in the survival of these increasingly rare habitats, which are under pressure from other sectors. With the increasing appreciation of the fragile nature and intrinsic value of wetlands, many wetland habitats are now considered just too vulnerable to be modified by the addition of nutrients in the form of wastewater. They are now protected under the Ramsar Convention, which was signed in 1971 and came into force in 1975. There are currently 171 countries that are contracting parties to the convention, with 2,412 wetlands of international importance listed covering 254.5 million hectares. Contracting parties to the convention are expected to designate suitable wetlands for the list of *Wetlands of International Importance* (known as the Ramsar List), ensure their effective management, cooperate internationally on transboundary wetlands, shared wetland systems, and shared species, and work towards the wise use of all their wetlands. So, while the major wetlands are now protected and managed, many smaller wetlands remain at risk from both direct and indirect pollution.

12.2 Natural Wetlands

Wetlands are habitats where the water table lies above or close to the surface of the rooting medium for a significant part of the year. Depending on their location, they can be flooded with either fresh or saline waters; the former is normally but not exclusively associated with high rainfall regions, whereas the latter is situated on the coast or along the fringes of inland salty lakes. Freshwater wetlands can be subdivided into those where the decomposition rate is low, resulting in a continuous accumulation of organic matter (peat) to form bogs and fens, or those where the negligible accumulation of organic matter occurs and the rooting medium is mainly mineral such as marshes and swamps. Fens, marshes and swamps are characterized by the presence of emergent aquatic species, and although the wetland communities of the tropics are floristically different from those of the temperate zones, the actual diversity of emergent species is not large. The primary productivity of wetlands, excluding bogs, is among the highest for any natural community, thus making them important carbon sinks in helping to sequester carbon dioxide (Westlake, 1963; Keddy, 2000; Mitsch and Gosselink, 2015; Finlayson *et al.*, 2018). Natural wetlands have been used as a design template for constructed wetlands that have been developed specifically for wastewater treatment purposes (Seidel, 1976; Small, 1976; Sloey *et al.*, 1978; Gersberg *et al.*, 1983; 1984; 1986) and these are increasingly used instead of existing natural wetlands.

Natural wetlands are normally dominated by a single species, which restricts the establishment and growth of other plants. Jones (1986) cites several examples, including the papyrus swamps of East and Central Africa that are almost monospecific communities of *Cyperus papyrus*, and the extensive stands of *Phragmites australis* or *Typha latifolia* that dominate many wetland areas in temperate regions, although these species are extending into much warmer climates. Although some wetlands can have a high species diversity, species-rich communities are normally less productive than wetlands with few species (Grime, 1979; Wheeler and Giller, 1982). Three genera of emergent macrophytes dominate wetlands — *Typha*, *Phragmites* and *Cyperus* — although other genera such as *Misacanthidium, Cladium* and *Echmochloa* can be important locally. Examples of dominant emergent species in tropical, subtropical and temperate zones are given in Table 12.1. Marine wetlands include mangrove swamps that are restricted between the latitudes 30°N and 30°S, whereas salt-tolerant grasses (*Spartina* spp.) are the primary vegetative cover elsewhere (Valiela and Vince, 1978; Whigham and Simpson, 1976). Cypress domes are a common palustrine wetland throughout the pine-palmetto woodlands of the Southern Atlantic and Gulf coastal plain. They are topographic depressions that vary in size from 1 to 10 ha, which are dominated by two species of tree — the tall pond cypress (*Taxodium distictium* var. *nutans*) and the smaller black gum (*Nyssa sylvatica* var. *biora*) (Monk and Brown, 1965; Coultas and Calhoun, 1975; Dierberg and Brezonik, 1983; Ewel, 1983; Keddy, 2000).

Table 12.1. Emergent species of some key wetland areas (Jones, 1986). Reproduced with permission from Elsevier Publishing.

Location	Dominants
Tropical, subtropical	
Nyumba ya Mungu reservoir, Tanzania	*Typha domingensis, Cyperus alopecuroides*
Lake Naivasha, Kenya	*Cyperus papyrus,* Cyperus immensus, Cyperus digitatus* spp. *auricomus, Cyperus alopecuroides*
Kashambya swamp, Kigezi, Uganda	*Cyperus papyrus, Cladium mariscus* (pH > 6.3), *Mischanthidium violaceum* (pH 4.5), *Cyperus latifolius, Typha* sp.
S.E. Lango, lowland West Nile, Uganda. On alkaline clay-like soils in valleys draining into Lake Kyoga (seasonal swamps)	*Echinochloa pyramidalis,* Oryza barthii* (sometimes codominant with *E. pyramidalis*), *Leersia hexandra*
River Amazon, Manaus, Brazil (seasonal swamp)	*Echinochloa polystachya,* Oryza perennis,* Paspalum *repens, Hymenachne amplexicaulis*
Namiro swamp, northen end Lake Victoria, Uganda, landward zone (disturbed?)	*Miscanthidium violaceum,* Loudetia phragmatoides;* (towards water's edge), *Cyperus papyrus,* Miscanthidium violaceum*
Northern Sudd, Sudan	*Cyperus papyrus, Phragmites karka, Vossia cuspidata, Typha domingensis*
Temperate	
Huntingdon Marsh, Lake St. Francis, P.Q., Canada	*Carex aquatilis,* Carex Lanuginosa,* Calamagrostis canadensis, Typha angustifolia*
Prairie glacial marsh, Eagle Lake, IA, US	*Typha glauca, Scirpus fluviatilis, Scirpus validus, Sparganium eurycarpum, Sagitaria latifolia*
Freshwater tidal marsh, Augustine Creek, GA, US	*Zizaniopsis miliacea,* Zizania aquatica, Peltandra virginica, Pontederia cordata*
Norfolk Broadland, UK	*Cladium mariscus,* Phragmites australis,* Calamagrostis canescens, Carex acutifomis, Carex lasiocarpa, Carex paniculata, Glyceria maxima, Juncus subnodulosus, Phalaris arundinacea, Typha angustifolia*

*Most common species.

12.3 Removal Processes

The purification process in wetlands is complex and the most obvious biological component of the habitat, the emergent macrophytes, play an important, albeit minor role in the treatment of wastewater (Stephenson *et al.*, 1980; Nichols, 1983). While the major removal mechanisms are bacterial and fungal transformations and physicochemical processes such as adsorption, precipitation and sedimentation (Chan *et al.*, 1982), the emergent plants have a number of important functions. The plant rhizomes provide a stable surface for heterotrophic growth as well as enhancing sedimentation of solids by ensuring flocculation and maintaining near quiescent conditions. Emergent macrophytes dominate wetlands due to the hostile root environment, which has a restricted oxygen supply allowing facultative and anaerobic bacteria to flourish (Armstrong, 1982). The plants are able to translocate oxygen from their shoots to the roots, making the rhizosphere (the root zone) an area where aerobic microorganisms can survive (Sec. 13.3). Heterotrophic and nitrifying bacteria flourish with nitrates diffusing into the oxygen-limited zones (anoxic) of the wetland where they are removed from the system by denitrification (Sherr and Payne, 1978; Iizumi *et al.*, 1980; Gersberg *et al.*, 1986). Nutrients are trapped in wetlands, which are often referred to as nutrient sinks, not only in the sediment but in the actual plant biomass by the nutrient uptake.

Intensification of agriculture and the increase in land taken over for food production has led to a global problem of nutrient enrichment of surface waters, which directly affect wetlands, and these nutrients and other pollutants have led to a widescale deterioration of all types of wetlands. The phosphorus in the wastewater does not move far in the soil as it is retained near the surface. The phosphate ions are chemically adsorbed onto the surface of hydrous oxides of iron and aluminum, and onto silicate clay minerals, where the phosphate binds to the aluminum atoms exposed at the edges of the clay particles. The chemical and physical adsorption of phosphate onto the surface of soil minerals is a rapid process, which in the laboratory occurs mostly within the first few minutes. In addition to this initial fast reaction, slower reactions continue to remove phosphate from the solution for periods of several days to months. Adsorption-precipitation by soils is not necessarily a permanent sink for the phosphorus in wastewater and is partially reversible by a reduction in the phosphate concentration in the water in contact with the soil, by plant uptake, or by flushing or dilution with low phosphate concentration water, which will release some phosphorus back into solution. The soil, therefore, regulates the concentration of phosphate in the solution.

It is interesting that soils that adsorb phosphorus least readily are those very same soils that typically release phosphorus most readily. Peaty material, made up of plant remains, can adsorb minerals by ion exchange and retain them indefinitely. Organic soils are typically low in phosphorus and consequently have a high C:P ratio. The phosphorus content of peat is commonly less than 0.05 to 0.10%. Microbial immobilization has been suggested as a mechanism for the retention of wastewater phosphorus, where the C:P ratio is high. An initial immobilization of phosphorus will occur in response to wastewater application but as the C:P ratio of wastewater is low compared to peat, the continued application would soon satisfy the microbial requirement for phosphorus. Therefore, immobilization by microorganisms is not likely to play a significant role in the long-term fixation of phosphorus by peat but will initially be important.

Nitrogen exists in a multitude of organic forms and as ammonia, nitrite and nitrate, as well as gaseous forms of ammonia, nitrogen and nitrogen oxides, and is converted from one form to another in wetlands (Boyt *et al.*, 1976) (Fig. 12.1). Denitrification occurs in the anoxic sediments, although it occurs much more slowly under acid compared with either neutral or alkaline conditions. Below pH 5, chemical rather than biochemical reactions converts nitrogen to gaseous forms. A major mechanism for nitrogen removal is through nitrification-denitrification, although this is controlled by the movement and velocity of the wastewater as it passes through the wetland. This was demonstrated by Itoh *et al.* (2011) who evaluated the movement of wastewater using an isotope analysis of nitrogen ($\delta^{15}N\text{-}NO_3^-$, $\delta^{18}O\text{-}NO_3^-$, and $\delta^{15}N\text{-}NH_4^+$) and water ($\delta^{18}O\text{-}H_2O$) from the central wastewater treatment plant in Ulaanbaatar (Mongolia), which is discharged to the Tuul River through a riparian wetland area (6.8 km^2).

Algae and bacteria associated with wetlands can fix atmospheric nitrogen into available forms that can reduce the overall removal efficiency of wastewater nitrogen of wetlands. The application to a wetland of secondary wastewater effluent, which typically has an N:P ratio of less than 10 may stimulate nitrogen fixation, although this is dependent on the amount of wastewater applied and other nutrient sources. The soil rather than the water is the major source of nutrients for emergent vegetation, non-rooted plants such as algae and

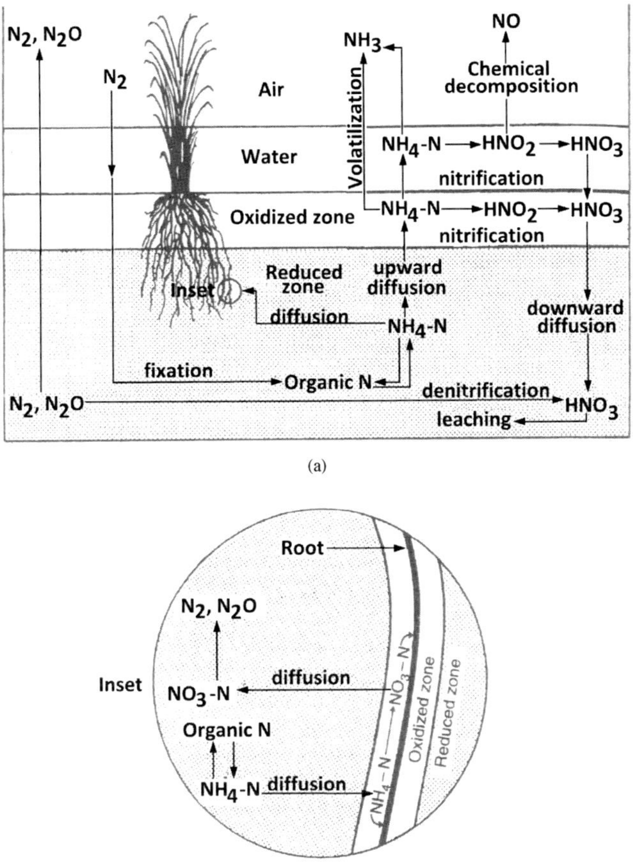

(b)

Fig. 12.1. The key steps of the nitrogen cycle in flooded soils and sediments. The major stock of sediment N is the organic-N component, and the rate of release of this nitrogen is largely dependent upon oxygen availability (Reddy and Patrick, 1984). Reproduced with permission from the Taylor & Francis Group.

duckweeds, which may also be present that obtain their nutrients directly from the water, and the subsequent incorporation of their detritus into the soil as a net transfer of nutrients from the water to the soil. Nutrient retention is greatest during periods of active vegetative growth and is low during the non-growing season. Unless the vegetation can be harvested, much of the stored nutrients will be released during the seasonal cycle of dieback and decomposition. Subsequently, there is often a release of nutrients from the decaying biomass in the winter followed by the rapid release into the water of 35 to 75% of the phosphorus from the plant tissue, although less of the nitrogen is released.

Natural wetlands effectively remove contaminants, especially heavy metals from surface waters, which come from a variety of point and diffuse sources within the catchment. Main sources are through industrial and domestic wastewaters, fluvial transport, and atmospheric deposition, and are closely linked with industrial and economic developments within the catchment. Tidal and marine wetlands are particularly at risk with metals accumulating in the sediments affecting the food chain of these ecosystems. Wetlands in Asia are particularly at risk due to unprecedented industrial development over the past 20 to 30 years (Wang *et al.*, 2017). Coastal swamps and marshes are widely used for nutrient assimilation from wastewater effluents in the south-eastern areas of the United States (US). For example, studies on cypress-tupelo swamps in the Mississippi delta have shown that wastewater nutrients can stimulate foliar growth but not root development, altering the shoot to root ratios and making the plants more vulnerable to damage during flooding, as well as causing a shift from forested to herbaceous marshes, allowing invasive species to colonize and dominate (Shifflett and Schubauer-Berigan, 2019). Metals present in the wastewater can become immobilized in the anaerobic mud, forming metallic sulfides. Release of carbon dioxide and methane from the anaerobic mud will also occur. Ma *et al.* (2021) studied the effects of heavy metals on the soil microbiota of a mangrove swamp and found that heavy metal pollution reduced the diversity and species-richness of soil microbiota, including methanotrophs, which utilize methane as a carbon source, reducing soil respiration and the release of carbon dioxide. In contrast, the methanogens were more tolerant to heavy metals, resulting in methane production increasing. Carbon dioxide has a global warming potential (GWP) of 1, while methane has an equivalent GWP of 21 over 100 years, or 56 over 20 years (Table 3.1), thus making methane a much more serious greenhouse gas. This shift in the respiratory emission ratio of CH_4 to CO_2 indicates how metal pollution can reduce the overall efficiency of wetlands as carbon sinks by increasing the release of methane.

Pharmaceutical and other micropollutants are effectively removed within wetlands, due to the stable environmental conditions such as temperature and pH, by a combination of soil microbial activity leading to degradation, adsorption and plant uptake. An example is given by Dalahmeh *et al.* (2020) who found that pharmaceuticals from domestic and industrial sources in Kampala (Uganda) were effectively removed by the Nakivubo riparian zones and wetlands as the river entered Lake Victoria, effectively removing the contaminants and protecting the lake's water quality. As with metals, pharmaceuticals

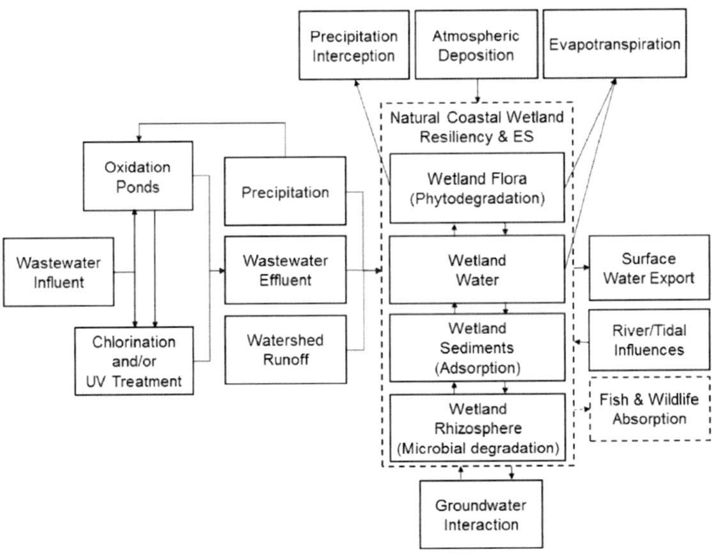

Fig. 12.2. The life cycle of pharmaceuticals and personal care products from wastewater treatment to their potential fate in coastal wetlands. Solid lines are direct relationships and dashed lines are indirect relationships (Shifflett and Schubauer-Berigan, 2019). Reproduced with permission from Elsevier Publishing.

and personal care products (PPCPs) in general can enter the food chain as well as affect the ecological balance within the wetland ecosystem. They have multiple fates within wetlands, including dilution and transportation within the liquid phase due to currents or tidal action, microbial breakdown by microorganisms in the soil or associated with the rhizosphere of plants, modified by plant uptake and phytodegradation or chemical alteration, and adsorption in both the water or soil phases. Shifflett and Schubauer-Berigan (2019) have studied the fate of PPCPs in coastal wetlands and the processes involved, which is summarized in Fig. 12.2. A long retention time is an important factor in many of the physicochemical processes and also ensures maximum removal of pathogens.

12.4 Natural Wetlands for Wastewater Treatment

Nearly all the water, both precipitation and surface water, entering the wetland leaves it by surface discharge with little entering the groundwater strata. The quantity of water leaving the wetland will vary seasonally, with little or no discharge in summer, but high in winter and spring. Most natural wetlands contain open channels to enhance the movement of water, and much of the water never comes into contact with either the soil or vegetation, whereas salt marshes may be regularly inundated with high tides. In a natural system, there will be periodic flushes of accumulated solids and nutrients from the area. It is clear, therefore, that a natural wetland must be managed to ensure that the wastewater comes into maximum contact with the soil and vegetation. During winter, the loading must be reduced as the emergent vegetation diesback and the rate of biological activity of the associated microorganisms in the rhizosphere is greatly reduced. Loading capacity varies, with a natural wetland able to treat the organic fraction in wastewater from about 100 people ha^{-1}, although the same area can remove the nitrogen produced by 125 people, but the phosphorus from only 25 (Dinges, 1982).

To ensure maximum removal, the wastewater must come into maximum contact with the soil and vegetation, which can be optimized by ensuring a shallow depth of <30 cm. In essence, this is the basic design objective of constructed wetlands that use long shallow channels or several ponds in series. Nichols (1983) has listed several natural wetlands which at the time received applications of secondary wastewater (Table 12.2). At low loading rates, wetlands can remove much of the phosphorus applied and continue to do so unless the loading rate is increased, when removal efficiency declines rapidly. It is not known how long a wetland can continue to remove phosphorus from wastewater but if sufficient nutrient is added, the adsorption capacity of the soil can become saturated. In addition, a wetland soil can release a proportion of the phosphorus previously adsorbed if the concentration of phosphorus in the water is reduced. The nitrogen pattern of wetlands is similar to that for phosphorus, with high removal efficiency, >70%, at low loading rates (<10 g N $m^{-2}y^{-1}$), rapidly declining as the loading rate increases. The hydraulic conditions in a wetland can also affect the removal of wastewater nutrients. Higher nitrogen and phosphorus loading rates are obtained by higher hydraulic loadings (Table 12.2), thus retention times in the wetland are reduced and less time for nutrient removal reactions is allowed. The morphology of wetlands is also important, for example, as the depth increases, the chance for reactions to occur between wastewater nutrients and the underlying soil decreases. Deeper waters will, however, provide longer retention times than shallower wetlands at the same hydraulic loading.

Spangler *et al.* (1976) evaluated a natural wetland in Wisconsin. Using a 156-ha portion of a marsh with a total surface area of some 18 km^2, the marsh was loaded with a secondary effluent. The biochemical oxygen demand (BOD) was reduced from 26.9 to 5.3 mg L^{-1} (80%), with an 86.2% reduction in total coliforms and a 51.3% and 13.4% removal of nitrate and total phosphorus, respectively. They observed that young shoots of emergent vegetation had the highest concentration of phosphorus, which decreased as the tissue aged. They concluded that the quality of phosphorus removal could be greatly increased by frequent harvesting as the tissue was not permitted to mature completely or subsequently die and decompose, thereby re-releasing the accumulated nutrients.

Table 12.2. The removal of total nitrogen and total phosphorus from wastewater applied to natural wetlands (Nicols, 1983). Reproduced with permission from John Wiley & Sons, Inc.

Types of wetland	Location	Size (ha)	Years wastewater applied	Hydraulic loading (cm y^{-1})		Nutrient loading (g $m^{-2}y^{-1}$)		Nutrient removal	
				Wastewater	Other	Total P	Total N	Total P	Total N
(1) Shrub sedge fen	Michigan	1[b]	1[c]	70[c]	—	1.7[C]	1.9[c]	95[c]	96[cd]
(2) Forest shrub fen	Michigan	18.2	1[e]	36.8[e]	—	0.9[e]	1.5[ed]	91[e]	75[ed]
			2[f]	74.1[f]	205[f]	2.6[f]	6.5[fd]	88[f]	80[fd]
			3[s]	65.2[g]	183[g]	1.7[s]	9.3[gd]	72[g]	80[gd]
			4[h]	55.7[h]	116[h]	1.8[h]	6.2[hd]	64[h]	77[hd]
			5[h]	57.3[h]	97[h]	1.7[h]	9.3[hd]	65[h]	75[hd]
(3) Blanket bog	Ireland	—	1	i	—	5.0	7.4[d]	96	82[d]
			2	i	—	13.1	15.4[d]	72	87[d]
			3	i	—	8.1	10.3[d]	43	68[d]
(4) Hardwood swamp	Florida	204	20	10.2	83	0.9	—	87	—
(5) Cattail marsh	Wisconsin	156	55	23.4	558	15.2	—	32	—
(6) Cattail marsh	Massachusetts	19.4	69	684	159	7.1	53.6	47	31
(7) Cattail	Massachusetts	2.4	69	5526	—	63.6	428	20	1
(8) Deepwater marsh	Ontario	162	55	231	5569	11.6	78.6	58[j]	41[j]
(9) Glycena marsh	Ontario	20	55	1870	—	77	404	24[j]	38[j]

[a]Secondary effluent; [b]Area affected by study, entire wetland is 710 ha; [c]May–September; [d]Inorganic N only, organic N not measured; [e]August–October; [f]March–November; [g]April–November; [h]June–November; [i]Chemical fertilizers, not wastewater applied; [j]Wastewater applied on a year-round basis but percent removal measured during the growing season only. Percent removed would likely have been more calculated on a year-round basis.

A natural littoral reed belt on the southern edge of Lake Borrevannet in south east Norway is being used to prevent nutrients and suspended solids in surface runoff from agricultural land entering the eutrophic lake. Over the monitoring period (1994–1996) the total phosphorus concentration fell by 65%, orthophosphate by 8%, total nitrogen by 44%, and suspended solids by 94%. The retention of suspended solids and phosphorus by the reed belt was unaffected by season, whereas the total nitrogen retention was higher in the summer (50–85%) than winter (30%) (Bratli *et al.*, 1999).

Nutrients, in particular phosphorus, can cause significant problems to wetlands, resulting in water eutrophication or changes in flora. In response, wastewater treatment plants discharging their final effluents into wetlands now generally install nutrient removal processes to reduce nitrogen and phosphorus loadings, often incorporating maturation ponds. Using a eutrophic-ecological model to study the effects of eutrophication of the Deepor Beel wetland in Assam (India), Dash *et al.* (2020) found that the rate of nutrient uptake by plants (water hyacinth) was relatively low, so harvesting the biomass was ineffective in controlling the negative effect of eutrophication (Fig. 12.3). They found that the only effective long-term control was the removal of nitrogen and phosphorus by treatment before entering

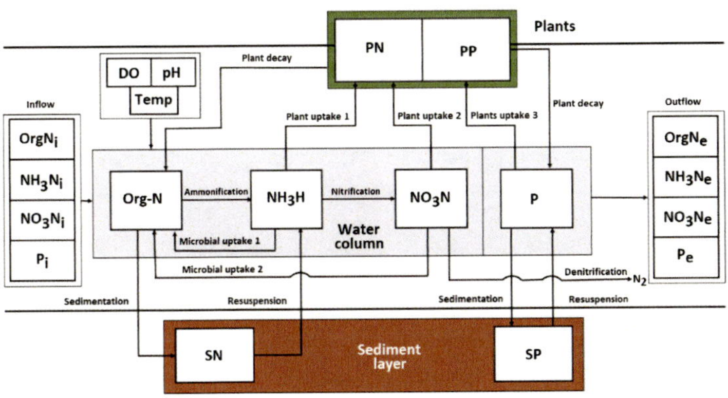

Fig. 12.3. Basis of the eutrophication-based ecological model for wetlands developed by Dash *et al.* (2020). Reproduced with permission from Elsevier Publishing.

the wetland. The study clearly demonstrates the importance of nutrient management plans in agriculture and nutrient removal at all point sources such as wastewater treatment plants, and that wetlands have a limited capacity for self-purification. Phosphorus can be specifically targeted for removal by chemical treatment or using adsorbents such as zeolites (Ning *et al.*, 2008). Adsorbents allow the recovered phosphorus to be easily reused as a fertilizer. A novel method to remove phosphorus from treated wastewater in Málaga, Spain, which discharges into the Fuente de Piedra, a Ramsar-listed wetland, was developed by Álvarez-Manzaneda *et al.* (2020). Using magnetic particles, which can remove up to 90% of the phosphorus present, the final effluent can be safely discharged to the wetland, causing nutrient enrichment. The particles are easily recovered from the final effluent before the discharge to the wetland and can be used up to four times at a ratio of 0.16 g of magnetic particles per mg of phosphorus.

12.5 Conclusions

Unlike constructed wetlands, it is not possible, or desirable in terms of conservation, to harvest natural wetlands to control nutrient release back into the solution. Gopal (1999), while recognizing the need to restore lost and degraded wetlands, particularly, river floodplains, lake littoral zones, and coastal wetlands that reduce pollution from non-point sources, warns of their over-exploitation. He stresses the problems of eutrophication and the pollution of wetlands and feels it is ironic that natural wetlands should have ever been considered as alternatives for wastewater treatment. The increasing disappearance rate of smaller wetlands, in particular, the habitat importance of wetlands and their fragility, which has resulted in extensive legislative protection, means that wetlands should only be considered for use for wastewater treatment, including tertiary treatment, as the last result. The use of natural wetlands must be subject to extensive environmental impact analysis, including water management assessment, and the removal of key contaminants (e.g. metals and trace organics, including pharmaceuticals) prior to the discharge of wastewaters to wetlands.

The loss of plant species due to either disease or environmental stressors is an important consideration in the design of constructed wetlands. In the past thirty years, cases of dieback of the common reed Phragmites australisin natural wetlands have become increasingly common in many central and western European countries and the US (van der Putten, 1997; Cerri and Reale, 2018). This indicates the need to build into any natural treatment design sufficient resilience, especially against pests and disease and pests, to sustain the long-term viability of treatment.

References

Álvarez-Manzaneda, I., Guerrero, F., Cruz-Pizarro, L., Rendón, M. and de Vicente, I. (2020) Magnetic particles as new adsorbents for the reduction of phosphate inputs from a wastewater treatment plant to a Mediterranean Ramsar wetland (Southern Spain). *Chemosphere*, 128640. https://doi.org/10.1016/j.chemosphere.2020.128640

Armstrong, W. (1982). Water-logged soils. In *Environment and Plant Ecology* (ed. J.R. Etherington). John Wiley, Chichester, pp. 290–330.

Boyt, F.L., Bayley, S.E. and Zoltec, J. (1976) Removal of nutrients from treated municipal wastewater by wetland vegetation. *Journal of the Water Pollution Control Federation*, **48**, 789–799.

Bratli, J.L., Skiple, A. and Mjelde, M. (1999) Restoration of Lake Borrevannet: Self-purification of nutrients and suspended matter through natural reed-belts. *Water Science and Technology*, **40**(3), 325–332.

Cerri, M. and Reale, L. (2018) *Claviceps arundinisidentification* and its role in the die-back syndrome of *Phramities australispopulations* in central Italy. *Plant Biosystems*, **152**, 818–824.

Chan, E., Bursztynsky, T.A., Hantzsche, N. and Litwin, T.J. (1982) The use of wetlands for water pollution control. Municipal Environmental Research Laboratory, US EPA 600/2-82-086. Environmental Protection Agency, Washington, D.C.

Coultas, C.L. and Calhoun, F.G. (1975). A toposequence of soils in and adjoining a cypress dome in North Florida. *Soil and Crop Society of Florida Proceedings*, **35**, 186–192.

Cui, S., Yu, T., Zhang, F., Fu, Q., Hough, R. An, L., Gao, S., Zhang, Z., Hu, P., Zhu, Q. and Pei, Z. (2020) Understanding the risks from diffuse pollution on wetland eco-systems: The effectiveness of water quality classification schemes. *Ecological Engineering*, **155**, 105929.

Dalahmeh, S., Björnberg, E., Elenström, A.-K., Niwagaba, C.B. and Komakech, A.J. (2020) Pharmaceutical pollution of water resources in

Nakivubo wetlands and Lake Victoria, Kampala, Uganda. *Science of the Total Environment*, **710**, 136347.

Dash, S., Borah, S.S. and Kalamdhad, A.S. (2020) Study of the limnology of wetlands through a one-dimensional model for assessing the eutrophication levels induced by various pollution sources. *Ecological Modelling*, **416**, 108907.

Dierberg, F.E. and Brezonik, P.L. (1983) Tertiary treatment of municipal wastewater by cyprus domes. *Water Research*, **17**, 1027–1040.

Dinges, R. (1982) *Natural Systems for Water Pollution Control.* Van Nostrand Reinhold, New York.

Dolan, T.J., Bayley, S.E., Zoltek, J. and Herman, A.J. (1981). Phosphorus dynamics of a Florida freshwater marsh receiving treated wastewater. *Journal of Applied Ecology*, **18**, 205–219.

Ewel, K.C. (1983). Effects of fire and sewage on understorey vegetation in cypress domes. In *Cypress Swamps* (eds. K.C. Ewel and H.T. Odum). University Presses of Florida, Gainesville.

Fetter, C.W., Sloey, W.E. and Sprangler, F.L. (1978) Use of a natural marsh for wastewater polishing. *Journal of the Water Pollution Control Federation*, **50**, 240–307.

Finlayson, C.M., Everard, M., Irvine, K., McInnes, R.J., Middleton, B.A., van Dam, A.A. and Davidson, N.C. (eds.) (2018) *The Wetland Book I: Structure and Function, Management, and Methods.* Springer, Dordrecht. https://doi.org/10.1007/978-90-481-9659-3

Gersberg, R.M., Elkins, B.V. and Goldmann, C.R. (1983) Nitrogen removal in artificial wetlands. *Water Research*, **17**, 1009–1014.

Gersberg, R.M., Elkins, B.V. and Goldmann, C.R. (1984) Wastewater treatment by artificial wetlands. *Water Science and Technology*, **17**, 443–450.

Gersberg, R.M., Elkins, B.V., Lyon, S.R. and Goldmann, C.R. (1986). Role of aquatic plants in wastewater treatment by artificial wetlands. *Water Research*, **20**, 363–368.

Good, R.E., Whigham, D.F. and Simpson, L. (eds.) (1978) *Freshwater Wetlands, Ecological Processes and Management Potential.* Academic Press, New York.

Gopal, B. (1999) Natural and constructed wetlands for wastewater treatment: Potentials and problems. *Water Science and Technology*, **40**(3), 27–36.

Grime, J.P. (1979) *Plant Strategies and Vegetation Processes.* John Wiley, Chichester.

Iizumi, H., Hattori, A. and McRoy, C.P. (1980) Nitrate and nitrite in interstitial waters of eelgrass beds in relation to the rhizosphere. *Journal of Experimental Marine Biology and Ecology*, **47**, 191–201.

Itoh, M., Takemon, Y., Makabe, A., Yoshimizu, C., Kohzu, A., Ohte, N., Tumurskh, D., Tayasu, I., Yoshida, N. and Nagata, T. (2011) Evaluation of wastewater nitrogen transformation in a natural wetland (Ulaanbaatar, Mongolia) using dual-isotope analysis of nitrate. *Science of the Total Environment*, **409**, 1530–1538.

Jones, M.B. (1986) Wetlands. In *Photosynthesis in Contrasting Environments* (eds. N.R. Baker and S.P. Long). Elsevier, Amsterdam, pp. 103–138.

Keddy, P.A. (2000) *Wetland Ecology*. Cambridge University Press, Cambridge.

Ma, J., Ullah, S., Niu, A., Liao, Z., Qin, Q., Xu, S. and Lin, C. (2021) Heavy metal pollution increases CH_4 and decreases CO_2 emissions due to soil microbial changes in a mangrove wetland: Microcosm experiment and field examination. *Chemosphere*, 128735. https://doi.org/10.1016/j.chemosphere.2020.128735

Mitsch, W.J. and Gosselink, J.G. (2015) *Wetlands, 5th ed.* John Wiley & Sons, New Jersey.

Monk, C.D. and Brown, T.W. (1965) Ecological consideration of cypress heads in north central Florida. *American Midland Naturalist*, **74**, 125–140.

Nichols, D.S. (1983) Capacity of natural wetlands to remove nutrients from wastewater. *Journal of the Water Pollution Control Federation*, **55**(5), 495–505.

Ning, P., Bart, H.J., Li, B., Lu, X. and Zhang, Y. (2008) Phosphate removal from wastewater by model-La(III) zeolite adsorbents. *Journal of Environmental Science*, **20**, 670–674. https://doi.org/10.1016/S1001-0742(08)62111-7

Reddy, K.R. and Patrick, W.H. (1984) Nitrogen transformations and loss in flooded soils and sediments. *CRC Critical Reviews in Environmental Control*, **13**, 273–309.

Seidel, K. (1976). Macrophytes and water purification. In *Biological Control of Water Pollution* (eds. J. Tourbier, R.W. Pierson and E.W. Furia). University of Pennsylvania Press, pp. 109–121.

Sherr, B.F. and Payne, W.J. (1978) Effect of *Spartina alterniflora* root-rhizome system on salt marsh soil denitrifying bacteria. *Applied Environmental Microbiology*, **35**, 724–729.

Shifflett, S.D. and Schubauer-Berigan, J. (2019) Assessing the risk of utilizing coastal wetlands for wastewater management. *Journal of Environmental Management*, **236**, 269–279.

Sloey, W.E., Sprangler, F.L. and Fetter, C.W. (1978) Management of freshwater wetland for nutrient assimilation. In *Freshwater Wetlands: Ecological Processes and Management Potential* (eds. R.E. Good,

D.F. Whigham and R.L. Simpson). Academic Press, New York, pp. 321–340.

Small, M. (1976) Marsh/pond sewage treatment plants. In *Freshwater Wetlands and Sewage Effluent Disposal* (eds. D.L. Tilton, R.H. Kadlec and C.J. Richardson). University of Michigan, Ann Arbor, pp. 197–213.

Sprangler, F., Sloey, W. and Fetter, C.W. (1976) Experimental use of emergent vegetation for the biological treatment of municipal wastewater in Wisconsin. In *Biological Control of Water Pollution* (eds. J. Tourbier, R.W. Pierson and E.W. Furia). University of Pennsylvania Press, pp. 161–171.

Stephenson, M., Turner, G., Pope, P., Colt, J., Knight, A. and Tchobanoglous, G. (1980) The environmental requirements of aquatic plants. Publication No. 65, Appendix A. The State Water Resources Control Board, Sacramento, California.

Tilton, D.L., Kadlec, R.H. and Richardson, C.J. (eds.) (1976) Freshwater wetlands and sewage effluent disposal. University of Michigan, Ann Arbor.

Tourbier, J., Pierson, R.W. and Furia, E.W. (eds.) (1976) *Biological Control of Water Pollution*. University of Pennsylvania Press.

Valiela, I. and Vince, S. (1976) Green borders of the sea. *Oceanus*, **19**, 10.

van der Putten, W. (1997) Die-back of *Phragmites australisin* European wetlands: An overview of the European Research Programme on Reed Die-back and Progression (1993–1994). *Aquatic Botany*, **59**, 263–275

Wang, J., Ye, S., Laws, E.A., Yuan, H., Ding, X. and Zhao, G. (2017) Surface sediment properties and heavy metal pollution assessment in the shallow sea wetland of the Liaodong Bay, China. *Marine Pollution Bulletin*, **120**, 347–354.

Westlake, D.F. (1963) Comparisons of plant productivity. *Biological Reviews of the Cambridge Philosophical Society*, **38**, 385–425.

Wheeler, B.D. and Giller, K.E. (1982) Species richness of herbaceous fen vegetation in Broadland, Norfolk, in relation to the quantity of above ground plant material. *Journal of Ecology*, **70**, 179–200.

Whigham, D.F. and Simpson, R.L. (1976) The potential use of tidal marshes in the management of water quality in the Delaware River. In *Biological Control of Water Pollution* (eds. J. Tourbier and R.R. Pierson). University of Pennsylvania Press, pp. 173–186.

13

Constructed Wetlands Incorporating Emergent Vegetation

13.1 Introduction

Although there are several hundred potential emergent wetland species, only a few have been selected as being potentially suitable for wastewater treatment, such as the bulrushes *Scripus lacustris*, *S. acutus* and *S. validus*; the reeds, the commonest being *Phragmites australis*, which is synonymous with *P. communis*; and the yellow flag (*Iris pseudocorus*). Reeds have underground rhizomes and jointed stems that root at the nodes, and both reeds and rushes multiply vegetatively from adventitious buds located on the rhizomes. Rushes grow between 2 and 3 m in height and are restricted to shallow water <0.5 m deep, whereas reeds can grow at depths of up to 2.5 m, with some species growing to heights of 5 m (Dinges, 1982) (Figs. 13.1 and 13.2). Emergent species are generally very hardy and resistant to both pests and diseases. They can also withstand cold and even freezing temperatures, making them ideal for more temperate climates. These plants are perennials, and with proper management, stands of rush or reeds can be maintained indefinitely.

As discussed at the beginning of Sec. 10.1, constructed wetlands are categorized as either free water surface (FWS) or subsurface flow (SSF) systems (Fig. 13.3). The former most resemble natural wetlands such as swamps and marshes and develop a complex and diverse ecology. In contrast, SSF systems are normally monocultures of reeds

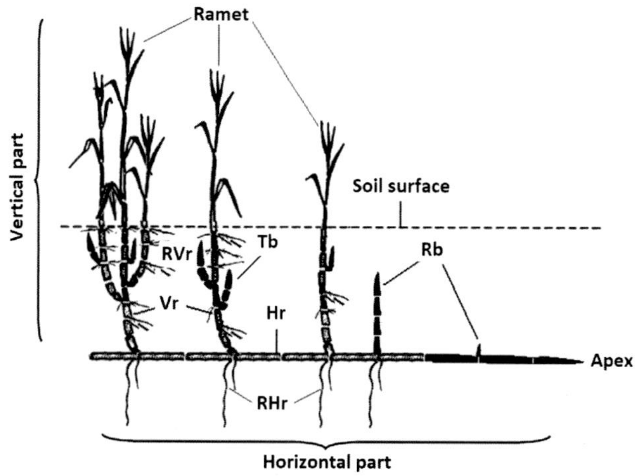

Fig. 13.1. Schematic representation of above and below-ground morphology of *Phragmites australis.* Where Rb are rhizomatous buds, Tb are tiller buds, Vr are vertical rhizomes, Hr are horizontal rhizomes, RVr are the roots on the vertical rhizomes, and RHr are the roots on the horizontal rhizomes (Liu *et al.,* 2014). The term "ramet" is applied to the individual members of a clone of plants produced by vegetative reproduction from a single parent plant. Reproduced with permission from Springer International Publishing.

that do not have permanent areas of open water, thus resembling a fen where the water movement is normally through the soil and not over the surface (Keddy, 2000) (Fig. 13.4). In contrast to natural wetlands, constructed systems are subject to close control and designed to maximize treatment efficiency and biomass production, with the shape of constructed wetlands made to facilitate harvesting. They can be used in all climates and provide valuable habitats for wildlife. In temperate climates, they function best in the warmer months. Therefore, there is a need to reduce organic loading during the winter and possibly even store effluent until the following spring. However, research in moderately cold climates, such as Ontario, Canada, has shown that they can be operated successfully all year round and produce good quality final effluents even when covered with ice (Wile *et al.,* 1982; Reed *et al.*, 1984). In terms of cost, they

Fig. 13.2. A *P. communis* reed bed treating 25 to 30 m^3 of wastewater per day in Dubai. Photograph by W. Sievert, reproduced with permission from the Sustainable Sanitation Alliance (SuSanA) Secretariat under the Creative Commons License CC BY 2.0 (https://creativecommons.org/licenses/by/2.0).

require a large area of land, a high degree of labor to construct the bed and plant the vegetation, and periodic harvesting. However, work in the United States (US) has shown that they are much more cost-effective on a long-term basis than natural wetlands (Kadlec and Knight, 1996) as well as creating new habitats rather than damaging existing wetlands (Sec. 12.5).

Constructed wetlands should be carefully integrated into the existing landscape, especially when used for the treatment of diffuse or

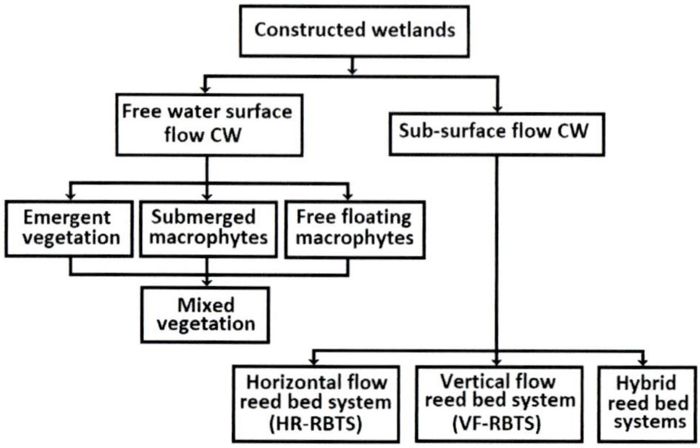

Fig. 13.3. Classification of constructed wetland designs.

Fig. 13.4. Anther Creek FWS constructed wetland, Shulin District, New Taipei, Taiwan. Image by Lien Yuan Lee, reproduced under the Creative Commons License CC BY 3.0 (https://creativecommons.org/licenses/by/3.0).

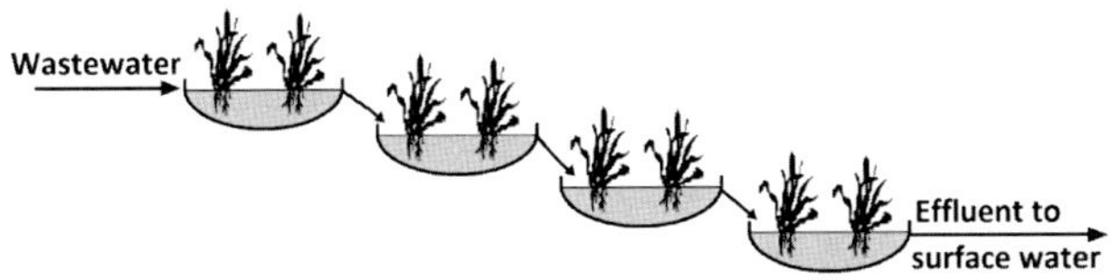

Fig. 13.5. Constructed wetlands can be used in almost any terrain, with designs following contours and boundaries. Where steep slopes are present, then wetlands are employed as a number of smaller ponds in series (Mitsch, 1992). Reproduced with permission from Elsevier Publishing.

non-point pollution. Mitsch (1992) gives many examples of possible designs including the construction of ponds in steep terrain (Fig. 13.5), using a bypass system to treat diffuse pollution in-stream while preventing overloading the pond hydraulically, and finally wetlands fed during the flood season (Fig. 13.6). Mitsch has also listed a number of factors to be considered when designing constructed wetlands:

(1) Wetlands should be designed for minimum maintenance.
(2) Design a system that utilizes natural energies, such as potential energy of streams, as natural subsidies to the system.
(3) Design the system with the landscape, not against it.
(4) Design the system with multiple objectives but identify at least one major objective and several secondary objectives.
(5) Design the system as an ecotone. This means including a buffer strip around the site, but it also means that the wetland site itself is often a buffer system between terrestrial and aquatic systems.
(6) Wetlands do not become functional overnight and several years may elapse before nutrient retention or wildlife enhancement is optimal.
(7) Design the system for function, not for form.
(8) Do not over-engineer wetland design with rectangular basins, rigid structures and channels, and regular morphology. Ecological engineering recognizes that natural systems should be mimicked to accommodate biological systems.

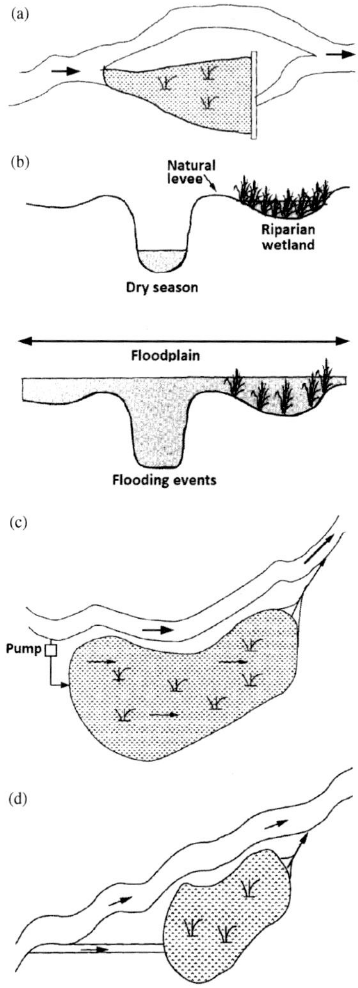

Fig. 13.6. Examples of how constructed wetlands can be used for treating diffuse pollution within water catchments, especially in agriculture, using good landscape design: (a) instream constructed wetland with part of the flow bypassing during high flow conditions, (b) natural or constructed riparian wetland only fed during flooding seasons, (c) riparian wetland fed by pump, and (d) riparian wetland fed by gravity from upstream (Mitsch, 1992). Reproduced with permission from Elsevier Publishing.

Constructed wetlands can be used to treat a wide variety of wastewaters from domestic sewage to acid mine drainage. They are extremely versatile, but careful consideration must be given at the design stage as to the most suitable plants for treating any specific wastewater. For example, saline wastewaters are becoming a more frequent problem, due primarily to saline intrusion into sewers. Many industrial, agricultural and aquaculture wastewaters can also contain high concentrations of salts as well as other contaminants such as metals. While normal constructed wetlands can be damaged by salinity, saline wastewaters can be successfully treated so long as suitable halophytes are selected for planting at the construction stage (Liang *et al.*, 2017) (Table 13.1).

Table 13.1. Commonly used haolophytes used for the treatment of saline wastewaters in constructed wetlands (Liang *et al.,* 2017). Reproduced with permission from Elsevier Publishing.

Type	Latin name	Locality
Emergent	*Typha angustifolia* Linn.	China, Philippines, Japan, the previous Soviet Union, etc.
	Phragmites australis (Cav.) Trin.	Around the world
	Lythrum salicaria Linn.	Asia, Europe, Algeria, North America, Australia, etc.
	Iris pseudacorus Linn.	Europe and China
	Vetiveria zizanioides Linn.	Southeast Asia, India, Africa, etc.
	Spartina alterniflora Lois.	Southeast coast of America, coastal cities of China except Hainan
Submerged	Canna *indica* Linn.	India, tropical America, etc.
	Ruppia maritima Linn.	Temperate and subtropical waters of the world
	Ceratophyllum demersum Linn.	Mongolia, North Korea, Japan, Malaysia, China, Europe, etc.
	Hydrilla verticillata (Linn. f.) Royle	Europe, Asia, Africa, Oceania, etc.
	Potamogeton crispus Linn.	Around the world
	Myriophyllum verticillatum Linn.	Around the world
Floating	*Eichhornia crassipes* (Mart.)	Around the world
	Pistia stratiotes Linn.	Tropical and subtropical regions, some regions of China, etc.
	Lemna minor Linn.	Around the world

13.2 Free Water Surface Treatment Wetlands

Designs of FWS-constructed wetlands are based on natural systems, which may include just a single species or many different groups of plants (e.g. submerged, floating, emergent, etc.), each comprising many different species (Fig. 13.7). Plant selection may not only be based on performance, but also aesthetics including flowering and ornamental plants.

FWS wetlands utilize emergent plants in two distinct ways. Most commonly, they are planted in shallow basins of soil or other mediums with the water level controlled to ensure that the sediment, leaf litter, and soil are always submerged and only the living and dead stems of the plants are above the water level. This is similar to a natural marsh with an oxygen gradient formed from the air-water interface with most of the medium anaerobic. Bulrushes (*Scripus* spp.), spike rush (*Eleocharis* spp.), sedges (*Cyperus* spp. and *Carex* spp.), and rushes (*Juncus* spp.) are most frequently used along with cattails (*Typha* spp.) and the reed (*Phragmites communis*). A shallow depth of

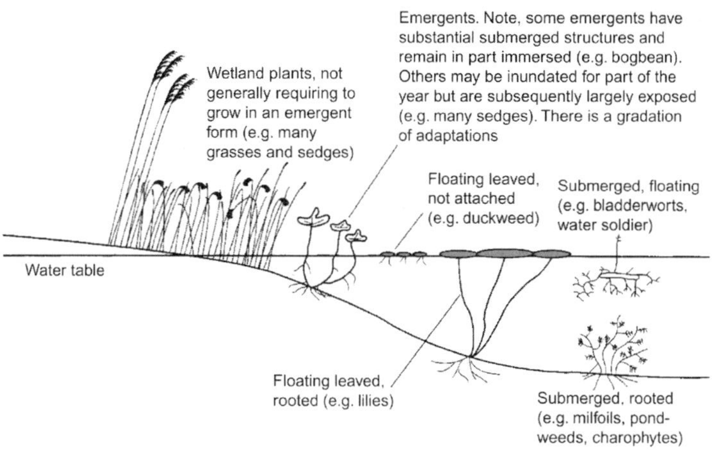

Fig. 13.7. Different classifications of aquatic macrophytes in a pond during summer (Jeffries and Mills, 1993). Reproduced with permission from Wiley-VCH Verlag GmbH & Co. KGaA.

the medium, penetrated by roots and rhizomes, and the liquid layer above the medium surface are the major pathways for the wastewater. Young shoots protrude up from the medium into the water. Therefore, a dense matrix of roots develops above and below the water-medium interface where a range of biochemical transformations occur (Khatiwada and Polprasert, 1999). These systems are particularly attractive to wildlife. The influent wastewater spreads slowly through the shallow water, with the suspended solids removed by settlement or entrapment by the vegetation and leaf litter. Attached and suspended microorganisms remove the soluble biochemical oxygen demand (BOD). Nitrification and denitrification occur in the aerobic and anoxic areas, respectively (Fig. 12.1), whereas phosphorus removal is at a low but sustained rate by various sorption, complexation and precipitation reactions (Fig. 12.2) (Morshiri, 1993; Kadlec and Knight, 1996). While emergent species take up nutrients from the substrate, submerged and floating macrophytes take nutrients directly from the water. In a series of mesocosm experiments, Tang *et al.* (2017) compared the uptake of nutrients by a selection of macrophytes grown in various substrates (i.e., clay, peaty-clay and peat). At low loadings (0.43 mg P m^2d^{-1}) 100% of phosphorus removal was solely by the plant uptake, but this rapidly declined as loadings increased with 50 to 90% of phosphorus adsorbed onto the substrate/sediment at loadings >21.4 mg P $m^{-2}d^{-1}$. Nitrogen removal was limited to 64% at a loading of 62 mg N $m^{-2}d^{-1}$, falling to 21% at 249 mg N $m^{-2}d^{-1}$. The authors concluded that macrophytes are best suited to polishing (tertiary treatment) rather than secondary treatment and that sediments, especially clay, are important sinks for phosphorus until they become saturated.

Constructed systems can be lined to prevent exfiltration from the wetland, with the vegetation planted in a shallow layer of gravel. Where the substrate is suitable, emergent vegetation can be planted directly into the soil. Where wetlands are also constructed for wildlife use, then it is important to include a diversity of habitats in the overall design such as areas of deep, as well as shallow water, islands and adequate areas of open water for wildfowl. Such a system has been constructed at Martinez in California, where the wetland was created primarily for wildlife. The system is 6.1 ha in area and has a retention time of 10 days. Some 60% of the wetland is open water with stands of *Scripus* and *Typha*. It has been necessary to use a further 2.1 ha around the edge of the existing wetland for growing plants, which the wildlife using the wetland can use as a food resource.

Constructed wetlands function in the same way as natural ones except that more nitrogen and phosphorus are lost from the system by harvesting. Such systems have been studied using 18.5 m × 3.5 m plastic-lined beds that are 0.76 m deep, with emergent macrophytes grown in the gravel (Gersberg *et al.*, 1986). Four beds were used to treat primary municipal wastewater from Santee, California. This is a particularly ideal climate for wetlands as the mean air temperature rarely falls below 12°C, and the plants continuously grow throughout the year. Three plants were compared for their ability to treat wastewater, *Scripus validus* (soft-stemmed bulrush), *P. communis* (common reed), and *Typha latifolia* (common cattail). During the experimental period, the mean ammonia concentration was reduced from 24.7 mg L^{-1} in the influent to 1.4 mg L^{-1} in the bulrush bed, 5.3 mg L^{-1} in the reed bed, 17.7 mg L^{-1} in the cattail bed, and only 22.1 mg L^{-1} in the control bed, which did not contain any emergent plants. The ability of the bulrush and reed to remove ammonia from water is attributed to the ability of these plants in being able to translocate oxygen from the shoots to the rhizosphere, which stimulated nitrification-denitrification (Armstrong, 1964). Similarly, BOD removal was significantly higher in the bulrush and reed beds at 5.3 and 22.2 mg L^{-1}, respectively, compared to the cattail bed at 30.4 mg L^{-1}, or the control bed at 36.4 mg L^{-1}, where the mean BOD of the influent was 118.3 mg L^{-1}. This showed that these two species were able to treat wastewater to an equal or even better quality than conventional secondary treatment and that the plants have a significant function in the removal of nitrogen separate to assimilation into plant tissue. Thus, the mean percentage removal of BOD was 96, 81, 74, and 69% for the bulrushes, reeds, cattails, and control beds, respectively. The poor performance of cattails can be attributed to the fact that it is less able to transport oxygen to the root zone, although this ability is seen in many similar plants such as

Spartina and *Zostera*, and has a shallow root zone with most of the root mass confined to the top 30 cm of the substrate compared to >60 cm for bulrushes and >75 cm for reeds. This provides less root surface area for the development of heterotrophic as well as nitrifying bacteria. In this study, a hydraulic loading rate of 4.7 cm d^{-1} produced a 10:10 (mg L^{-1} BOD:mg L^{-1} Suspended Solids) final effluent. At this loading rate, 5 ha of wetlands would be able to treat 2,300 m^3 of preliminary-treated wastewater each day.

In Flevoland (the Netherlands), an experimental rush pond of 1 ha in area and 0.4 m in depth was constructed to cope with the seasonal influx of tourists at an isolated camping site (de Jong, 1976). The square experimental pond had a star-shaped flow pattern to ensure maximum utilization of the available area. To minimize maintenance, this design was later modified to long narrow channels that were 375 m in length, which could be maintained mechanically. Rushes were rooted in the mud of the unlined basin and the results showed that the BOD of the raw wastewater was reduced by 95.7%, COD by 80.7%, and fecal coliforms (MPN) by 99.9%. Although the removal of nitrogen and phosphorus was initially rapid, the removal efficiency declined as the supply of nutrients exceeded the uptake rate, with mature rush stands removing between 300 and 500 kg N $ha^{-1}y^{-1}$ and 50–75 kg P $ha^{-1}y^{-1}$. De Jong (1976) observed that purification was achieved not only by the bacteria associated with plant roots but also by infiltration of the wastewater into the soil. In the channels, the performance of reeds and rushes was compared, and while there was little difference in their ability to purify wastewater, the reeds required less maintenance. Reeds are harvested annually during the summer, whereas rushes can be gathered every 2nd or 3rd year. However, if left alone, reeds can grow to >5 m in constructed wetlands after four years and be more difficult to harvest, so regular cutting is recommended. Stem densities are about the same for both groups at 120 to 150 m^{-2} but reeds have little commercial value, unlike rushes, which are highly sought after as wicker for basket and furniture manufacture and can also be used for feeding livestock (Pomoell, 1976). The total biomass of rushes produced in a treatment pond can reach between 5 and 10 tons $ha^{-1}y^{-1}$.

Spangler *et al.* (1976) compared the efficiency of *Iris versicolor* (the blue flag iris), *Scripus validus* and *S. acutus* (hard stemmed bulrush) in bench-scale ponds over a winter period. They were planted in pea-sized gravel 70 mm deep in plastic-lined basins (800 × 900 mm). Primary effluent was fed into the basins each day from Monday to Friday with the retention times of 3, 5 and 15 days studied. They found that there was no difference between the various ponds, including the control pond without the plants, in reducing biochemical oxygen demand (BOD) or chemical oxygen demand (COD). However, the presence of bulrushes made a significant difference in the removal of both the total phosphorus and orthophosphate, being more effective than the control or Iris ponds. The removal efficiency increased with the longer retention times, with total phosphorus removal reaching 98% on occasion (average >80%).

The efficiency of four emergent species *Typha domingensis*, *T. orientalis*, *Phragmites australis (≈communis)*, and *Scripus validus* were compared to treat the effluent from a poultry abattoir (Finlayson and Chick, 1983). The plants were grown in a gravel substrate in plastic-lined trenches. Using a retention time of between 2.7 and 3.6 d^{-1}, the effluent was allowed to percolate through the trenches. All three genera were able to significantly reduce the suspended solids (83–89%) and turbidity (58–67%) of the effluent, whereas *Phragmites* and *Scripus* were both able to oxygenate the anaerobic inflow. *Scripus* was more effective than the other genera in reducing the concentrations of total nitrogen and phosphorus, and also reduced the concentrations of sodium, potassium and chloride. The genus *Typha* comprises nearly 20 species with three species (i.e., *Typha latifolia*, *T. domingensis* and *T. orientalis)* used for treatment studies (Chick *et al.*, 1983).

Although emergent plants are an inexpensive treatment system for many types of wastewaters, some stronger effluents appear to be phytotoxic. Piggery effluents for example can severely affect the vigor of emergent plants, especially *T. domingensis* (Finlayson and Mitchell, 1982; Bowmer, 1985).

Increasingly, FWS treatment wetland employs a range of macrophytes that form floating mats. Cattails, giant sweet manner grass

(*Glyceria maxima*), pennywort (*Hydrocotyle umbellata)* and the common reed can all colonize mats of dead plant debris that can become quite thick (>200 mm) and float on the surface of the water. The mat or raft is stabilized by the rhizomes or roots that intertwine and grow into the water phase (van Oostrom, 1995). Free water surface ponds are widely used in combined systems with other constructed wetlands, normally SSF beds planted with emergent macrophytes. Wastewater from a restaurant and swimming pool resort in Nairobi, Kenya, is treated by an SSF horizontal flow reed bed planted with *Typha* followed by three FWS ponds planted with indigenous reeds (*Cyperus alternifolius* and *Cyperus latifolius*), ornamental plants (Arundo donax varigata, *Pontederia* spp. and *Sagittaria* spp.), and flowering ornamentals (*Ajuga remota* and *Aspilia mossambicensis*). Treating a population equivalent (PE) of 1,200, the system achieves high removal rates of 98% for BOD, 85% for SS, 96% for COD, 99% for fecal coliforms, 92% for ammonia, and 88% for orthophosphate. The constructed wetland, as well as being aesthetically pleasing, has attracted over 120 bird species and amphibians (Nyakang'o and van Bruggen, 1999). An integrated FWS system was designed for the Beijing Zoo for wastewater treatment (Jiang and Zhu, 1998). The design created numerous niches and comprised floating macrophytes (*Eichhornia crassipes* and *Azolla filiculoides*), submerged macrophytes (*Potamogeton crispus, Vallisneria spiralis*, and *Myriophyllum spicatum*), floating leaf plants (*Nelumbo nucifera*), and emergent species (*Ipomoea aquatica* and *Alternanthera philoveroides*). The ponds also contained bacteria, microalgae, zooplankton as well as fish (herbivorous species such as grass carp and triangular bream, phytoplanktonic-feeding fish including silver and big-head carp, and other fish species that are detritivores). Waterfowl and a range of mammals were also introduced creating a complex community structure (Fig. 13.8).

Mohan *et al.* (2010) evaluated a pilot-scale-constructed wetland system that they called an ecologically engineered system (EES). This consisted of three tanks in series. Tank 1 was a floating macrophyte system, followed by Tank 2, which contained submerged and emergent macrophytes, and Tank 3, which was a filter-feeding system

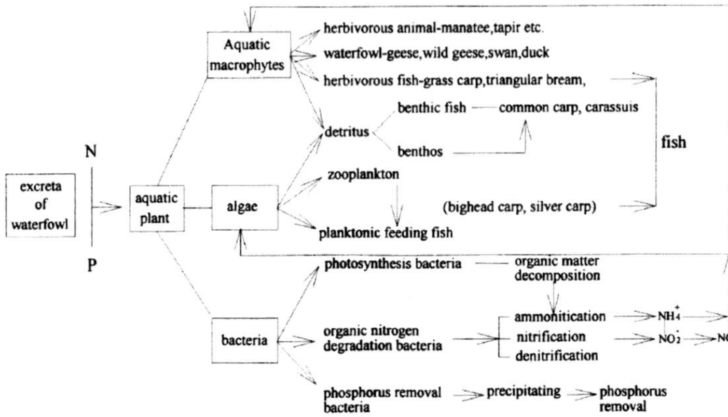

Fig. 13.8. Simplified food web of FWS-constructed wetland used to treat wastewater at the Beijing Zoo (Jiang and Zhu, 1998). Reproduced with permission from Elsevier Publishing.

containing algae, duckweed and fish. When treating settled sewage, removal rates were 68.1% for COD, 22.4% for nitrate, and 59.8% for turbidity. The ecology of the system is summarized in Table 13.2, which gives the details of the function of each flora or faunal group within each tank.

13.3 Subsurface Flow Treatment Wetlands

It is in the construction of subsurface flow (SSF) treatment wetlands that emergent macrophytes are most widely used. The original concept was developed by Seidal and his colleagues at the Max Planck Institute in Germany between 1960 and 1980 (Seidal, 1976). Originally known as the root zone method, the system is now universally known as the reed bed treatment system (RBTS). Reed beds are water-tight basins designed to generate either a horizontal flow (HF) or vertical flow (VF) of wastewater through the rooting medium (Brix, 1994; Reed *et al.*, 1995; Vymazal *et al.*, 1998; Kadlec *et al.*, 2000). The main application of the RBTS is the secondary treatment of domestic wastewater from small communities (PE <500) or

Table 13.2. The biota and their contribution to the purification of domestic wastewater in a three-tank ecologically engineered system incorporating floating macrophytes (Tank 1), submerged and emergent vegetation (Tank 2), and a filter-feeding-based system (Tank 3) (Mohan *et al.*, 2010). Reproduced with permission from Elsevier Publishing.

Tank	Biota and function	Survival*
Floating macrophyte based system (Tank 1)	***Eichhornia crassipes***	+++++
	• Extensive filamentous roots with high specific area has good filtration and bacterial support potential • Creates low oxygen conditions beneath the mat • Can also remove toxic elements from the wastewater	
	***Cyphoma gibbosum* (Snails)**	+++++
	• Removes algae from the walls of the tanks • Reduces suspended organic matter	
	Nematodes	++++
	• Reduces organic matter	
	Bacteria	
	• Nitrification/denitrification • Immobilization and conversion of minerals and nutrients	
	Algae	++++
	• Releases oxygen into water and thus, helps to stabilize organic waste	
Submerged and emergent macrophyte based integrated system (Tank 2)	***Asparagus***	+++
	• Provides higher root surface area • Absorbs high amount of nutrients • Can grow under stress conditions	
	***Bryophillum pinnatum*, *Lycopersicum esculentum* (Tomato), *Coriandrum sativum* (Coriander), *Capsicum annum* (Capsicum)**	+
	• Roots provide surface area for the attachment of microbes	

(*Continued*)

Table 13.2. (*Continued*)

Tank	Biota and function	Survival*
	Ferns • Roots provide surface area for the attachment of microbes	+++
	Rice (*Oriza sativa*) • Roots provide surface area for the attachment of microbes	++++
	Hydrilla verticillata* (Hydrilla) and *Myriophyllum • Improve DO levels • Provide good habitat for aerobic bacteria	+++
	***Cyphoma gibbosum* (Snails)** • Removes algae from the walls of the tanks • Reduces suspended organic matter	+++++
	***Tagetes erecta* (Marigold)** • Provides higher root surface area	+++
	Bacteria • Nitrification/denitrification • Immobilizeation and conversion of minerals and nutrients	
	Algae • Releases oxygen into water and thus, helps to stabilize organic waste	+++++
Filter-feeding-based system (Tank 3)	***Gambusia affinis* (Mosquito fish)** • Removes mosquito larva • Feeds on algae and other suspended matter • Can grow in polluted water	++++ (survived for 7 months)
	***Cyprinus carpio* (Common carp)** • Feeds on algae and other suspended matter • Can grow in polluted water, robust in nature and can grow under varied temperatures	++++ (survived for 3 months)
	Myriophyllum • Oxygenates water which helps for fish survival	++++

Table 13.2. (*Continued*)

Tank	Biota and function	Survival*
	***Cyphoma gibbosum* (Snails)**	++++
	• Removes algae from the walls of the tanks (tank maintenance) • Reduces suspended organic matter	
	***Limna gibba* (Duckweed)**	+++++
	• Reduces BOD, COD and suspended solids • Reduces bacterial pathogens of wastewater to a level that is safe for disposal • Removes toxic elements for the wastewater	
	Bacteria	
	• Nitrification/denitrification • Immobilization and conversion of minerals and nutrients	
	Algae	++++
	• Releases oxygen into water and thus, helps to stabilize organic waste	

*Overgrowth (+++++): Sharp improvement in growth (doubled in 4 days); Good (++++): Growth doubled (approximately) within 15 days of time; Moderate (+++): Growth doubled (approximately) within 30 days of time; Low (+): Insignificant growth within 30 days.

individual houses (PE <10). Very few European systems treat loads of PE >1,000 (Vymazal *et al.*, 1998). By 1998, there were over 5,000 systems in operation in Europe, with 3,500 in Germany alone (Börner *et al.*, 1998), and they have continued to grow in popularity especially for on-site applications. Design and treatment guidelines have been published by Copper *et al.* (1996) and Gschlößl and Stuible (2000).

13.3.1 *Design of horizontal flow reed bed treatment systems*

In horizontal flow reed bed treatment systems (HF-RBTSs), emergent macrophytes are grown in beds of graded soil or gravel retained by an impermeable subsurface barrier (clay or synthetic liner). In a review, Wu *et al.* (2015) list 28 different substrates including natural, industrial and artificial materials. The base of the bed has a slope of

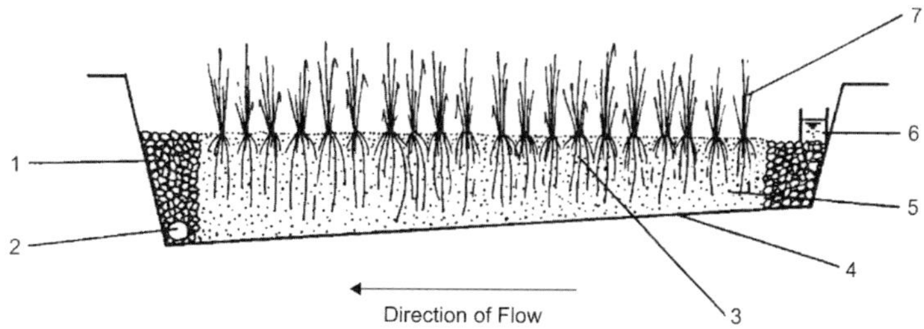

Fig. 13.9. Typical design of a horizontal flow reed bed treatment system: (1) drainage zone comprising large rocks, (2) water level control and drainage collection for treated effluent, (3) roots and rhizomes (root zone), (4) impervious liner, (5) substrate (e.g. gravel or soil), (6) influent wastewater distribution system, and (7) emergent vegetation.

2 to 8% to encourage settled sewage to pass horizontally through the gravel (Fig. 13.9). The outlet control pipework is designed to allow the water level in the bed to be adjusted to permit near saturation of the substrate during the establishment of the reeds while lowering the level as the reeds mature. The depth of the bed depends on the root penetration of the plants used, for example, cattails (300 mm), reeds (600 mm), and bulrushes (750 mm). The most widely employed emergent aquatic macrophytes are *Phragmites australis* (Packer *et al.*, 2017) and *Typha latifola,* both native to Europe and with global distributions. *Glyceria maxima* and the reed canary grass (*Phalaris arundinacea*), being native, are also used in Europe, normally in combination with the *Phragmites* spp., whereas in the US, the *Scripus* spp. are popular. However, many local plants are also employed (Fig. 13.10). The substrate and roots provide a large potential surface area for the growth of heterotrophic microorganisms (Table 13.3). The long roots of the vegetation penetrate deep into the substrate. The internal gas spaces of the plant, an adaptation to its partially submerged existence, transport oxygen to the saturated rhizomes and roots (i.e., aerenchyma tissue). Thus, an aerobic zone is created within otherwise water-logged sediment. Within the rhizosphere, there are also anoxic and reduced areas (Fig. 13.11). This combination allows both aerobic and anaerobic bacteria to

Fig. 13.10. Growth of papyrus for the treatment of wastewater in a constructed wetland. Image by Heike Hoffmann. Reproduced with permission from the Sustainable Sanitation Alliance (SuSanA) Secretariat under the Creative Commons License CC BY 2.0 (https://creativecommons.org/licenses/by/2.0).

survive, allowing carbonaceous oxidation, nitrification, denitrification, and anaerobic degradation to occur. Ciliate protozoa also play an important role in SSF RBTS. In a study of four reed bed systems, Decamp *et al.* (1999) found 22 different ciliate species with the abundance being greatest in the first third of the bed and decreasing towards the outlet. The main function of ciliates is to graze the bacteria present. They estimated that grazing rates in the gravel media were in the order of 49 bacteria per ciliate per hour. Assuming a mean abundance of 600 ciliates per ml of interstitial liquid, they estimated that ciliates by their predatory activities were capable of removing up to 2.35×10^5 *Escherichia coli* per ml of liquid per day in the first third of the bed. Other bacterivores, including rotifers, nematodes and amoebae, are also known to remove *E. coli* (Decamp and Warren, 1998).

Table 13.3. Predominant removal mechanisms in reed bed treatment systems.

Wasterwater constituent	Removal mechanism
Suspended solids	Sedimentation Filtration
Soluble organics	Aerobic microbial degradation Anaerobic microbial degradation
Nitrogen	Ammonification/Nitrification Denitrification Plant uptake Matrix adsorption Ammonial volatilization
Phosphorus	Matrix sorption Plant uptake
Metals	Adsorption and cation exchange Complexation Precipitation Plant uptake Microbial oxidation/reduction
Pathogens	Sedimentation Filtration Natural die-off Predation UV irradiation Root excretion of antibiotics

The performance data of European HF systems is given by Coombes (1990), Schierup *et al.* (1990) and Brix (1994), and North American systems by Kadlec and Knight (1996) (Table 13.4). The plants take up nutrients and store them in their tissues, which are removed from the system when harvested. The rooting substrate also has a sorption potential to remove metals from the solution. Removal potential exceeds 90% for most metals and these become immobilized in the anaerobic mud layer in the base of the bed as metal sulfides (Table 13.5), although excessive accumulation will eventually inhibit the process. During winter, the emergent part of the vegetation die back and treatment is restricted to the root zone alone, but in practice, performance is not severely affected. Reed beds do not require

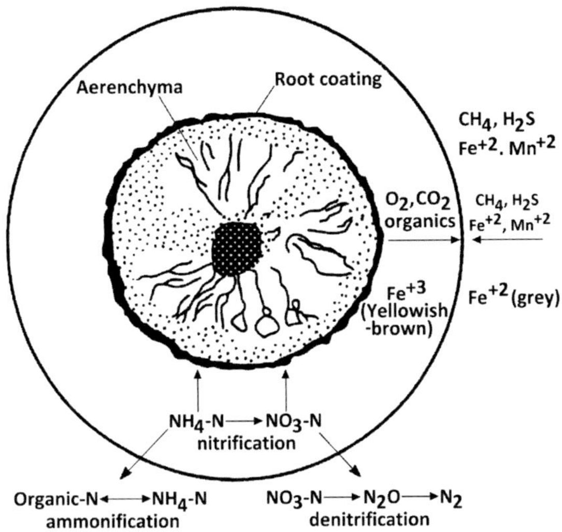

Fig. 13.11. Cross-section of an oxidizing root growing in reduced sediment. The oxidized rhizosphere is depicted, along with some processes that occur as a result of this aerobic-anaerobic interface (Meddelssohn and Postek, 1982). Reproduced with permission from John Wiley & Sons.

Table 13.4. Mean influent and effluent concentrations, and mass loading rates, in gravel-based subsurface horizontal flow constructed reed beds in the Czech Republic (Vymazal *et al.*, 1998). Reproduced with permission from John Wiley & Sons.

		Influent		Effluent	
Parameter	*n*	Mean	SD	Mean	SD
Concentrations (mg L^{-1})					
SS	37	71.9	47.2	10.8	7.1
BOD_5	39	87.4	65.7	11.9	11.4
TN	26	46.1	18.5	27.6	9.7
TP	27	6.4	3.8	3.1	2.1
Mass loading rates (g $m^{-2}d^{-1}$)					
SS	31	3.34	3.11	0.44	0.42
BOD_5	35	3.36	2.86	0.53	0.67
TN	26	1.39	0.91	0.80	0.16
TP	24	0.30	0.18	0.18	0.16

Table 13.5. Metal removal mechanisms in constructed reed beds.

Vegetation:	Ion uptake/translocation Adsorption Organic decomposition
Water:	Evaporation Dilution Complex formation Decomposition Microbial oxidation/reduction Precipitation

Substrate:	Microbial oxidation/reduction
	Ion exchange:
	  Precipitation
	  Adsorption
	  Chelation
	  Chemical (inorganic) decomposition

Inorganic substrate and sediment:	Microbial oxidation/reduction Precipitation Adsorption

daily attention, although, during the first year of operation, beds require weeding, either by hand or flooding, before the reeds are fully established.

A minimum hydraulic conductance of 10^{-3} to 10^{-4} m s^{-1} is required to prevent surface flow and channeling from occurring, therefore the selection of the correct substrate or medium is vital. Once the reeds are established and the rhizomes have penetrated the entire root zone, the hydraulic conductance is secured by the network of micro-channels left after dead rhizomes have decayed (5 to 15 mm in diameter). Therefore, porosity increases with the age of the bed. The best results are achieved using gravel (5 to 32 mm in diameter) rather than soils, although this medium has a low potential for the removal of heavy metals and phosphorus. Different substrate compositions have different qualities. For example, substrates with a low nutrient content encourage direct uptake of nutrients from wastewater by the plants rather than from the substrate itself. A high Al or Fe content provides effective phosphorus adsorption (Davies and Cottingham, 1993). This is achieved by ligand exchange reactions in which phosphate displaces water or hydroxyl ions from the surface of Fe and Al hydrous oxides. Other specific phosphorus-sorbing media are available (Mann and Bavor, 1993; Drizo *et al.*, 1997; Zhu *et al.*, 1997). As normal gravel media does not contain Fe, Al or Ca in any great quantities, removal of phosphorus is generally low (Cooke, 1992; Kadlec *et al.*, 2000). A high organic or clay content ensures a good removal of heavy metals. Crushed limestone is used in the treatment of acidic wastewaters and also gives enhanced phosphorus removal. Due to their ability to withstand a wide range of operating conditions and their high potential for wildlife conservation, reed beds have become popular with the industrial sector as part of an overall environmental management strategy. In an exhaustive review, Vymazal (2014) examined 138 installations in 33 different counties, detailing performance data for 25 different industry types, including the abattoir, aquaculture, brewing and distilling, chemical, dairy, paper pulp, potato, refinery, steel sugar, tannery, textile, and winery sectors. Most industries use RBTS (HF, VF and hybrid systems). Those with more degradable wastewaters also use FWS systems, although often integrated with an RBTS. However, the main application of the HF-RBTS remains the treatment of domestic sewage from isolated developments such as hotels, schools, housing estates, etc. Due to the relatively low loading rates that can be applied, constructed wetlands require large areas of land compared to conventional processes, and thus are normally restricted to treating domestic wastewater from small communities. Wastewaters require preliminary treatment (i.e., screening, grit, and, where necessary, fat removal) and primary settlement before being applied to a reed bed. The average design loading for domestic wastewaters is 5 m^2 per capita. The precise area of an HF reed bed required is calculated using Eq. (13.1):

$$A = \frac{Q\left(\text{In}\,C_o - \text{In}\,C_i\right)}{k_1}, \tag{13.1}$$

where A is the area of the bed (m^2), Q is the influent flow rate (m^{3}d^{-1}), C_o is the influent BOD concentration (mg L^{-1}), C_1 is the required effluent BOD concentration (mg L^{-1}) and k_1 is the BOD reaction rate constant. Cooper *et al.* (1996) recommend that a k_1 value of 0.1 can be used in most European (temperate) situations. The equation tends to produce an area of 5 m^2 PE^{-1} for the secondary treatment HF beds loaded with settled sewage. This will produce a 20:20 effluent but little ammonia reduction. This equation can also be used for HF beds used for tertiary treatment. The area can be reduced to 0.5 to 0.7 m^2 PE^{-1} and will often achieve full nitrification (Cooper, 1999). Severn and Trent Water have installed tertiary treatment SSF wetlands at many of their small treatment plants (Green and Upton, 1995). Using a design figure of 1.0 m^2 PE^{-1}, a final effluent of <5 mg BOD L^{-1} and 10 mg TSS L^{-1} has been obtained. This figure has now been revised downward to 0.7 m^2 PE^{-1}. Performance data for a tertiary treatment HF-RBTS at Leek Wooton in England is given in Table 13.6.

During the growing season, between 1.5 and 1.8 m^3 of water is lost by evapotranspiration per m^2 of fully developed *Phragmites*, significantly reducing the effluent volume at a time when receiving waters are at their most vulnerable to organic and nutrient enrichment. Reed beds can be designed to operate with the wastewater flowing in an either horizontal or vertical direction, with the latter achieving better overall nitrogen removal.

There has been much interest in the use of RBTSs to cultivate commercial plants, such as *Chrysopogon zizanioides* (formally *Vetiveria*

Table 13.6. Mean annual performance data for Leek Wootton's (UK) tertiary treatment horizontal flow reed bed treatment system (Kadlec *et al.*, 2000). Reproduced with permission from IWA Publishing.

	BOD_5 (mg L^{-1})		**COD (mg L^{-1})**		**TSS (mg L^{-1})**		**NH4-N (mg L^{-1})**		**TON (mg L^{-1})**	
Year	**In**	**Out**	**In**	**Out**	**In**	**Out**	**In**	**Out**	**In**	**Out**
1990/1991	11.6	4.8	75.7	32.1	27.6	6.1	7.6	5.8	32.8	23.4
1991/1992	11.9	2.0	76.7	34.0	19.1	3.7	5.4	1.9	29.7	20.8
1992/1993	15.4	2.7	109.0	55.5	24.2	5.3	7.0	2.8	20.4	8.7
1993/1994	9.1	1.5	93.8	48.3	16.3	4.4	7.2	3.0	25.6	16.8
1994/1995	9.1	1.0	82.1	46.6	18.4	4.5	6.6	1.9	25.7	18.4

Abbreviation: TON, total oxidized nitrogen (NO_2-N + NO_3-N).

zizanioides), as part of the treatment process. This sedge species, which has an infertile domestic variety, is incredibly resilient, with a dense root system that can reach 3 m in depth. It is particularly useful in the prevention of soil erosion and has been used for water conservation. It has high nutrient removal capacities and is ideal for tertiary treatment using bankside infiltration of the RBTS. It is invasive, although susceptible to glyphosate, and so requires careful management using the infertile variety only where it is not native. It has a wide variety of uses for its 2.5-m-long tough leaves — oil production, pharmacological uses, and in the food and cosmetic industries. It is widely used for basket making, thatch, rope and string manufacture, animal fodder, and insect repellent. This species is also very tolerant to contaminants and has been successfully used to phytotransform textile dyes into no-toxic metabolites when grown as a floating mat or in constructed narrow beds by the enhanced production dye degrading enzymes (i.e., lignin peroxidase, veratryl alcohol oxidase, laccase, and azo reductase) (Chandanshive *et al.*, 2020).

At an activated sludge biological nutrient treatment plant in South Africa, trials were carried out by Badejo *et al.* (2017) to see how constructed wetlands planted with *C. zizanioides* in a layer of granite chips overlaid with sludge from the plant, could be integrated into the system. Using four 1,000 L pilot units in parallel to treat the effluent from each of the anaerobic, anoxic and aerobic zones (Fig. 13.12), achieving removal efficiencies of 96.6, 96.9 and 97.2% (COD); 33.3, 85.7 and 92.5% (NO_2 and NO_3); 53.5, 46.5 and 88.8% (SO_4); and 98.3, 99.7 and 99.6% (TSS) from each zone, respectively (Table 13.7).

13.3.2 *Design of vertical flow reed bed treatment systems*

The vertical flow reed bed treatment systems (VF-RBTSs) were also developed by Seidal, and the process takes its original name from the institute where the research was done, the Max Planck Institute Process (MPIP). Vertical flow beds are made up of a number of graded layers of gravel media (Fig. 13.13), through which the wastewater percolates. The surface of the bed must be level and is normally topped with sharp sand. The bed is loaded intermittently with

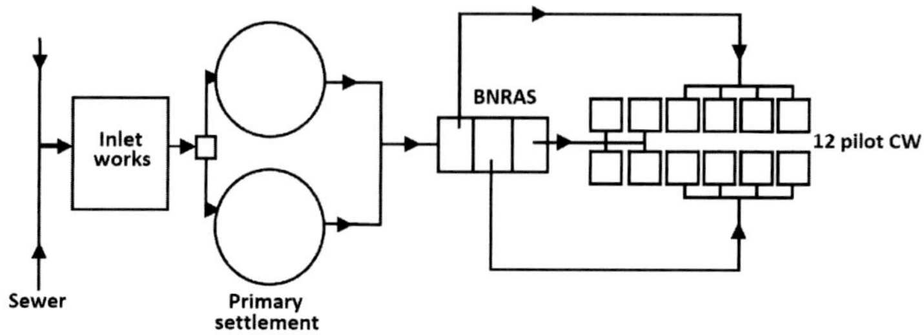

Fig. 13.12. Schematic layout of a BNR-activated sludge plant used for assessing the efficiency of *Chrysopogon zizanioides*-constructed wetlands to treat the effluent from the anaerobic, anoxic and aerobic zones (Badejo *et al.*, 2017). Reproduced with permission from Elsevier Publishing.

wastewater that floods the surface and slowly drains vertically through the media to a drainage system at the base. The bed is allowed to drain completely to allow the interstices to refill with air. The next dose of wastewater traps this air, ensuring good oxygen transfer into the microbial biofilm that develops on the roots and media (Gray, 2021). Although the reeds transfer some oxygen into the root zone via the rhizomes, this is not as important as in HF beds, with the majority of oxygen transfer in VF beds provided by the dosing system. Tertiary treatment HF systems produce well-nitrified effluents (Green, 1997), whereas secondary treatment HF systems do not, due to their limited oxygen transfer capability. In contrast, VF systems provide sufficient aerobic conditions to ensure good BOD oxidation and nitrification. They are considerably more compact (1–2 m^2 PE^{-1}) than HF systems (5–10 m^2 PE^{-1}) for secondary treatment. Vertical flow beds are normally planted with the common reed, although cattails and bulrushes are occasionally used. The precise area of the VF reed bed required for small systems (<100 PE) treating septic tank effluent can be calculated using Eq. (13.2):

$$A_1 = 3.6P^{0.35} + 0.6P, \tag{13.2}$$

where A_1 is the area of the first VF bed (m^2), and P is the PE. Treatment requires two VF beds to be used in series, with the area of the second bed (A_2) being 50% of A_1. This gives an A_1 design value

Table 13.7. Chemical removal efficiency of the *C. zizanioides*-constructed wetlands in treating the effluent from the anaerobic, anoxic and aerobic zones of an activated Biological Nutrient Removal plant over time (Badejo *et al.*, 2017). Reproduced with permission from Elsevier Publishing.

Parameters*	Before treatment	Period of treatment (days)				
Anaerobic compartment*		6	12	18	24	30
pH	7.56	6.99	7.05	6.49	7.13	7.30
COD (mg/L^{-1})	2046	236	101	93	85	68
Free & saline ammonia (mg/L^{-1})	25.76	5.57	4.81	4.67	4.17	3.84
TKN (mg/L^{-1})	97.90	19.40	17.22	8.45	8.30	3.51
TSS (mg/L^{-1})	675	45	22.20	21.20	20.80	11.20
Anoxic compartment						
pH	7.45	7.23	6.83	7.25	7.13	7.18
COD (mg/L^{-1})	2118	146	142	108	85	65
Free & saline ammonia (mg/L^{-1})	10.67	1.2 1	4.67	0.35	0.46	0.38
TKN (mg/L^{-1})	139.40	7.93	5.96	2.37	8.30	3.06
TSS (mg/L^{-1})	1718	36.40	22.20	12	8.80	4.80
Aerobic compartment						
pH	7.40	6.92	7.21	7.11	7.11	7.23
COD (mg/L^{-1})	2076	163	140	107	86	61
Free & saline ammonia (mg/L^{-1})	8.90	7.10	5.26	5.30	4.07	0.68
TKN (mg/L^{-1})	132	51.95	9.20	9.99	0.87	2.12
TSS (mg/L^{-1})	1998	35.40	27.20	25.20	21.30	9.60

*Effluent from the primary sedimentation tank mixed with return sludge from the secondary sedimentation tank.

ranging from 2 m^2 PE^{-1} at a PE of 4, falling to 0.78 m^2 PE^{-1} at a PE of 100.

A VF-RBTS was used to treat the domestic wastewater from an isolated farmhouse in Schillhuber, Austria. After settlement, the wastewater was pulsed four times a day onto a 40-m VF bed designed at 5 m^2 PE^{-1}. The bed was 6.5 m × 6.5 m in area and 0.8 m deep,

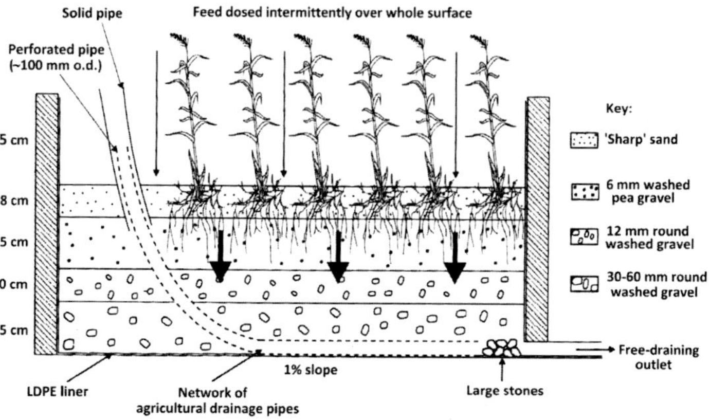

Fig. 13.13. Typical design of a vertical flow reed bed treatment system (Cooper *et al.*, 1996). Reproduced with permission from WRc Publications.

Table 13.8. Mean annual performance of a vertical flow reed bed treatment system treating the wastewater from a farmhouse in Wolfern Schillhuber, Upper Austria (Kadlec *et al.*, 2000). Reproduced with permission from IWA Publishing.

	Hydraulic load	BOD5			COD			NH_4-N			TP		
		In	Out	Rem.	In	Out	Rem.	In	Out	Rem.	In	Out	Rem.
	(mm d^{-1})	(mg L^{-1})		(%)	(mg L^{-1})		(%)	(mg L^{-1})		(%)	(mg L^{-1})		(%)
1992	24	143	11	92	378	54	86	48	8.6	82	10.8	3.0	72
1993	32	186	9	95	533	47	91	88	16.4	81	12.5	3.6	71
1994	27	139	3	98	366	36	90	71	1.3	98	11.5	6.1	47
1995	30	120	3	98	383	30	92	63	4.9	92	14.0	7.0	51
1996	37	157	<3	99	436	30	93	49	1.5	97	9.4	6.2	34
1997	30	278	<3	99.6	549	25	95	59	5.5	91	9.6	4.9	49
Average	30	171	5	97	441	37	91	63	6.4	90	11.3	5.1	54

made water-tight with a 2-mm polyethylene liner and filled with gravel topped with sharp sand. It was planted with *P. communis* and the wastewater was distributed over the surface via a fixed distribution network of PVC pipes with 8-mm holes and splash plates. Removal efficiencies were 97% for BOD, 91% for COD, 90% for NH_4, and 54% for TP (Table 13.8). Phosphorus removal decreased from 72 to <50%

over the five-year study due to the limited adsorption potential of the medium (Kadlec *et al.*, 2000). The progressive loss in phosphorus removal is commonly seen in all constructed wetlands. As the net plant storage capacity becomes stabilized and the sorption capacity of the media is saturated, retention of the phosphorus decreases. Continuing retention and accumulation within the bed becomes increasingly linked to the retention of organic and inorganic matter from the wastewater, or accumulation of organic matter produced by the macrophytes. This provides new sorption sites for phosphorus. Therefore, future removal is largely determined by the ability of the system to retain organic matter.

13.3.3 *Design of hybrid reed bed treatment systems*

Hybrid reed bed systems combine the use of VF and HF beds (Cooper and Maesseneer, 1996; Cooper, 1999). Vertical flow beds are primarily used to provide nitrification and are used in series to provide a high level of BOD and COD removal, followed by nitrification. A HF bed normally follows as a final step to remove suspended solids and pathogens. Although VF beds are not efficient at retaining fine solids, they are employed primarily for denitrification, and extra BOD and phosphorus removal. An example of a multistage system using both VF and HF reed beds is at Oaklands in Gloucestershire, United Kingdom (Burka and Lawrence, 1990). Serving a PE of 65, the flow rate was approximately 9.8 m^3d^{-1}. After settlement in a septic tank, the wastewater flows through two VF stages in series, followed by two HF stages and finally a pond (Fig. 13.14). Stage 1 consists of six parallel VF beds (each 8 m^2) planted with *Phragmites australis*, followed by Stage 2, comprising three parallel VF beds (each 5 m^2) planted with a mixture of *Phragmites*, *Iris* and *Schoenoplectus*. The beds are rotated with only one bed in operation in each stage; the others are rested. The flow then passes through a single HF bed of 8 m^2 (Stage 3) planted with *Iris*, and then another HF bed of 20 m^2 (Stage 4) planted with *Accorus*, *Carex*, *Schoenoplectus*, and *Sparganium*. The final stage is a pond. Gravel media is used on all the beds with the VF beds graded from the top as 8 cm of sand, 15 cm of 6 mm of washed gravel, 10 cm of 12 mm of round-washed gravel, and 15 cm

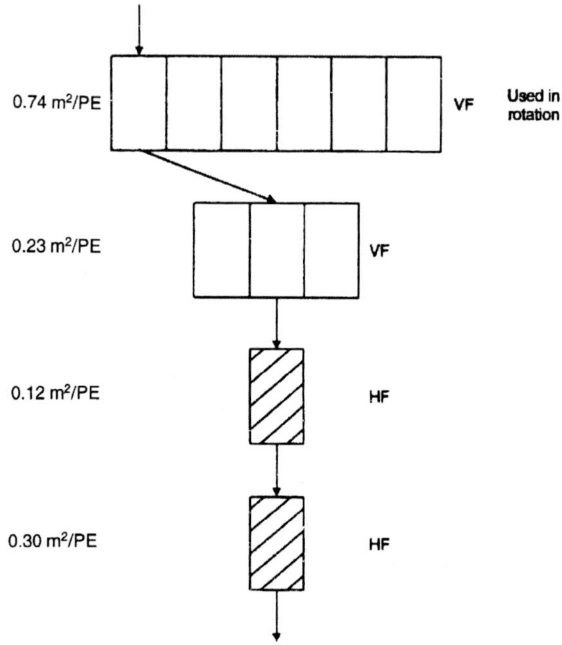

Fig. 13.14. Layout and design loadings of the reed bed treatment system used at Oaklands Park, United Kingdom (Burka and Lawrence, 1990). Reproduced with permission from Elsevier Publishing.

of 30 to 60 mm of round-washed gravel overlying drainage pipes. Details of the performance are given in Table 13.9. The VF stages achieve good BOD removal and partial nitrification, and the HF stages significantly denitrify the cation, reducing the total oxidized nitrogen from 22 to 7.2 mg L^{-1}.

The design and operation of RBTSs are reviewed by Wu *et al.* (2015) who suggested a number of areas where increased performance and sustainability can be achieved and where more research is required (Table 13.10). In design terms, careful selection of both macrophytes and substrate are required as they play critical roles in nutrient and contaminant uptake. Plant tolerance to contaminants,

Table 13.9. Mean performance of different stages of the reed bed wastewater treatment system at Oaklands Park, United Kingdom (Burka and Lawrence, 1990). Reproduced with permission from Elsevier Publishing.

	Concentration (mg L^{-1})					
		Effluents				
	Influent	Stage 1	Stage 2	Stage 3	Stage 4	Stage 5
BOD_5	285	57	14	15	7	11
TSS	169	53	17	11	9	21
NH_4-N	50.5	29.2	14.0	15.4	11.1	8.1
TON	1.7	10.2	22.5	10.0	7.2	2.3
PO_4-P	22.7	18.3	16.9	14.5	11.9	11.2

Table 13.10. Optimum design and operational criteria for the operation of constructed wetlands as outlined by Wu *et al.* (2015). Reproduced with permission from Elsevier Publishing.

	Design criteria	
Parameter	FWS CWs	SSF CWs
Bed size (m^2)	Larger if available	<2,500
Length to width ratio	3:1–5:1	<3:1
Water depth (m)	0.3–0.5	0.4–1.6
Hydraulic slope (%)	<0.5	0.5–1
Hydraulic loading rate (m/day)	<0.1	<0.5
Hydraulic retention time (day)	5–30	2–5
Media	Natural media and industrial byproduct preferred, porosity of 0.3–0.5, particle size <20 mm (50–200 mm for the inflow and outflow)	
Vegetation	Native species preferred, plant density 80% coverage	

extreme environmental conditions, nutrient requirement, and biomass production are important selection criteria for specific wastewaters and the location. The development of new substrates with high adsorptive capacities from industrial or agricultural wastes is also required. Proper maintenance of constructed wetlands ensures continuous optimal operation. To ensure nutrients and contaminants are not lost due to plant dieback, suitable harvest strategies are required, with the ultimate goal of the plant biomass as well as the final effluent being reused. The optimum performance of constructed wetlands is dependent on a wide range of environmental, hydraulic and operating conditions, which all need to be fully understood and controlled. Hydraulic and organic loading regimes, in particular, should be appropriate to the environmental conditions and season. Likewise, contaminant loading should be closely linked to the ability of the substrate and plants to remove them. New advances in operation are constantly being reported, including supplementary aeration, bioaugmentation, and step feeding, while new designs such as hybrid and baffled constructed wetlands are all improving performance for complete wastewater treatment and the treatment of stronger wastewaters.

13.4 Pathogen Removal

There is an overall reduction in viruses, bacteria, helminth eggs, and protozoan cysts, as screened or settled wastewater passes through wetland systems (Herskowitz 1986; Gersberg *et al.*, 1989; Rivera *et al.*, 1995). While little research has been done on pathogen removal in the different types of wetlands in use, they provide excellent physical, chemical and biological conditions for their removal. These include exposure to sunlight/UV, sedimentation, mechanical filtration, chemical oxidation, adsorption to organic matter and the substrate, exposure to plant secreted biocides, predation by micro and macrofauna, attack by lytic bacteria, natural antibiotics, and normal die-off due to prolonged retention (Gersberg *et al.*, 1989). Macrophytes are known to produce biocides that have a negative effect on some pathogens (Alufasi *et al.*, 2017), with this effect more pronounced in SSF systems compared to FWS wetlands, although the precise mechanisms are unknown. However, the removal of pathogens appears unpredictable as it is influenced by both environmental and operational factors. This includes the type of macrophytes used in constructed wetlands, their life cycles, and their habits. For example, shading can significantly reduce the inactivation effect of UV exposure from sunlight. Wetlands offering good filtration (e.g. SSF and hybrid systems) are more effective in removing bacterial pathogens and especially helminth eggs and protozoan cysts, although adsorption onto the substrate is also an important removal mechanism. Mandi *et al.* (1996) compared three HF reed beds of differing lengths (30, 40 and 50 m) planted with *P. communis.* They found excellent removal rates of helminth eggs ranging from 88% in the 39-m bed to 93% in the 50-m bed at short retention times of 1 to 4 hours.

The larger the wetland is in area, especially if they incorporate FWS areas, the greater the risk of pathogen contamination from visiting, resident or adjacent warm-blooded wildlife, especially birds. This can result in an increase in the pathogen load in the final effluent after treatment. The most common pathogens include fecal coliforms, streptococci and *Salmonella*, and zoonotic parasites such as microsporidia coccidia (*Encephalitozoon bieneusi, E. hellem*), protozoan oocysts, and cysts (*Cryptosporidium meleagridisa, C. muris, C. parvum, Giardia duodenalis*), nematode eggs *(Ascaris lumbricoides, A. suum, Toxocara canis*) and cestode eggs (*Taenia saginata*) (Shingare *et al.*, 2019). Wildfowl, in particular, are known vectors of several important human viruses, while surface runoff is another source of contamination. Natural wetlands often have significant background concentrations of fecal coliforms. For example, the wetland at Bryon Bay, Australia, has background fecal coliforms of up to 645 per 100 ml (Kadlec *et al.*, 2000).

The variability in design and construction of constructed wetlands makes the interpretation of existing studies difficult (Kadlec and Knight, 1996). However, while removal of fecal coliforms follows first-order reaction kinetics in gravel beds, longer retention times are required in those planted with emergent vegetation (i.e., reeds, rushes) than unvegetated beds to achieve the same degree of removal (Bavor *et al.,* 1989). Similar results are reported by Qunonez-Diaz *et al.* (2001). Kadlec and Knight (1996) have reported similar removal rate constants (k) for fecal coliforms in the US, as found elsewhere ranging from 50 to 300 m yr^{-1} with an average k value of around 70 m yr^{-1}. Shingare *et al.* (2019) have produced an extensive review on pathogen removal efficiencies in constructed wetlands, which they summarized in a range of tables detailing case studies for different designs. The authors warn about using fecal coliforms as a generic indicator for the removal of other pathogen groups in constructed wetlands, and emphasize the need for effluents when reused from wetlands, to be effectively monitored, especially for the *Cryptosporidium* spp., *Giardia* sp., helminths and human-virulent microsporidia.

References

Alufasi, R., Gere, J., Chakauya, E., Lebea, P., Parawira, W. and Chingwaru, W. (2017) Mechanisms of pathogen removal by macrophytes in constructed wetlands. *Environmental Technology Reviews*, **6**, 135–144.

Armstrong, W. (1964) Oxygen diffusion from the roots of some British bog plants. *Nature (London)*, **204**, 801–802.

Badejo, A.A., O. Omole, D.O., Ndambuki, J.M. and Kupolati, W.K. (2017) Municipal wastewater treatment using sequential activated sludge reactor and vegetated submerged bed constructed wetland planted with *Vetiveria zizanioides. Ecological Engineering*, **90**, 525–529. https://doi.org/10.1016/j.ecoleng.2016.11.012

Bavor, H.J., Roser, D.J., Fisher, P.J. and Smalls, I.C. (1989) Performance of solid matrix wetland systems, viewed as fixed film reactors. In *Constructed Wetlands for Wastewater Treatment: Municipal, Industrial and Agricultural* (ed. D.A. Hammer). Lewis, Chelsea, MI, pp. 646–656.

Börner, T., von Felde, K., Geschlössl, E., Geschlössl, T., Kunst, S. and Wissing, F.W. (1998) Germany. In *Constructed Wetlands for Wastewater Treatment in Europe* (eds. J. Vymazal, H. Brix, P.F. Cooper, M.B. Green and R. Haberl). Backhuys Publishers, Leiden, pp. 169–190.

Bowmer, K.H. (1985) Detoxification of effluents in a macrophyte treatment system. *Water Research*, **19**, 57–62.

Brix, H. (1994) Use of constructed wetlands in water pollution control: Historical development, present status and future perspectives. *Water Science and Technology*, **30**(8), 209–223.

Burka, U. and Lawrence, P. (1990) A new community approach to wastewater treatment with higher water plants. In *Constructed Wetlands in Water Pollution Control* (eds. P.F. Cooper and B.C. Findlater). Pergamon Press, Oxford, pp. 359–371.

Chandanshive, V., Kadam, S., Rane, N., Jeon, B.-H., Jadhav, J. and Govindwar, S. (2020) In situ textile wastewater treatment in high-rate transpiration system furrows planted with aquatic macrophytes and floating phytobeds. *Chemosphere*, **252**, 126513.

Chick, A.J., Finlayson, C.M. and Swarbrick, J.T. (1983) Typhaceae — The cumbungis or bulrushes. *Australian Weeds*, **23**(2), 24–27.

Cooke, J.G. (1992) Phosphorus removal processes in a wetland after a decade of receiving a sewage effluent. *Journal of Environmental Quality*, **21**, 733–739.

Coombes, C. (1990) Reed bed treatment systems in Anglian Water. In *Constructed Wetlands in Water Pollution Control* (eds. P.F. Cooper and B.C. Findlater). Pergamon Press, Oxford, pp. 223–234.

Cooper, P.F. (1999) A review of the design and performance of vertical-flow and hybrid reed bed treatment systems. *Water Science and Technology*, **40**(3), 1–9.

Cooper, P.F. and De Maeseneer, A. (1996) Hybrid systems — What is the best way to arrange the vertical and horizontal flow stages. *IAWQ Specialist Group on the Use of Macrophytes in Water Pollution Control Newsletter*, **15**, 8–13.

Cooper, P.F., Job, G.D., Green, M.B. and Shutes, R.B.E. (1996) Reed beds and constructed wetlands for wastewater treatment. WRc Publications, Marlow, UK.

Davies, T.H. and Cottingham, P.D. (1993) Phosphorus removal from wastewater in a constructed wetland. In *Constructed Wetlands for Water Quality Improvement* (ed. G.A. Moshiri). Lewis Publishers, Bocta Raton, FL, pp. 315–320.

De Jong, J. (1976) The purification of wastewater with the aid of rush or reed ponds. In *Biological Control of Water Pollution* (eds. J. Tourbier, R.W. Pierson and E.W. Furia). University of Pennsylvania Press, pp. 133–139.

Decamp, O. and Warren, A. (1998) Bacterivory in ciliates isolated from constructed wetlands (reed beds) used for wastewater treatment. *Water Research*, **32**, 1989–1996.

Decamp, O., Warren, A. and Sanchez, R. (1999) The role of ciliated protozoa in subsurface flow wetlands and their potential as bioindicators. *Water Science and Technology*, **40**(3), 91–98.

Dinges, R. (1982) *Natural Systems for Water Pollution Control*. Van Nostrand Reinhold, New York.

Drizo, A., Frost, C.A., Smith, K.A. and Grace, J. (1997) Phosphate and ammonium removal by constructed wetlands with horizontal subsurface flow, using shale as a substrate. *Water Science and Technology,* **35**(5), 95–102.

Finlayson, C.M. and Chick, A.J. (1983) Testing the potential of aquatic plants to treat abbatoir effluent. *Water Research,* **17**, 415–422.

Finlayson, C.M. and Mitchell, D.S. (1982) Treatment of rural wastewaters in Australia with aquatic plants: A summary. *Der Tropenlandwirt,* **83**, 153–165.

Gersberg, R.M., Gearhart, R.A. and Ives, M. (1989) Pathogen removal in constructed wetlands. In *Constructed Wetlands for Wastewater Treatment: Municipal, Industrial, and Agricultural* (ed. D.A. Hammer). Lewis, Chelsea, MI, pp. 431–446.

Gersberg, R.M., Elkins, B.V., Lyon, S.R. and Goldmann, C.R. (1986) Role of aquatic plants in wastewater treatment by artificial wetlands. *Water Research,* **20**, 363–368.

Gray, N.F. (2021) *Fixed-Film Reactors in Wastewater Treatment.* World Scientific Publishing, Singapore.

Green, M.B. (1997) Experience with establishment and operation of reed bed treatment for small communities in the UK. *Wetlands Ecology and Management,* **4**, 147–158.

Green, M.B. and Upton, J. (1995) Constructed reed bed: Appropriate technology for small communities. *Water Science and Technology,* **32**(3), 339–348.

Gschlößl, T. and Stuible, H. (2000) Reed bed systems: Design, performance and maintainablility. *Water Science and Technology,* **41**(1), 73–76.

Herskowitz, J. (1986) Listowel artificial marsh project report. Water Resources Branch, Ontario Ministry of the Environment, Toronto, Ontario.

Jefferies, M. and Mills, D.H. (1993) *Freshwater Ecology: Principles and Applications.* Belhaven Press, London.

Jiang, Z. and Zhu, X. (1998) Treatment and utilization of wastewater in the Beijing Zoo by an aquatic macrophyte system. *Ecological Engineering,* **11**, 101–110.

Kadlec, R.H. and Knight, R.L. (1996) *Treatment Wetlands.* CRC Press, Boca Raton, FL.

Kadlec, R.H., Knight, R.L., Vymazal, J., Brix, H., Cooper, P. and Haberl, R. (2000) Constructed wetlands for pollution control: Processes, performance, design and operation. Scientific and Technical Report No. 8. International Water Association, IWA Publishing, London.

Keddy, P.A. (2000) *Wetland Ecology.* Cambridge University Press, Cambridge.

Khatiwada, N.R. and Polprasert, C. (1999) Assessment of effective specific surface area for free water surface constructed wetlands. *Water Science and Technology,* **40**(3), 83–90.

Liang, Y., Zhu, H., Bañuelos, G., Yan, B., Zhou, Q., Yu, X. and Cheng, X. (2017) Constructed wetlands for saline wastewater treatment: A review. *Ecological Engineering*, **98**, 275–285. http://dx.doi.org/10.1016/j.ecoleng.2016.11.005

Liu, B., Liu, Z., Wang, L. and Wang, Z. (2014) Responses of rhizomatous grass *Phragmites communis* to wind erosion: Effects on biomass allocation. *Plant Soil,* **380,** 389–398. https://doi.org/10.1007/s11104-014-2104-y

Mandi, L., Houhoum, B., Asmama, S. and Schwartzbrod, J. (1996) Wastewater treatment by reed beds an experimental approach. *Water Research*, **30**, 2009–2016.

Mann, R.A. and Bavor, H.J. (1993) Phosphorus removal in constructed wetlands using gravel and industrial waste substrata. *Water Science and Technology,* **27**(1), 107–113.

Mendelssohn, I.A. and Postek, M.T. (1982) Elemental analysis of the deposits on the roots of *Spartina alterniflora* Loisel. *American Journal of Botany*, **69**, 904–912.

Mitsch, W.J. (1992) Landscape design and the role of created, restored, and natural riparian wetlands in controlling nonpoint source pollution. *Ecological Engineering*, **1**, 27–47.

Mohan, S.V., Mohanakrishna, G., Chiranjeevi, P., Peri, D. and Sarma, P.N. (2010) Ecologically engineered system (EES) designed to integrate floating, emergent and submerged macrophytes for the treatment of domestic sewage and acid rich fermented-distillery wastewater: Evaluation of long term performance. *Bioresource Technology*, **101**, 3363–3370.

Moshiri, G.A. (1993) *Constructed Wetlands for Water Quality Improvement.* Lewis Publishers, Boca Raton, FL.

Nyakang'o, J.B. and van Bruggen, J.J.A. (1999) Combination of a well functioning constructed wetland with a pleasing landscape design in Nairobi, Kenya. *Water Science and Technology,* **40**(3), 249–256.

Packer, J.G., Meyerson, L.A., Skálová, H., Pyšek, P. and Kueffer, C. (2017) Biological flora of the British Isles: *Phragmites australis. Journal of Ecology,* **105**(4), 1123–1162.

Pomoell, B. (1976) The use of bulrushes for livestock feed. In *Biological Control Of Water Pollution* (eds. J. Tourbier and R.W. Pierson). University of Pennsylvania Press, pp. 187–189.

Quinonez-Diaz, M.D., Karpiscak, M.M., Ellman, E.D. and Gerba, C.P. (2001) Removal of pathogenic and indicator micro-organisms by a constructed wetland receiving untreated domestic wastewater. *Journal of Environmental Science and Health: Part A. Toxic/Hazardous Substances and Environmental Engineering*, **36**, 1311–1320.

Reed, S., Bastian, R., Black, S. and Khettry, R. (1984) Wetlands for wastewater treatment in cold climates. *Proceedings of the Water Reuse Symposium, III*. American Water Works Association Research Foundation, Denver, CO.

Reed, S.C., Middlebrooks, E.J. and Crites, R.W. (1995) *Natural Systems for Waste Management and Treatment, 2nd ed.* McGraw-Hill, New York.

Rivera, F., Warren, A., Ramirez, E., Decamp, O., Bonilla, P., Gallegos, E., Calderon, A. and Sanchez, J.T. (1995) Removal of pathogens from wastewaters by the root zone method (RZM). *Water Science and Technology*, **32**(3), 211–218.

Schierup, H.H., Brix, H. and Lorenzen, B. (1990) Wastewater treatment in constructed reed beds in Denmark — State of the art. In *Proceedings of the International Conference on the Use of Constructed Wetlands in Water Pollution Control* (eds. P.F. Cooper and B.C. Findlater). Pergamon Press, Oxford, pp. 495–504.

Seidel, K. (1976) Macrophytes and water purification. In *Biological Control of Water Pollution* (eds. J. Tourbier, R.W. Pierson and E.W. Furia) University of Pennsylvania Press, pp. 109–121.

Shingare, R.P., Thawale, P.A., Raghunathan, K., Mishra, A. and Kumar, S. (2019) Constructed wetlands for wastewater reuse: Role and efficiency in removal of enteric pathogens. *Journal of Environmental Management*, **246**, 444–461.

Spangler, F., Sloey, W. and Fetter, C.W. (1976) Experimental use of emergent vegetation for the biological treatment of municipal wastewater in Wisconsin. In *Biological Control of Water Pollution* (eds. J. Tourbier, R.W. Pierson and E.W. Furia). University of Pennsylvania Press, pp. 161–171.

Tang, Y., Harpenslager, S.F., van Kempen, M.M.L., Verbaarschot, E.J.H., Loeffen, L.M., Roelofs, J.G.M., Smolders, A.J.P. and Lamers, L.P.M. (2017) Aquatic macrophytes can be used for wastewater polishing but not for purification in constructed wetlands. *Biogeosciences*, **14**, 755–766. https://doi.org/10.5194/bg-14-755-2017

van Oostrom, A.J. (1995) Nitrogen removal in constructed wetlands treating nitrified meat processing effluent. *Water Science and Technology*, **32** (3), 137–148.

Vymazal, J. (2014) Constructed wetlands for treatment of industrial wastewaters: A review. *Ecological Engineering*, **73**, 724–751. http://dx.doi.org/10.1016/j.ecoleng.2014.09.034

Vymazal, J., Brix, H., Cooper, P.F., Green, M.B. and Harberl, R. (eds.) (1998) *Constructed Wetlands for Wastewater Treatment in Europe*. Backhuys Publishers, Leiden.

Wile, I., Miller, G. and Black, S. (1982) Design and use of artificial wetlands. In *Proceedings of the Workshop on Ecological Considerations in Wetlands Treatment of Municipal Wastewaters*. University of Massachusetts, Amherst.

Wu, H., Zhang, J., Ngo, H.H., Guo, W., Hu, G., Liang, S., Fan, J. and Lin, H. (2015) A review on the sustainability of constructed wetlands for wastewater treatment: Design and operation. *Bioresource Technology*, **175**, 594–601.

Zhu, T., Jenssen, P.D., Maelum, T. and Krogstag, T. (1997) Phosphorus sorption and chemical characteristics of lightweight aggregates (LWA) — Potential filter media in treatment wetlands. *Water Science and Technology*, **35**(5), 103–108.

Part IV
Waste Stabilization Ponds

14

Introduction to Waste Stabilization Ponds and Lagoons

14.1 Development of Waste Stabilization Ponds (WSPs)

Ponds have been widely used as a method of sewage disposal since ancient times. It is possible that castle moats, which received excrement directly from garderobes built into the walls, as well as night soil thrown from the battlements, although primarily defensive devices were also effective oxidation ponds (Porges and Mackenthun, 1963). Because such ponds were highly productive, they were often also utilized for fish culture. The development of waste stabilization ponds (WSPs) as a secondary treatment process has been largely accidental, with ponds initially constructed as simple sedimentation basins or as emergency holding tanks at treatment plants. It is only relatively recently that the design and operational criteria needed to successfully operate ponds have been established (USEPA, 1983). They are now accepted as a major treatment process and are used throughout the world, serving populations ranging from <1,000 to >100,000 (Gloyna, 1971; WHO, 1987; USEPA, 1983; Mara and Pearson, 1998; Pearson *et al.*, 2000; USEPA, 2011). The development of WSPs has been reviewed by Ho and Goethals (2020) (Table 14.1).

Ponds are a popular alternative to other biological treatment systems in countries where there is plenty of sunshine, and where land is both cheap and readily available. They are particularly favored in

Table 14.1. The development of waste stabilization ponds over the past 120 years (Ho and Goethals, 2020). Reproduced with permission from Elsevier Publishing.

• Accidentally dicovered the purification ability of pond systems • Combined with water storage or growing crops	• First favorable experiences promoted the presence of ponds in US, Europe, and South Africa • Removal mechanisms of pond technology was unexplored • The presence of algae was not appreciated, being referred as 'unattractive green coloration of the effluent	• Rapid increase in presence of WSPs in treating both municipal and industrial wastewaters • W.J. Oswald pioneered the use of algae in wastewater treatment and the use of high rate algal ponds • Developing model-based guidelines for pond design and operation	• International and national recognition of importance of pond technology in wastewater treatment • Establishment of the IWA specialist group on WSPs • Advance the understanding on biochemical removal processes • Start applying the chemical reactor concept to represent the mixing in ponds	• Introduction of advanced integrated wastewater pond systems (AIWPS) and aerated ponds • First CFD models of Wood *et al.* (1995) and integrated models of Sah *et al.* (2011) • Advance the understanding on pathogen and nutrient removal mechanisms • Increasing the research on new issues such as energy, gas emissions, emerging pollutants
1901	1920s	1950s	1970s	1990s – 2020

Australia and New Zealand (Parker *et al.*, 1950; 1959; Park and Craggs, 2007), Africa (Stander and Meiring, 1965; Marais, 1970; Mara and Pearson, 1998), India (Arceivala *et al.*, 1970), Israel (Watson, 1962), the United States (US) (Oswald, 1963; Porges and Mackenthun, 1963), and Canada (Townshend and Knoll, 1987). Since the 1980s, they have become widely employed in Europe where they are used primarily for treating domestic wastewater from small communities (<2,000 PE [population equivalent]), whereas larger-sized plants are used for tourist areas where there is a large increase in population during the summer (Vuillot and Boutin, 1987; Gomes de Sousa, 1987). In France, there are currently >2,500 systems, with 1,925 systems serving populations <1,000 (Racault *et al.*, 1995). The largest facility in France covering an area of 40 ha is at Rochefort sur Mer, which serves a summer population of 50,000. In the south of France, it is not unusual for WSP systems to serve populations 2 to 20 times greater in July and August than those served in the winter (Drakides and Calignon, 1983). Over 20 WSP systems have been constructed to serve tourist areas in the Algarve region of Portugal (Rodrigues, 1997), while there are currently 3,000 WSPs in Germany used for domestic wastewater treatment (Ho *et al.*, 2017). They are especially popular in rural areas of the US, with 90% of the communities served having a PE of less than 10,000. In Canada, there are over 1,000 stabilization ponds in operation, representing approximately half of the wastewater treatment systems in the country. Dependence on ponds increases to over 70% in Alberta, Manitoba, Saskatchewan, the Northwest Territories, and the Yukon Territory (Townshend and Knoll, 1987). WSPs are used as far north as the arctic, being widely used for treating the wastewater from remote villages. The climatic conditions are very extreme with short cool summers with intense sunshine and long cold winters. Ice cover can occur from mid-September to late June due in part to permanent permafrost. These lagoon systems tend to store solids in the winter with a subsequent breakdown in the summer where Biological Oxygen Demand (BOD) removals of 80% can be achieved, although suspended solids (SS) can be higher than normal. Specific design criteria are required for such conditions (Smith, 1986; Heinke *et al.*, 1991), usually with a 365-day retention time and effluent discharge limited to the summer months only. Facultative pond designs are generally used but with depths of up to 6 m and organic loading rates of 11 to 22 kg BOD $ha^{-1}d^{-1}$. Biological activity may be limited to just 60 days per year corresponding with the effluent discharge period (Ragush *et al.*, 2015).

The terminology and classification of ponds used in the literature are somewhat confusing. Hawkes (1983) classifies a waste stabilization pond as: "any natural, or more commonly, artificial lentic (i.e., standing) body of water in which organic wastes (either crude sewage, settled sewage, organic and oxidizable industrial effluents, or oxidized sewage effluents) are treated by natural biological, biochemical and physical processes commonly referred to as 'self-purification' or 'stabilization'". Waste stabilization ponds are most conveniently classified by the type of biological activity that occurs, for example, whether the breakdown is primarily anaerobic or aerobic, and the importance of algae in the process (Fig. 14.1). Except for anaerobic lagoons, they are similar to the activated sludge process but differ in that ponds have: (i) a much longer retention period, (ii) lower loading rates, (iii) a less active microbial biomass, (iv) may incorporate algae in their operation, (v) require less aeration due to a lower oxygen demand

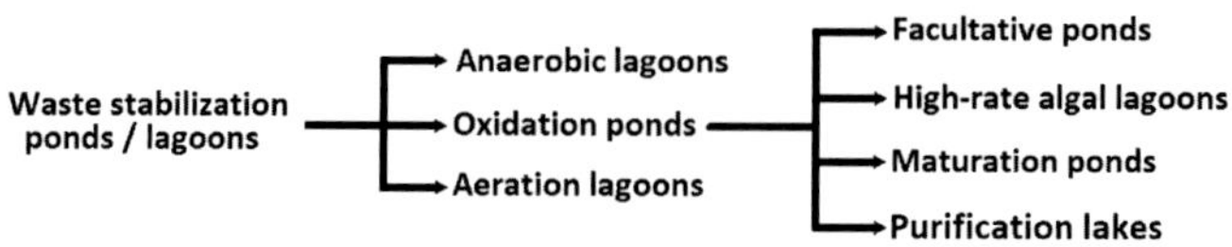

Fig. 14.1. The major types of waste stabilization ponds.

from the biomass, and (vi) lower aeration results in less mixing and agitation with the result that particulate solids settle and form a sludge layer in which anaerobic breakdown occurs.

Where appropriate, modern WSPs have significant advantages over conventional systems. They are simple to construct, operate and maintain and are extremely robust. Due to long hydraulic retention times, WSPs are able to withstand both organic and hydraulic shock loadings better than any other secondary wastewater treatment process. Their simplicity is reflected by low capital costs, and with no aeration or mixing required they have very low energy requirements resulting in low operational costs. Land requirements for WSPs are, however, significantly higher than more conventional systems. Well-designed aerobic WSPs can produce high-quality effluents with particularly good removal of all pathogens, including viruses, bacteria, protozoan cysts, and helminth eggs (Feachem *et al.,* 1983). Waste stabilization ponds are rarely used on their own, rather in a series of anaerobic, facultative and maturation ponds (Fig. 14.2). Configurations vary widely, although there are predominant WSP system designs employed by some European Countries (Gratziou and Chalatsi, 2013) (Table 14.2). Anaerobic ponds are used primarily for BOD and solids removal and are so efficient that they reduce the area of other ponds required (Pfeffer *et al.,* 1970) (Chap. 15). Facultative ponds also remove BOD as well as nutrients (Chap. 16), although aeration lagoons, which are hybrid activated sludge systems (Chap. 18), are also used as a secondary treatment step (Sec. 14.2.2). Maturation ponds are employed mainly for pathogen removal, although they have a polishing effect on the treated wastewater removing residual BOD, nutrients and SS (Chap. 17).

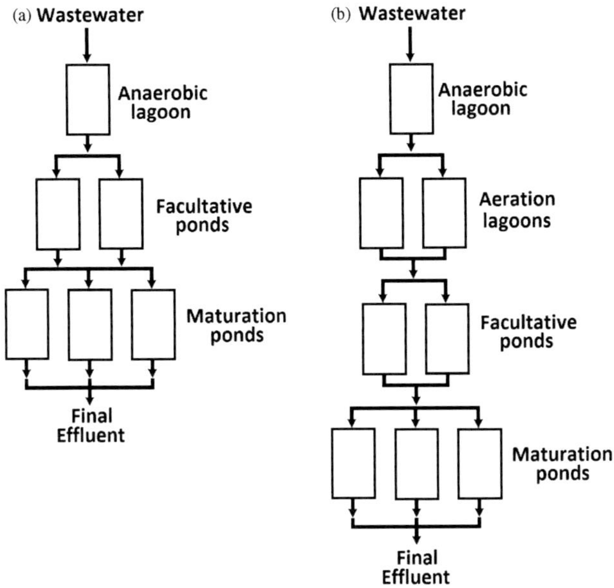

Fig. 14.2. Typical design of waste stabilization pond systems: (a) high sustainability design widely used for domestic wastewater, and (b) incorporating aeration lagoons that required energy for aeration, used primarily for very strong organic wastewaters.

Table 14.2. Predominant waste stabilization pond system design in some European countries (Gratziou and Chalatsi, 2013). Reproduced with permission from the Taylor & Francis Group.

Region	Succession of ponds in series
France	Anaerobic-facultative-2 maturation (French system)
Portugal	Anaerobic-facultative-2 maturation (French system)
Spain	Anaerobic or facultative-2 maturation
Italy	Facultative-2 maturation or French system
Greece	Older system: facultative-2 maturation; New system: French system
Cyprus	Aerated ponds

14.2 Case Studies

14.2.1 *Medium to large scale WSP treatment system*

With an annual flow rate of 5.5×10^6 m^3, the Haspolat Wastewater Treatment Plant serves the northern and central areas of the capital of Cyprus, Nicosia, using waste stabilization ponds. The plant has a design hydraulic and BOD loading of 20,000 m^3 d^{-1} and 11,300 kg d^{-1}, respectively. The wastewater treatment plant (WWTP) comprised aerated, facultative and maturation ponds in series, but was upgraded in 2001 to include a pretreatment step comprising five anaerobic lagoons. The general design of the WWTP is shown in Fig. 14.3. Its dimensions are shown in Table 14.3. The average BOD of the influent was 572 mg L^{-1} (sd [standardard deviation] 108) and total suspended solids (TSS) 576 mg L^{-1} (sd 174) measured over a four-year-period immediately prior to the installation of the anaerobic lagoons with the final effluent 82.0 mg L^{-1} (sd 37.5) and 133 mg L^{-1} (sd 63.5) being achieved. Over a similar four-year sampling period after the anaerobic lagoons were incorporated into the design, the mean BOD of the influent was similar at 573 mg L^{-1} (sd 158) and the mean TSS 491 mg L^{-1} (sd 193) while the mean final effluent was reduced to 32.9 mg L^{-1} (sd 15.2) and 75.5 mg L^{-1} (sd 60.1),

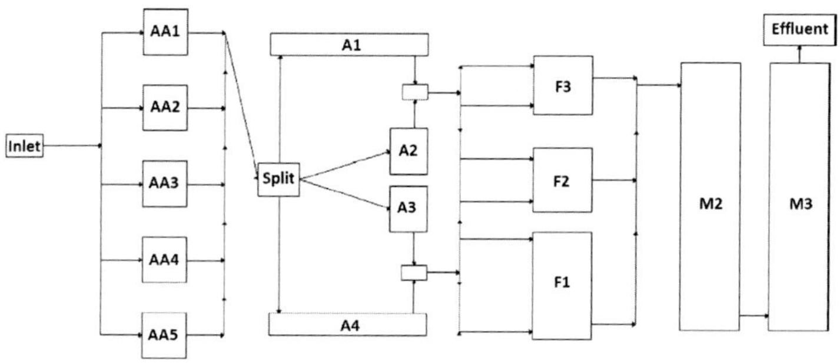

Fig. 14.3. Schematic layout of the Haspolat Wastewater Treatment Plant in Cyprus comprising anaerobic lagoons (AA), aerated lagoons (A), facultative ponds (F), and maturation ponds (M) in series (Türker *et al.*, 2009). Reproduced with permission from Elsevier Publishing.

Table 14.3. Lagoon dimensions at the Haspolat Wastewater Treatment Plant (Türker *et al.*, 2009). Reproduced with permission from Elsevier Publishing.

	Anaerobic	Aerated	Facultative	Maturation
Number of lagoons	5	4	3	4
Total surface area (m^2)	23,805	16,000	390,348	244,670
Total volume (m^3)	120,900	40,000	510,390	293,604
Depth of lagoons (m)	5	2.5	1.3	1.2

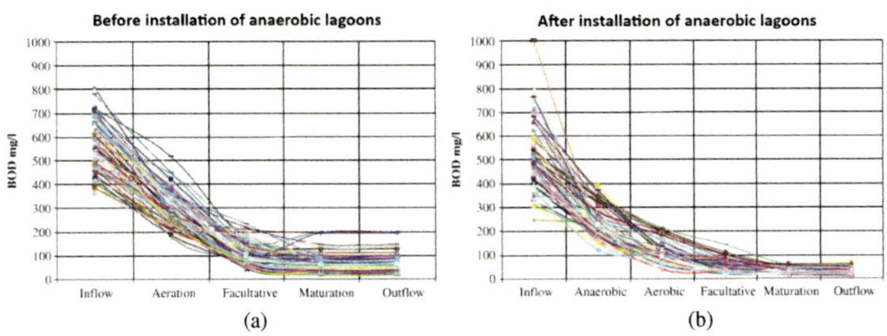

Fig. 14.4. Comparison of BOD removal from individual samplings as wastewater passes through the Haspolat Wastewater Treatment Plant before and after the installation of anaerobic lagoons (Türker *et al.*, 2009). Reproduced with permission from Elsevier Publishing.

respectively. The anaerobic lagoons not only improved the BOD and TSS removal by 60 and 42%, respectively (Fig. 14.4), they also significantly reduced the seasonal variations in the performance seen in the plant before their installation (Türker *et al.*, 2009).

Facultative ponds are often operated in series, creating a substrate gradient between ponds for the secondary treatment phase. An example of this is seen in the treatment of municipal wastewater from the City of Arak in Iran, which is treated by two WSP systems in parallel named M1 and M2, commissioned in 1993 and 2006, respectively. M1 serves a PE of 25,000 and M2 a PE of 80,000, with the latter incorporating tertiary facultative ponds (Fig. 14.5 and Table 14.4). The performance data shows that solids are lost from the anaerobic ponds and that BOD values are high due to the presence of algae

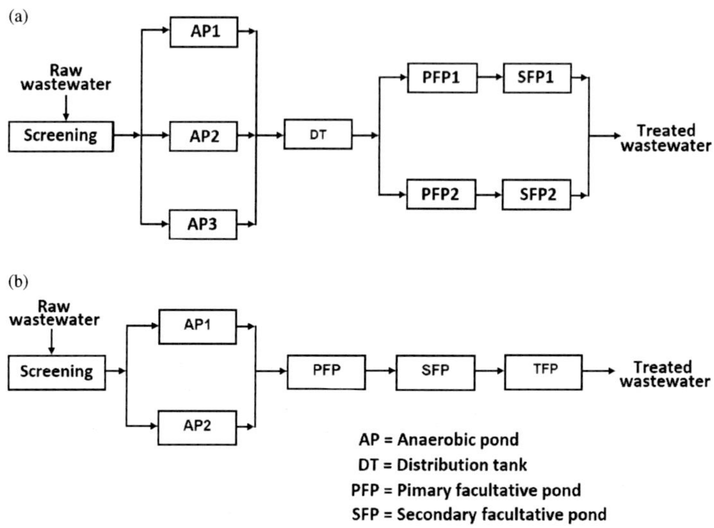

Fig. 14.5. Schematic layout of two waste stabilization pond systems (M1 and M2) operated in parallel at the Wastewater Treatment Plant, Arak, Iran (Naddafi *et al.*, 2009).

(Naddefi *et al.*, 2009) (Table 14.5). The authors suggest that tertiary facultative ponds are not advantageous and consideration should be given to increasing the capacity of the primary facultative ponds.

One of the largest WSP systems in the world is the Western Treatment Plant situated at Werribee in the State of Victoria, Australia. The plant treats approximately 50% of Melbourne's wastewater, which is equivalent to 485 mL d^{-1} using a series of anaerobic and aerated lagoons covering 10,500 ha, with a 30 to 35-day retention time (Hodgson and Paspaliaris, 1997). The anaerobic lagoons are covered in order to collect the biogas that makes the plant carbon neutral in terms of energy, but also for odor control. Aerated lagoons are used to achieve full nitrification followed by settlement. As water is at a premium in Australia, the final effluent can be reused with 40 billion liters of recycled water produced annually. The effluent can be used directly for on-site agricultural irrigation (23% of the total

Table 14.4. Physical characteristics of lagoons and ponds used at the two parallel waste stabilization pond treatment systems (M1 and M2) at Arak, Iran (Naddafi *et al.*, 2009).

M1 treatment stream

Component	**Area (m^2)**	**Depth (m)**	**Volume (m^3)**	**Hydraulic retention time (day)**
Anaerobic pond 1 (AP1)	1,768	9.25	5,200	2.9
Anaerobic pond 2 (AP2)	1,768	4.25	5,200	2.9
Anaerobic pond 3 (AP3)	1,768	4.25	5,200	2.9
Total anaerobic	5,304	—	15,600	2.9
Primary facultative pond 1 (PFP1)	20,800	3.5	33,280	11.3
Primary facultative pond 2 (PFP2)	20,800	3.5	33,280	11.3
Secondary facultative pond 1 (SFP1)	20,800	3.5	33,280	11.3
Secondary facultative pond 2 (SFP2)	20,800	3.5	33,280	11.3
Total facultative	83,200	—	133,120	22.6
Total	**88,504**	—	**148,720**	**25.5**

Daily treated flow: 5,500 m^3

M2 treatment stream

Component	**Area (m^2)**	**Depth (m)**	**Volume (m^3)**	**Hydraulic retention time (day)**
Anaerobic pond 1 (AP1)	10,000	4.1	32,000	2.73
Anaerobic pond 2 (AP2)	10,000	4.1	32,000	2.73
Total anaerobic	20,000	—	64,000	2.73
Primary facultative pond (PFP)	147,000	2.3	215,000	12.2
Secondary facultative pond (SFP)	77,600	2.3	113,000	6.42
Tertiary facultative pond (TFP)	7,800	2.3	113,500	6.5
Total facultative	302,600	—	441,500	25.12
Total	**322,600**	—	**505,500**	**27.85**

Daily treated flow: 17,600 m^3

flow), usually for pasture grazing, or be disinfected using a mixture of UV and chlorine to bring it up to class A quality where it is used for irrigating parks and gardens (18%). If necessary, excess salt can be

Table 14.5. Average BOD, COD and SS removal after different pond/lagoons at the two parallel waste stabilization pond treatment systems (M1 and M2) at Arak, Iran (Naddafi *et al.*, 2009).

M1 treatment stream (n = 18)				
Pond type		**BOD_5 (mg L^{-1})**	**COD (mg L^{-1})**	**SS (mg L^{-1})**
AP1	Influent	242.2	524.9	127.8
	Effluent	145.7	279.5	140.4
	% Removal	40	47	−10
AP2	Influent	242.2	524.9	127.8
	Effluent	158	249	142
	% Removal	32	43	−11
AP3	Influent	242.2	524.9	127.8
	Effluent	171.2	318.3	144.2
	% Removal	29	39	−13
PFP1	Influent	158	299	142
	Effluent	90.6	166.3	111
	% Removal	43	44	22
PFP2	Influent	158	299	142
	Effluent	92.5	172.4	116.7
	% Removal	42	42	18
SFP1	Influent	90.6	166.3	111
	Effluent	73.3	148.2	117.7
	% Removal	19	11	−6
SFP2	Influent	92.5	172.4	116.7
	Effluent	65.7	153.4	128.2
	% Removal	29	11	−10
Total	**% Removal**	71	71	4

M2 treatment stream (n = 18)				
Pond type		**BOD_5 (mg L^{-1})**	**COD (mg L^{-1})**	**SS (mg L^{-1})**
AP1	Influent	242	525	128
	Effluent	133	287	188
	% Removal	45	45	−46

Table 14.5. (*Continued*)

M2 treatment stream (n = 18)				
Pond type		**BOD_5 (mg L^{-1})**	**COD (mg L^{-1})**	**SS (mg L^{-1})**
AP2	Influent	242	525	128
	Effluent	126	294	168
	% Removal	48	44	–32
PFP	Influent	130	290	178
	Effluent	107	151	84
	% Removal	18	48	53
SFP	Influent	107	151	84
	Effluent	69	101	77
	% Removal	36	33	8
Total (without TFP)	**% Removal**	71	**80**	**40**
TFP	Influent	69	101	77
	Effluent	76	127	78
	% Removal	–10	–25	–2
Total (with the TFP)	**% Removal**	69	76	39

removed to supplement the household water supply. The remaining final effluent is released into Port Philip Bay (Melbourne Water, 2020).

14.2.2 *Small scale WSP treatment system*

Throughout Mediterranean Europe, WSPs are widely used, especially at coastal holiday resorts where the population may increase by a factor of 20 during the summer (Mara and Pearson, 1998). Of the 2,500 of the WSPs employed in France during the late 1990s, nearly 80% served small rural communities. The standard design uses three ponds in series: a primary facultative pond that is 1.0 to 1.5 m deep with an area equivalent to 6 m^2 per person followed by two maturation ponds in series and of similar depth, each with an area equivalent to 2.5 m^2 per person (Table 14.2). This is equivalent to an organic loading of 83 kg $ha^{-1}d^{-1}$ employing a total pond area of 11 m^2 per person (CEMAGREF *et al.*, 1997). Typical performance data is given in Table 14.6. In a comparison of the performance of WSPs in Mediterranean countries with similar climates, Gratziou and Chalatsi (2013) directly compared WSPs from Greece with those in Cyprus, France, Italy, Portugal, and Spain (Tables 14.7 and 14.8). They also briefly reviewed the design and reuse of pond effluent for each country studied. Detailed performance data for three WSPs treating domestic wastewater in Greece is presented by

Table 14.6. The mean performance of 178 waste stabilization pond systems in France (Racult *et al.*, 1995). Reproduced with permission from IWA Publishing.

Parameter	Raw wastewater (mg L^{-1})	WSP effluent (mg L^{-1})
BOD	277	23[a]
COD	657	99[a]
SS	256	60
Total Kjeldahl nitrogen	70	22
Ammoniacal nitrogen	48	14
Total phosphorus	21	8.5

[a]Filtered. Unfiltered values: BOD, 43; COD, 162.

Table 14.7. The average percentage removal efficiencies of waste stabilization pond systems in the Mediterranean countries studied by Gratziou and Chalatsi (2013). Reproduced with permission from the Taylor & Francis Group.

Region	BOD_5	COD	TSS	TKN	NH_4-N	TP	TC/100 ml	FC/100 ml
Greece	95.2	85.0	77.3	75.0	81.5	45.2	94.99	99.99
Cyprus	92.5	89.5	81.8	65.1	84.5	45.1	N.D	N.D
France	91.7	84.9	76.6	68.6	70.8	59.5	N.D	N.D
Spain	70.1	N.D	61.5	67.0	N.D	52.0	N.D	99.80
Portugal	80.0	N.D	N.D	N.D	N.D	N.D	N.D	N.D

N.D: Not determined.

Table 14.8. The average influent and effluent quality for WSPs in Northern Greece, South of France, and Catalonia in Spain (Gratziou and Chalatsi, 2013). Reproduced with permission from the Taylor & Francis Group.

Region parameter	North Greece (n = 10)		South France (n = 178)		Catalonia-Spain (n = 7)	
	Influent	**Effluent**	**Influent**	**Effluent**	**Influent**	**Effluent**
BOD (mg L^{-1})	316	15	277	23	280	67
COD (mg L^{-1})	647	97	657	99	—	—
SS (mg L^{-1})	269	61	256	60	260	100
TN (mg L^{-1})	78	19.5	70	22	100	33
NH_4 (mg L^{-1})	65	12	48	14	—	—
TP (mg L^{-1})	14.6	11.7	21	8.5	15	7.2
F.Coliforms (col/100 ml)	7.0×10^6	1.18×10^3	—	—	2.0×10^7	4.0×10^4

Chalatsi and Gratziou (2015). These plants all use rock filters after facultative and maturation ponds in series for the removal of algae.

In the United Kingdom (UK), waste stabilization pond systems are becoming increasingly popular to treat raw domestic wastewater with many more used solely for tertiary treatment. Mara *et al.* (1998) describe a three-pond system at Burwarton Estate in Shropshire, UK. It comprises a primary facultative pond and two maturation ponds in series (Fig. 14.6). The facultative pond is 1.5 m deep with an area of 1,380 m^2, followed by two maturation ponds in series, both 1.5 m deep and 1,057 m^2 in area. The surface BOD loading rate for the facultative pond is 55 kg ha^{-1}d^{-1}. The final effluent is discharged from the final pond into two constructed wetland ditches before entering a small stream. The system achieves the maximum discharge limits of 25 mg L^{-1} unfiltered BOD, 45 mg L^{-1} SS, and 10 mg L^{-1} ammonia.

The Bertix Wastewater Treatment Plant in Belgium primarily treats domestic wastewater from a rural community of 7,500 PE. After preliminary treatment (bar screens, grit separation, rotary drum screen), the wastewater is treated by a series of two anaerobic lagoons,

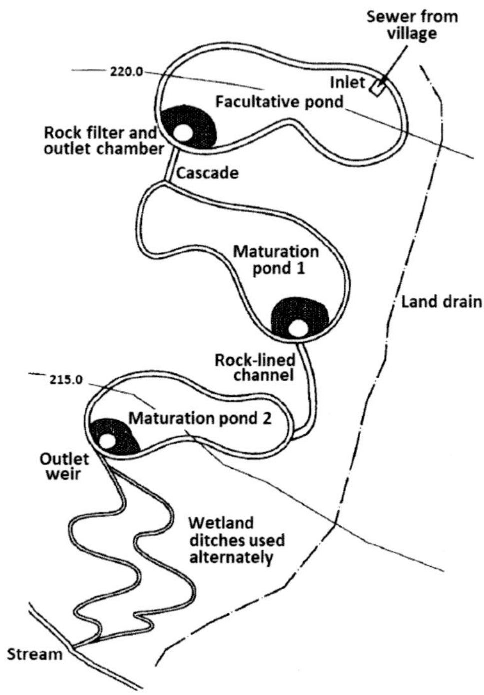

Fig. 14.6. Layout of the waste stabilization ponds used to treat domestic sewage at the Burwarton Estate, UK (Mara *et al.*, 1998). Reproduced with permission from the Chartered Institution of Water and Environmental Management.

four facultative ponds, and one maturation pond. Although unfiltered effluent values are within EU guidelines for BOD, COD and SS, the removal of nitrogen is similar to a conventional non-nitrifying system with the average removal of phosphorus approaching zero (Nameche *et al.*, 2000).

14.3 Design

Waste stabilization ponds can be designed for either BOD reduction or pathogen removal. The design of individual lagoons or ponds has been discussed in the following relevant chapters and is further discussed by Mara and Pearson (1998) and the United States Environmental Protection Agency (USEPA, 2011). An example is given below. Many authors feel that the general design approaches based largely on fixed-loading rates, regression equations or first-order kinetic models, could be improved by focussing more on the hydraulic characteristics of ponds and lagoons, which may include the use of baffles, repositioning of the inlet and outlets, or different pond configurations, as well as better biokinetic evaluation and modeling (Wood *et al.*, 1995; Shilton and Harrison, 2003). The adoption of integrated mechanistic models has led to improved performance, more accurate sizing of ponds, and a better understanding of sludge accumulation (Sah *et al.,* 2011). The different approaches to the design of anaerobic, facultative and maturation ponds have been reviewed by Ho *et al.* (2017), particularly the evolution and application of computational fluid dynamics (CPD) models now widely in use. They also list the original rule of thumb designs based solely on loading rates, regression or empirical models used to predict performance as well as design, and finally kinetic values from past studies.

Design of a waste stabilization pond system for slaughterhouse wastewater

In this example a WSP system using anaerobic, facultative and maturations ponds is designed to treat the wastewater from a medium-sized slaughterhouse in Brazil. The BOD of the influent wastewater is 1,210 g m^{-3}, the flow rate is 240 m^3d^{-1}, and the temperature estimated at a mean reactor temperature of 21°C. The required depths of the units are 4.0, 1.5 and 1.5 m for the anaerobic lagoon, facultative and maturation ponds, respectively. Note that the BOD can be expressed as either mg L^{-1} or g m^{-3}.

(i) Anaerobic lagoon

Using Table 15.3 to calculate S_v, the maximum permissible organic loading rate (g BOD $m^{-3}d^{-1}$) based on mean reactor temperature (21°C), both V_a, the volume of the lagoon (m^3), and A_a, the area of the lagoon, can be calculated as well as the hydraulic retention time (Φ_a), where L_i is the influent wastewater BOD (mg L^{-1}), Q is the flow rate ($m^3\ d^{-1}$), and D_a is the depth of the lagoon (m).

$$Va = \frac{Li\,Q}{Sv} = \frac{1210 \times 240}{300} = 968 \text{ m}^3,$$

$$Aa = \frac{Va}{Da} = \frac{968}{4.0} = 242 \text{ m}^2,$$

$$\Phi a = \frac{Va}{Q} = \frac{968}{240} = 4 \text{ days.}$$

From Table 18.1, the expected BOD removal at 21°C is 62%. Therefore, the effluent BOD concentration is 460 g m^{-3} (Chap. 15).

(ii) Facultative pond

Table 16.3 is used to calculate $\boldsymbol{S_s}$, the maximum permissible surface loading rate (kg BOD $ha^{-1}d^{-1}$) based on the mean reactor temperature (21°C), while both $\boldsymbol{V_s}$, the volume of the lagoon (m^3), and $\boldsymbol{A_f}$, the area of the pond (m_2), can be calculated as well as the hydraulic retention time (Φ_f), where $\boldsymbol{L_i}$ is the influent wastewater BOD (mg L^{-1}), $\boldsymbol{Q}$ is the flow rate ($m^3\,d^{-1}$), and $\boldsymbol{D_f}$ is the depth of the pond (m).

$$A_f = 10\frac{Li\,Q}{Ss} = 10\frac{460 \times 240}{272} = 4{,}059 \text{ m}^2,$$

$$\Phi_f = \frac{Af\,Df}{Q} = \frac{4059 \times 1.5}{240} = 25.4 \text{ days.}$$

Facultative ponds can achieve BOD removals of 90 to 95%, so the expected BOD in the effluent will be between 23 and 46 mg L^{-1} or g m^{-3} (Chap. 16).

(iii) Maturation pond

Assuming a hydraulic retention (Φ_m) time of 12 days and depth ($\boldsymbol{D_f}$) of 1.5 m.

$$\text{Volume} = \Phi\text{m Q} = 12 \times 240 = 2{,}880 \text{ m}^3,$$

$$\text{Area}\left(A_f\right) = \frac{\Phi\text{m Q}}{Dm} = \frac{12 \times 240}{1.5} = 1{,}920 \text{ m}^2 .$$

The final effluent BOD should be in the range of 10 to 20 mg L^{-1} or g m^{-3} with good nitrogen and phosphorus removals (Chap. 17).

References

Arceivala, S.J., Lakshminarayana, J.S.S., Alagarsamy, S.R. and Sastry, C.A. (1970) Waste stabilization ponds: Design, construction and operation in India. Central Public Health Engineering Research Institute, Nagpur, India.

CEMAGREF, SATESE, Ecole National de la Santé Publique and Agences de l'Eau (1997) *Le Lagunage Naturel: Les leçons Tirées de 15 Ans de Pratique en France*. Centre national du Machinisme Agricole, du Génie Rural, des Eaux et des Forêts, Lyon, France.

Chalatsi, M. and Gratziou, M. (2015) Efficiency and design parameters of waste stabilization ponds in north-east Greece. *Desalination and Water Treatment*, **56**(5), 1240–1247. doi:10.1080/19443994.2014.981677

Drakides, C. and Calignon, M. (1983) *Lagunes à Charge Estivale*. Agence de Bassin Rhône-Mediterranée-Corse. Pierre-Bénite, France.

Feachem, R.G., Bradley, D.J. Gavelick, H. and Mara, D.D. (1983) *Sanitation and Disease: Health Aspects of Excreta and Wastewater Management*. John Wiley & Sons, Chichester, UK.

Gloyna, E.F. (1971) Waste stabilization ponds. WHO Monograph Series, 60. World Health Organization, Geneva, Switzerland.

Gomes de Sousa, J.M. (1987). Wastewater stabilization lagoon design criteria for Portugual. *Water Science and Technology*, **19**(12), 7–16.

Gratziou, M. and Chalatsi, M. (2013) Use of waste stabilization ponds' systems in Mediterranean Europe. *Desalination and Water Treatment*, **51**, 3018–3025. doi:10.1080/19443994.2012.748451

Hawkes, H.A. (1983). Stabilization ponds. In *Ecological Aspects of Used Water Treatment, Vol. 2. Biological Activities and Treatment Processes* (eds. C.R. Curds and H.A. Hawkes). Academic Press, London, pp. 163–217.

Heinke, G., Smith, D. and Finch, G. (1991) Guidelines for the planning and design of wastewater lagoon systems in cold climates. *Canadian Journal of Civil Engineering*, **18**(4), 556–567. http://dx.doi.org/10.1139/l91-068

Ho, L.T. and Goethals, P.L.M. (2020) Municipal wastewater treatment with pond technology: Historical review and future outlook. *Ecological Engineering*, **148**, 105791. https://doi.org/10.1016/j.ecoleng.2020.105791

Ho, L.T., Van Echelpoel, W. and Goethals, P.L.M. (2017) Design of waste stabilization and systems: A review. *Water Research*, **123**, 236–248. http://dx.doi.org/10.1016/j.watres.2017.06.071

Hodgson, B. and Paspaliaris, P. (1996) Melbourne water's wastewater treatment lagoons: Design modifications to reduce odours and enhance nutrient removal. *Water Science and Technology*, **37**(7), 157–164.

Mara, D.D. and Pearson, H.W. (1998) Design manual for waste stabilization ponds in Mediterranean countries. Lagoon Technology International, Leeds, UK.

Mara, D.D., Cogman, C.A., Simkins, P. and Schembri, M.C.A. (1998) Performance of the Burwarton Estate waste stabilization ponds. *Water and Environment Journal of the Chatered Institution of Water and Environmental Management*, **12**(4), 260–264.

Marais, G.v.R. (1970). Dynamic behaviour of oxidation ponds. In *Proceedings of the Second International Symposium on Water Treatment Lagoons* (ed. R.E. McKinney). University of Kansas, Laurence, KS, USA, pp. 15–46.

Melbourne Water. (2020) Western treatment plant. Melbourne Water, Melbourne. https://www.melbournewater.com.au/water-data-and-education/learning-resources/water-and-sewage-treatment-plants/western-treatment

Naddafi, K., Hassanvand, M.S., Dehghanifard, E., Faezi Razi, D., Mostofi, S., Kasaee, N., Nabizadeh, R. and Heidari, M. (2009) Performance evaluation of wastewater stabilization ponds in Arak-Iran. *Iranian Journal of Environmental Health Science and Engineering*, **6**(1), 41–46.

Nameche, T., Dufayt, O., Ouarghi, H. El. and Vassel, J.L. (2000) Performance of a system combining aerated lagoons and waste stabilization ponds in a temperate climate. *Water Science and Technology*, **42**(10–11), 23–34.

Oswald, W.J. (1963) Fundamental factors in wastewater stabilization pond design. In *Advances in Biological Waste Treatment* (eds. W.W. Eckenfelder and J. McCabe). Pergamon, Oxford, pp. 357–393.

Park, J.B.K. and Craggs, R.J. (2007) Biogas production from anaerobic waste stabilization ponds treating dairy and piggery wastewater in New Zealand. *Water Science and Technology*, **55**(11), 257–264.

Parker, C.D., Jones, H.L. and Greene, N.C. (1959) Performance of large sewage lagoons at Melbourne, Australia. *Sewage and Industrial Waste*, **31**, 133–152.

Parker, C.D., Jones, H.L. and Taylor, W.S. (1950) Purification of sewage in lagoons. *Sewage and Industrial Waste*, **22**, 760–775.

Pearson, H.W., Mara, D.D. and Azov, Y. (eds.) (2000) Waste stabilization ponds: Technology and the environment. *Water Science and Technology*, **42**(10–11), 1–408.

Pfeffer, J.T. (1970) Anaerobic lagoons: Theoretical considerations. In *Proceedings of the Second International Symposium on Water Treatment Lagoons* (ed. R.E. McKinney). University of Kansas, Laurence, KS, pp. 310–320.

Porges, R. and Mackenthun, K.M. (1963) Waste stabilization ponds: Use, function and biota. *Biotechnology and Bioengineering*, **5**, 255–273.

Racault, Y., Boutin, C. and Seguin, A. (1995) Waste stabilization ponds in France: A report on fifteen years experience. *Water Science and Technology*, **31**(12), 91–101.

Ragush, C.M., Schmidt, J.J., Krkosek, W.H., Gagnon, G.A., Truelstrup-Hansen, L. and Jamieson, R.C. (2015) Performance of municipal waste stabilization ponds in the Canadian Arctic. *Ecological Engineering*, **83**, 413–421. http://dx.doi.org/10.1016/j.ecoleng.2015.07.008

Rodrigues, A. (1997) *Relatório de Estágio Formal em Engenharia Civil (Engenharia do Ambiente)*. Ordem dos Engenheiros, Lisbon.

Sah, L., Rousseau, D.P.L., Hooijmans, C.M. and Lens, P.N.L. (2011) 3D model for a secondary facultative pond. *Ecological Modelling*, **222**(9), 1592–1603.

Shilton, A.N. and Harrison, J. (2003) Guidelines for the hydraulic design of waste stabilisation ponds. Institute of Technology and Engineering, Massey University, Wellington.

Smith, D. (1986) Cold climate utilities manual. Environment Canada, Montreal.

Stander, G.J. and Meiring, P.G.J. (1965) Employing oxidation ponds for low-cost sanitation. *Journal of the Water Pollution Control Federation*, **37**, 1025–1033.

Townshend, A.R. and Knoll, H. (1987) Cold climate sewage lagoons. Report EPS 3/NR/1. Environment Canada, Ottawa.

Türker, U., Okaygün, M. and Almaqadma, S.J. (2009) Impact of anaerobic lagoons on the performance of BOD and TSS removals at the Haspolat (Mia Milia) Wastewater Treatment Plant. *Desalination*, **249**, 403–410. https://doi.org/10.1016/j.desal.2009.06.044

USEPA. (1983) Design manual: Municipal wastewater stabilization ponds. EPA-625/1-83-015. Office of Water, US Environmental Protection Agency, Washington, DC.

USEPA. (2011) Principles of design and operation of wastewater treatment pond systems for plant operators, engineers and managers. EPA/600/R-10/088. Office of Research and Development, US Environmental Protection Agency, Cincinnati, OH.

Vuillot, M. and Boutin, C. (1987) Waste stabilization ponds in Europe: A state of the art review. *Water Science and Technology*, **19**(12), 1–6.

Watson, J.L.A. (1962) Oxidation ponds and use of effluent in Israel. *Proceedings of the Institute of Civil Engineers*, **22**, 21–40.

WHO. (1987) Wastewater stabilization ponds: Principles of planning and practice, Technical Publication: 10. Regional Office for the Eastern Mediterranean, World Health Organization, Alexandria.

Wood, M.G., Greenfield, P.F., Howes, T., Johns, M.R. and Keller, J. (1995) Computational fluid dynamic modelling of wastewater ponds to improve design. *Water Science and Technology*, **31**(12), 111–118.

15
Anaerobic Lagoons

15.1 The Process

Treatment in lagoons, which are designed to be predominantly anaerobic, relies on the development of a biologically active sludge layer. The sludge takes several months to build up before the maximum biological activity is reached. The design criteria for anaerobic lagoons are different, compared with other stabilization pond systems; the main difference being depth. Oxygen transfer through the air–water interface is undesirable in anaerobic ponds and is minimized by employing deep basins that are 2 to 5 m in depth. This reduces the surface area to volume ratio, effectively minimizing both reaeration and heat loss. Construction of lagoons of this depth requires specialist engineering training and will first be required to be lined with sand and then with several layers of welded butyl rubber or PVC sheeting to prevent any seepage to the unsaturated zone below that may ultimately result in groundwater pollution. Anaerobic lagoons need to be desludged occasionally, which requires the use of heavy machinery working from the edge of the lagoon, so simple earthen banks may not always be suitable. Smaller lagoons can be desludged using pumps. Sludge accumulation rates, sludge characteristics, and disposal are reviewed by Owusu-Twum and Sharara (2020).

Three identifiable zones are observed in lagoons: the scum, the supernatant that contains about 0.1% volatile solids, and the sludge with 3 to 4% volatile solids (Fig. 15.1). The scum or grease layer, which forms on the surface of the lagoon can become 40 to 60 mm

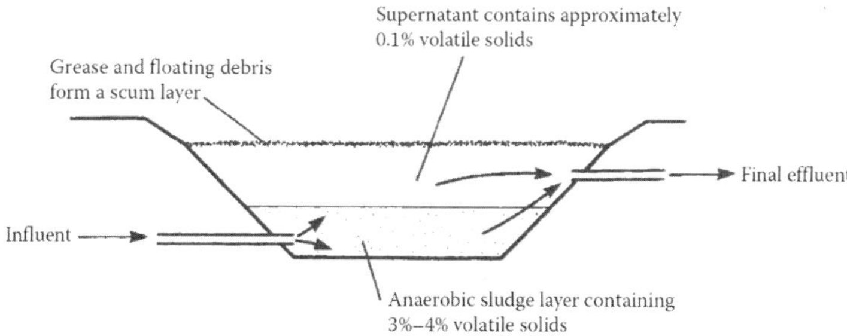

Fig. 15.1. The basic design of an anaerobic lagoon used for treating strong organic wastewaters.

thick and has a number of important functions. It insulates the lagoon, thus preventing heat loss, suppressing odors, but most importantly, maintains anaerobic conditions in the supernatant by eliminating oxygen transfer between the air–water interface. If the grease content of the wastewater is low (e.g. vegetable-processing wastewaters), then the lagoon surface can be covered using high-density polyethylene (HDPE) plastic balls with a 50 to 100-mm diameter. To be as effective as a grease layer, at least 90% of the surface must be covered. Some animal wastewaters can also be problematic. While cattle slurry rapidly forms a crust due to the high fiber content of their feces, pig slurry does not. The scum layer can become a breeding ground for a variety of flies and in warmer climates, this can become a serious nuisance. The operation of anaerobic lagoons in series is not recommended as it is difficult to maintain an adequate scum cover on the second lagoon. The reduced loading results in the surface layer of the lagoon becoming aerobic, thus the system becomes facultative rather than strictly anaerobic (Sec. 16.1). Under-loaded lagoons will also tend to become facultative ponds.

Although the inlet pipe can be at the surface (Hawkes, 1983), anaerobic lagoons function most efficiently if the influent wastewater enters near the bottom of the basin. This prevents short-circuiting of the liquid and ensures that it mixes with the microbial solids in the active anaerobic sludge layer. The flow pattern of the wastewater and the constant release of methane and carbon dioxide resulting from the degradation of organic matter ensures that some sludge particles remain in suspension within the supernatant. In this way, decomposition of organic matter continues both in the liquid phase and sludge blanket, thus a high mixing efficiency results in a high rate of biochemical oxygen demand (BOD) removal (Parker *et al.*, 1950). The discharge pipe is located just below the scum layer at the opposite end of the basin to ensure an upward flow pattern. This encourages the bacterial flocs to settle out of suspension, ensuring a clarified effluent, and also eliminating the need for sludge return by retaining the microbial biomass in the lagoon. Further design details are given by the United States Environmental Protection Agency (USEPA, 1983; 2011).

15.2 *Microbiology*

Anaerobic heterotrophic bacteria are either obligate — and unable to grow in the presence of oxygen — or facultative and can adapt to environments either with or without oxygen. The latter forms a bridge between obligate aerobes and obligate anaerobic species. Growth and metabolism of obligate anaerobes are inhibited by oxygen, with oxygen toxicity depending on the redox potential (*Eh*), partial pressure, the composition of the substrate, growth rate, and cell density (Hughes, 1980).

Anaerobic lagoons have a widely diverse population of bacteria. In a study of four anaerobic lagoons treating swine wastewater, Ducey and Hunt (2013) identified 15,682 phylotypes that could be classified into 839 individual genera with 223 genera common to all four lagoons. These belonged primarily to four major phyla: the Firmicutes (54.1%), Proteobacteria (15.8%), Actinobacteria (7.2%), and the Bacteroidetes (4.1%) (Fig. 15.2). These anaerobes are responsible for a wide range of functions including fermentation, sulfate-reduction, odor production, and both nitrogen and sulfur cycling. Twenty-five genera accounted for 20% of the operational taxonomic units (OTUs) measured in the four lagoons (Table 15.1). Each lagoon had a unique community structure controlled by key environmental parameters

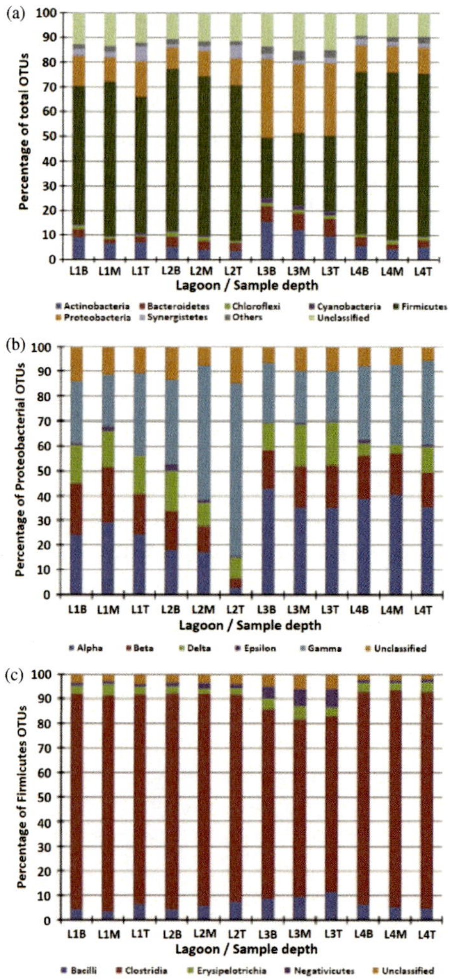

Fig. 15.2. Relative abundance of bacterial 16S rDNA genes from swine wastewater anaerobic lagoons expressed as taxonomic units (OTUs) at (a) the phylum level, (b) breakdown of proteobacteria, and (c) Firmicutes (Ducey and Hunt, 2013). Reproduced with permission from Elsevier Publishing.

Table 15.1. The relative abundance (RA) (%) and the number of operational taxonomic units (OTUs) of the 25 most common genera isolated from four anaerobic lagoons treating piggery wastewater (Ducey and Hunt, 2013). Reproduced with permission from Elsevier Publishing.

		Lagoon #1		Lagoon #2		Lagoon #3		Lagoon #4	
Phylum	**Genus**	**Ra**	**# OTUs**	**Ra**	**# OTUs**	**Ra**	**# OTUs**	**Ra**	**# OTU**
Firmicutes	*Anaerovorax*	2.58%	96	2.45%	126	1.94%	73	2.78%	85
Proteobacteria	*Thiolamprovum*	1.59%	59	3.81%	196	1.38%	52	1.14%	35
Firmicutes	*Clostridium* sp. cluster XI	3.14%	117	0.70%	36	0.40%	15	2.16%	66
TM7	TM7	0.89%	33	1.24%	64	1.04%	39	1.34%	41
Firmicutes	*Clostridium* (sensu stricto)	1.53%	57	0.64%	33	0.48%	18	1.63%	50
Firmicutes	*Oscillibacter*	0.43%	16	1.13%	58	0.85%	32	0.98%	30
Actinobacteria	*Leucobacter*	1.67%	62	0.45%	23	0.37%	14	0.78%	24
Proteobacteria	*Methylocystis*	0.11%	4	0.10%	5	2.77%	104	0.00%	0
Firmicutes	*Tissierella*	0.67%	25	0.86%	44	0.05%	2	1.18%	36
Firmicutes	*Acetanaerobacterium*	1.05%	39	0.56%	29	0.16%	6	1.05%	32
Firmicutes	*Turicibacter*	1.42%	53	0.40%	19	0.13%	5	0.85%	26
Firmicutes	*Saccharofermentans*	0.67%	25	0.70%	36	0.67%	25	0.33%	10
Firmicutes	*Lachnospiracea*	0.38%	14	0.80%	41	0.56%	21	0.62%	19
Actinobacteria	*Corynebacterium*	0.73%	27	0.52%	27	0.59%	22	0.56%	17
Firmicutes	*Sporobacter*	0.54%	20	0.76%	39	0.05%	2	0.75%	23
Synergistetes	*Aminobacterium*	0.73%	27	0.51%	26	0.24%	9	0.26%	8
Firmicutes	*Lactobacillus*	0.38%	14	0.45%	23	0.64%	24	0.29%	9
Proteobacteria	*Desulfomonile*	0.38%	14	0.06%	3	1.23%	46	0.20%	6
Proteobacteria	*Acinetobacter*	0.40%	15	0.45%	23	0.19%	7	0.59%	18
Actinobacteria	*Mycobacterium*	0.13%	5	0.08%	4	1.25%	47	0.03%	1
Cyanobacteria	Group IIa	0.00%	0	0.00%	0	1.52%	57	0.00%	0
Actinobacteria	*Actinomyces*	0.70%	26	0.25%	13	0.11%	4	0.39%	12
Proteobacteria	*Thioflavicoccus*	0.00%	0	0.00%	0	1.46%	55	0.00%	0
Bacteroidetes	*Prevotella*	0.16%	6	0.27%	14	0.53%	20	0.42%	13
Actinobacteria	*Klugiella*	0.02%	1	0.00%	0	1.09%	41	0.07%	2
Total		20.30%	755	17.19%	882	19.70%	740	18.40%	563

such as Kjeldhal nitrogen, chemical oxygen demand (COD), orthophosphate, total suspended solids, and dissolved oxygen (DO) (Ducey and Hunt, 2013). The microbial community structure changes seasonally, which may also affect odor production (Cook *et al.*, 2010).

In anaerobic lagoons, the organic substrate is degraded in the absence of oxygen to carbon dioxide and methane with only a small amount of bacterial growth. Approximately 90% of the available chemical energy, in the form of organic material, is retained as methane production (McInnery *et al.*, 1980). Apart from the economic value of the methane gas produced, anaerobic treatment has many advantages over aerobic treatment processes, such as less biomass produced per unit of substrate utilized (the lower biomass production means a lower requirement for nitrogen, phosphorus, and other nutrient and growth factors), higher organic loadings (as anaerobic processes are not limited by oxygen transfer rates), and lower constructional and operational costs. However, the major disadvantage in temperate and colder climates is the elevated temperatures required to maintain microbial activity at a reasonable level.

The process of anaerobic digestion or fermentation is considered to be a four-step or phase process comprising hydrolysis, acid formation, acetate formation, and, finally, methane production. From the perspective of microbial ecology, it is convenient to think of these phases as different trophic levels, and although each phase comprises a different consortium of microorganisms, each is metabolically dependent on the other for survival (Fig. 15.3).

Hydrolysis phase: Hydrolytic bacteria convert complex organic material, especially high-molecular-weight carbohydrates, fats and proteins, which are often insoluble, by enzymatic action into simple alcohols and acids, primarily amino acids, glucose, fatty acid, and glycerols (Brockett, 1976; Pearson, 2005; Paterson and Curtis, 2005).

Acid production phase: Acidogenic bacteria convert these soluble polymers into a range of organic acids (i.e., propionic, acetic and butyric acids), alcohols, hydrogen, and carbon dioxide. For example:

$$\underset{\text{(glucose)}}{C_6H_{12}O_6} \rightarrow \underset{\text{(acetic acid)}}{3CH_3COOH}. \tag{15.1}$$

As only acetic acid can be converted directly to methane, a third stage is required to convert the remaining organic acids to acetate.

Acetogenic phase: Acetogenic bacteria convert organic acids and alcohols into acetate (acetic acid), ammonia, hydrogen, and carbon dioxide. For example:

$$\underset{\text{(propionic acid)}}{CH_3CH_2COOH} + 0.5H_2O \rightarrow \underset{\text{(acetic acid)}}{CH_3COOH} + 0.25CO_2 + \underset{\text{(methane)}}{0.75CH_4}. \tag{15.2}$$

Although methane can also be produced from hydrogen and carbon dioxide, about 70% is from acetic acid.

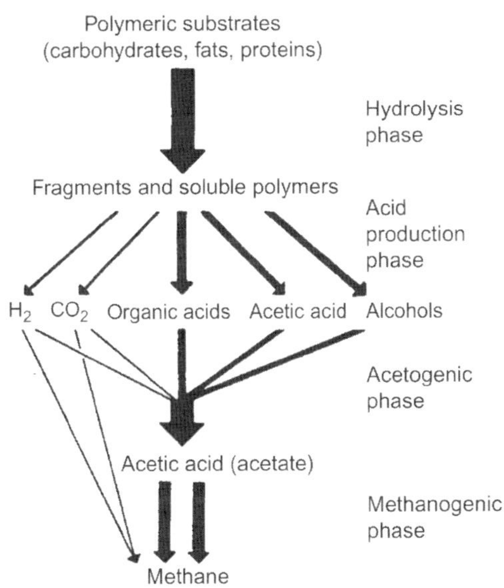

Fig. 15.3. Major steps in anaerobic fermentation.

Methanogenic phase: Methanogenic bacteria convert acetic acid further to methane and carbon dioxide (Gallert and Winter, 2005) (Fig. 15.3):

$$\underset{\text{(acetic acid)}}{CH_3COOH} \rightarrow \underset{\text{(methane)}}{CH_4} + CO_2. \tag{15.3}$$

A summary of the key bacterial groups involved in anaerobic digestion is given in Table 15.2 and is discussed in detail by Gray (2021).

As the methanogenic phase is dependent on the production of acetic acid/acetate, then, in theory, the acetogenic phase is rate-limiting in the fermentation process. In practice, as methanogenic bacteria are more sensitive and slower growing than the other groups of bacteria involved in the previous steps, it is the last step that is considered to be rate-limiting in the overall reaction, with the methanogenic phase used for modeling purposes, where the rate of methane production is normally expressed as a linear function of temperature

Table 15.2. Enumeration and identification of anaerobic bacterial populations in sewage sludge fermentation (Zeikus, 1980). Reproduced with permission from Elsevier Publishing.

Group	Numbers (per mL)	Generic identity
Hydrolytic bacteria		Majority unidentified
Total	10^8–10^9	Gram-negative rods
Proteolytic	10^7	*Eubacterium*
Cellulolytic	10^5	*Clostridium*
Hydrogen-producing acetogenic bacteria	10^6	Unidentified Gram-negative rods
Homoacetogenic bacteria	10^5–10^6	*Clostridium* *Acetobacterium*
Methanogens	10^6–10^8	*Methanobacterium* *Methanospirillum* *Methanococcus* *Methanosarcina* *'Methanothrix'*
Sulfate reducers	10^4	*Desulfovibrio* *Desulfotomaculum*

(Wu and Chen, 2011). Volatile acids not only act as a substrate for methanogenic bacteria, but they can also be inhibitory at high concentrations so that substrate concentration is not related to a specific growth rate (i.e., the Monod function). This is replaced with an inhibition function so that:

$$\mu = \left(\frac{\mu_m}{1 + \frac{K_s}{s} + \frac{s}{K_i}} \right), \tag{15.4}$$

where μ is the specific growth rate of the biomass, μ_m is the maximum specific growth rate at saturation concentration of the growth-limiting substrate (d^{-1}), s is the substrate concentration (mg L^{-1}), K_s is the saturation constant (mg L^{-1}), and K_i is the inhibition function. Methane production is explored in greater detail in Sec. 15.5.1.

When operating under ideal conditions, the rate of volatile acid production equals the rate of conversion to methane. However, methanogenic activity is adversely affected by fluctuations in organic loading, temperature and pH. So, if the methanogens are inhibited, then there is an immediate reduction in the conversion of volatile organic acids, leading to an increase in their concentration and a rapid reduction in the pH, causing further inhibition of methanogenic activity, eventually leading to process failure. Optimum (and maximum) operating conditions for methane formation are at a temperature of 30 to 35°C (25–40°C), a pH of 6.8 to 7.4 (6.2–7.8), redox (or oxidation-reduction potential [ORP]) of −520 to −530 mV (−490 to −550 mV), volatile organic acids of 50 to 500 mg acetic acid L^{-1} (50–2,000 mg L^{-1}), and an alkalinity of 2,000 to 3,000 mg $CaCO_3$ L^{-1} (1,000–5,000 mg L^{-1}) (USEPA, 2011).

15.3 Application

Anaerobic lagoons are generally used as a preliminary treatment for strong organic wastes, with an 80% reduction in the BOD and a 90% reduction in suspended solids (SS) possible. This means that the lagoon only partially stabilizes the wastewater, and secondary aerobic treatment is required before the effluent can be discharged to natural waters, either in the form of facultative ponds in series or as activated sludge. However, anaerobic treatment can considerably reduce the organic loading to secondary treatment units, thus reducing the secondary treatment capacity required. If the subsequent aerobic treatment process is a facultative pond, then the sludge will be less likely to rise to the surface in warm weather, as the sludge has already undergone anaerobic breakdown. Most industrial ponds operate as anaerobic lagoons and the process is widely used for the treatment of slaughterhouse waste, animal production, sugar beet, paper pulp, potato, and other food processing wastewaters (Porges and Brackney, 1962; Barnes *et al.,* 1984; Vuillot and Boutin, 1987). Wastewaters most amenable to anaerobic breakdown in lagoons have high organic strength, are rich in fats and protein, have a relatively high temperature, are free from toxic materials, especially heavy metals, and contain sufficient biological nutrients.

Slaughterhouse and meat processing wastewaters are particularly suitable for treatment by anaerobic lagoons as they have a high BOD (~1,400 mg L^{-1}), grease content (~500 mg L^{-1}), temperature (~28°C), and a neutral pH. Minimum pretreatment is required prior to discharge to an anaerobic lagoon, which should normally be limited to blood recovery, screening to remove coarse solids, and dissolved air flotation or skimming to remove excessive grease. If pretreatment is too extensive, then insufficient grease may remain in the wastewater to form a scum when discharged to the lagoon. Domestic sewage is not suitable for treatment by this method due to a relatively low BOD concentration and grease content. When sewage has a high industrial content, anaerobic treatment often leads to serious odor problems due to lagoons becoming facultative instead of strictly anaerobic.

Anaerobic lagoons used for vegetable processing wastewaters tend to be shallower in design and are widely used to treat sugar beet and potato wastewaters. The organic strength of sugar beet-processing wastewaters can be very high and although a 60 to 70% reduction of BOD and an 80 to 90% reduction in SS is possible, the final effluent is still far too strong (BOD >500 mg L^{-1}) for direct discharge to surface waters, requiring further aerobic biological treatment. These wastewaters are difficult to treat as they are seasonal with <100 days wastewater production per year. This can leave insufficient time for active anaerobic biomass to accumulate, resulting in partial anaerobic breakdown and the release of odors (Sec. 15.5.3). The shallow lagoons can also result in partial reaeration, while low temperatures and a high C:N ratio, common with vegetable wastewaters, all increase the inhibition of anaerobic activity.

In arid and semi-arid areas, anaerobic-lagoon-treated effluent may be used for irrigation, although it is more commonly further treated. In the United States (US), new regulations governing wastewater irrigation have been introduced, which include the use of nutrient (nitrogen and phosphorus) management plans commonly used in Europe (Sharpley *et al.*, 2011). Microbial standards for wastewater reuse on agricultural land are not included, but the World Health Organization (WHO) has set standards based on fecal coliform counts of $\leq 1 \times 10^5$ 100 mL^{-1} (WHO, 2006). However, little attention has been paid to the transfer of fecal bacteria and zoonotic pathogens from anaerobic lagoons treating animal rearing and slaughterhouse wastewaters. In a study of water quality from an anaerobic lagoon treating piggery wastewater, McLaughlin *et al.* (2012) studied the seasonal variations in both nutrients and pathogens, recording higher nitrogen and carbon losses at temperatures above 19°C, with the N:P ratio falling from 9.7 to 2.8 during the warmer months when irrigation was used. Enterococci and zoonotic pathogens were present throughout the year but decreased at the higher summer temperatures while water-soluble metals including Cu and Zn increased. Temperature is a key factor in pathogen inactivation and removal, with aerobic systems more effective than anaerobic lagoons (Pandey *et al.*, 2016). A key nutrient in eutrophication is phosphorus and the effluent from a piggery waste anaerobic lagoon is rich in this nutrient, with N:P ratios normally between 2:1 and 4:1. In order to comply with the crop requirements of a ratio of 4:1 to 9:1, either additional nitrogen must be added or the phosphorus concentration reduced. As both NH_4+ and Mg^{2-} are common in the lagoon effluent, Nelson *et al.* (2003) have proposed to remove phosphorus by struvite (magnesium ammonium phosphate hexahydrate $MgNH_4PO_4 \cdot 6H_2O$)

precipitation, which can reduce effluent PO_4-P concentration to 2 mg L^{-1} as well as also reducing the ammonium concentration and creating a useful fertilizer.

15.4 Design of Anaerobic Lagoons

The design loadings for stabilization ponds are normally measured in grams of BOD per unit area (g BOD $m^{-2}d^{-1}$), but as light is not an important factor in anaerobic lagoons, loading is expressed as BOD per unit volume (g BOD $m^{-3}d^{-1}$) or as the HRT ($m^{-3}h^{-1}$). The rate of treatment is temperature-dependent (Table 15.3), so in temperate or cold climates, lagoons may have very reduced anaerobic activity in the

Table 15.3. Design values calculated at various temperatures for the organic loading rate and percentage BOD removal in anaerobic lagoons.

Temperature (°C)	Organic loading rate (g BOD m^{-3} day^{-1})	% BOD removal
≤10	100	40
11	120	42
12	140	44
13	160	46
14	180	48
15	200	50
16	220	52
17	240	54
18	260	56
19	280	58
20	300	60
21	300	62
22	300	64
23	300	66
24	300	68
≥25	300	70

winter, acting primarily as settlement ponds. BOD reduction is a function of both HRT and temperature. The WHO (1987) has given different BOD reduction values at temperatures >20°C based on an HRT of 1 day for 50%, 2.5 days for 60%, and 5 days for 70% reductions. Hammer (1977) gives a mean organic loading value of 320 g BOD $m^{-3}d^{-1}$ at a minimum temperature of 25°C to achieve 75% BOD removal at a minimum retention time of 4 days. Similar figures are given by Gloyna (1971), who cites examples where longer retention times of up to 19 days are needed for strong organic industrial wastewaters. The volume of the lagoon can be calculated as:

$$\lambda_v = L_i Q / V_a, \tag{15.5}$$

where λ_v is the organic loading rate (g BOD $m^{-3}d^{-1}$), L_i is the influent wastewater BOD (mg L^{-1} or g m^{-3}), Q is the influent flow rate (m^2d^{-1}), and V_a is the volume of the lagoon (m^3) (Mara and Pearson, 1998). The value λ_v increases with temperature but lies between 100 and 350 g $m^{-3}d^{-1}$, the former to maintain anaerobic conditions and the latter to avoid odor release (Table 15.3). The mean temperature of the lagoon is used to select λ_v from Table 15.3, from which the lagoon volume (V_a) can be calculated using Eq. (15.5). The mean hydraulic retention time (θ_a) is calculated as:

$$\theta_a = V_a/Q \text{ days.} \tag{15.6}$$

An example of how to design an anaerobic lagoon is given in Table 15.4.

The major advantages of anaerobic systems are that less sludge is produced compared with an aerobic system and no aeration equipment or power supply is required. The main operating limitations of the process are the temperature and strength of the wastewater. The temperature of the lagoon controls the BOD removal. Below 15°C, purification will be due to physical settlement only, with a limited anaerobic breakdown of organic matter occurring. At higher temperatures, anaerobic digestion proceeds rapidly with biogas (70% CH_4, 30% CO_2) bubbling to the surface. The same bacterial groups are involved as with other anaerobic systems. Thus, the process is inhibited by the same conditions (e.g. pH <6.2). In temperate climates,

Table 15.4. An example of the design calculation for an anaerobic lagoon.

Design an anaerobic lagoon to treat piggery (swine) wastewater with a BOD of 3,500 mg L^{-1} and a flow rate of 80 m^3d^{-1}. The average temperature is 14°C.

Step 1 Calculate the organic loading rate (λ_v) using Table 15.3:

At 14°C, λ_v is 180 g BOD $m^{-3}d^{-1}$

Step 2 Use Eq. (15.1) to calculate the required volume of the lagoon (V_a):

$V_a = L_i Q/\lambda_v$

$V_a = (3{,}500 \times 80)/180 = \mathbf{1{,}556\ m^3}$

Step 3 Calculate the area of lagoon (A), assuming a depth of 4 m:

$A = 1{,}556/4 = \mathbf{389\ m^2}$

Step 4 Calculate the HRT (θ_a):

$\theta_a = Va/Q = 1{,}556/80 = \mathbf{19.45\ days}$

Step 5 Calculate the final effluent using Table 15.3:

At 14°C, the expected BOD removal is 48%, so the final effluent (L_e) is:

$L_e = [(3{,}500/100) \times (100 - 48)] = \mathbf{1{,}820\ mg\ L^{-1}}$

anaerobic lagoons act as settling tanks during the winter months and the accumulated sludge is subsequently degraded during the warmer summer months. This has been demonstrated by Parker *et al.* (1950), who recorded a 65 to 80% BOD removal in lagoons during the summer with a retention period of 1.2 days, which fell 45 to 65% in the winter, even after 5 days of retention. Therefore, special attention should be paid to those factors affecting the temperature of the lagoon, such as ambient and wastewater temperature and insufficient scum development for insulation. A well-designed anaerobic lagoon will achieve, on average, 40% BOD removal at 10°C, rising to 60% at 20°C. The required HRT is a function of temperature and wastewater strength. Low BOD and grease content of wastewater will also reduce scum development; where no scum formation occurs, then a thin surface film of the algae *Chlamydomonas* often develops. Protection from wind is important in preventing mixing and the resultant break-up of surface scum.

The large depth of anaerobic lagoons used to conserve heat and maintain anaerobicity can lead to stratification of key nutrients, trace

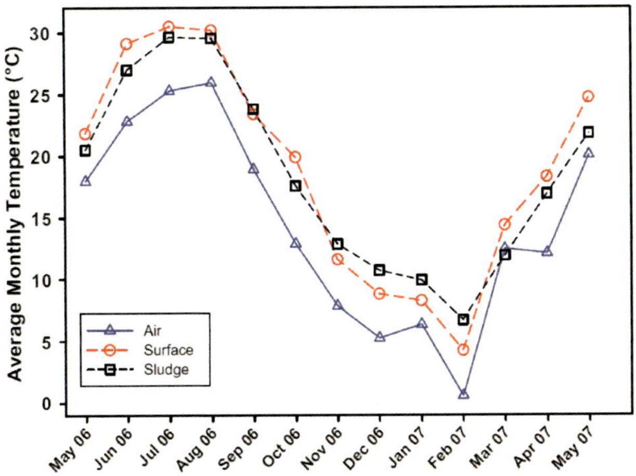

Fig. 15.4. The temperature profile of an anaerobic lagoon (0.4 ha) treating piggery wastewater (approximately 2,000 sows) (Lovanh *et al.*, 2009). Reproduced with permission from Elsevier Publishing.

elements, and the microbial community (Lovanh *et al.*, 2009; Cook *et al.*, 2010). These effects are both seasonal as well as temperature-dependent. Lagoons have been recorded as being up to 9 m in depth, resulting in some becoming vertically thermally stratified, which can result in the sludge layer remaining colder or warmer than the upper water phase, depending on the season. When open, lagoons are also affected by wind action, which causes mixing deep within the lagoon, resulting in the settled solids being mixed with the water phase. An example of the temperature variation within lagoons is illustrated in Fig. 15.4.

15.5 Gas Emissions from Lagoons

15.5.1 *Methane and carbon dioxide*

The final phase in anaerobic digestion is methane fermentation, where the end-products of acid fermentation are converted to gases, mainly methane and carbon dioxide, but also ammonia and hydrogen sulfide, by several different species of obligate anaerobic bacteria, see Eq. (15.7). In this stage, complete stabilization of the substrate occurs and since the end-products are only gases, it is more efficient than complete aerobic stabilization.

$$\text{Organic matter} \rightarrow CH_4 + CO_2 + H_2 + NH_3 + H_2S. \quad (15.7)$$

Methane is an ideal end-product as it is: (i) non-toxic, although as it is heavier than oxygen it will displace the latter, resulting in the possibility of asphyxiation if the oxygen levels fall from 21% to below 16%, (ii) not very soluble, (iii) inert under anaerobic conditions, and (iv) can be readily collected and used as an energy source. In the overall anaerobic fermentation of carbohydrates to carbon dioxide and methane, equal volumes of each gas are produced.

$$(C_6H_{12}O_6)_x + xH_2O = xC_6H_{12}O_6 + 3xCH_4 + 3xCO_2. \quad (15.8)$$

The carbon dioxide evolved only partially and escapes as gas because unlike methane, it is relatively soluble in water. It also reacts with any hydroxide ions (OH^-) in the system to produce bicarbonate ions (HCO_3^-). The evolution of carbon dioxide gas is therefore a function of factors such as pH, bicarbonate concentration, temperature, and substrate composition. The biodegradable protein is delaminated to produce ammonia, which reacts with water:

$$NH_3 + HOH \leftrightarrow NH_4^+ + OH^-. \quad (15.9)$$

This is the major source of hydroxide ions, which react with the carbon dioxide evolved during methanogenesis to form bicarbonate ions:

$$CO_2 + HOH \leftrightarrow H_2CO_3 \leftrightarrow H^+ + HCO_3, \quad (15.10)$$

$$H_2CO_3 + OH^- \leftrightarrow HCO_3^- + HOH. \quad (15.11)$$

Therefore, the protein content of the wastewater substrate will significantly affect the quantity of carbon dioxide actually released from the solution as well as the buffering capacity of the system in terms of bicarbonate. The portion of carbon dioxide incorporated in the bicarbonate ion is eventually removed from the reactor in the liquid rather than in the gas phase (Pfeffer, 1980).

Anaerobic fermentation occurs in natural environments, including the digestive tract of most animals, sediments of lakes and rivers, and estuaries, swamps, marshes, and peat bogs. The bacteria responsible for methanogenesis in anaerobic lagoons are similar to those found in other environments and are, in fact, Archaea belonging to the phylum Euryarchaeota. There are five common orders: Methanobacterialis, Methanococcales, Methanomicrobiales, Methanosarcinales, and Methanopyrales, which belong to the genera *Methanobacterium*, *Methanosarcina*, *Methanospirillum*, and *Methanococcus*. They are limited to the catabolism of either one carbon compound (e.g. H_2/CO_2, CH_3OH, CO, HCOOH, CH_3NH_2) or two carbon compounds (e.g. CH_3COOH). Methanogens are unusual in that they are composed of many species with very different cell morphology. They require a strict anaerobic environment for growth with a redox potential of below −300 mV. They have simple nutritional requirements: carbon dioxide, ammonia and sulfide. Ammonia is the essential nitrogen source for growth, with sulfide the most common source of sulfur, although some species can use cysteine instead. Methanogens contain a number of unique co-enzymes, for example, co-enzyme 420, which is involved in electron transfer instead of the usual ferredoxin, co-enzyme M, which is used in methyl transfer reactions, and factor B, which is required for the enzyme formation of methane from methyl co-enzyme M. The synthesis of adenosine 5′-triphosphate (ATP) appears to be via electron transport linked to phosphorylation. The ultrastructure and cellular composition of methanogens make them very different from typical aerobic and facultative bacteria and it has now been established that they are, in fact, a primitive group of bacteria. The group does not contain muramic acid in their cell walls and this, linked with their unique co-enzymes and the unique oligonucleotide sequences of the 16S ribosomal RNA molecule, has been used to re-classify the microorganisms of the group (Bapteste *et al.,* 2005; Liu, 2010).

There is a close relationship between gas production and the numbers of methanogenic bacteria present, generally being present in

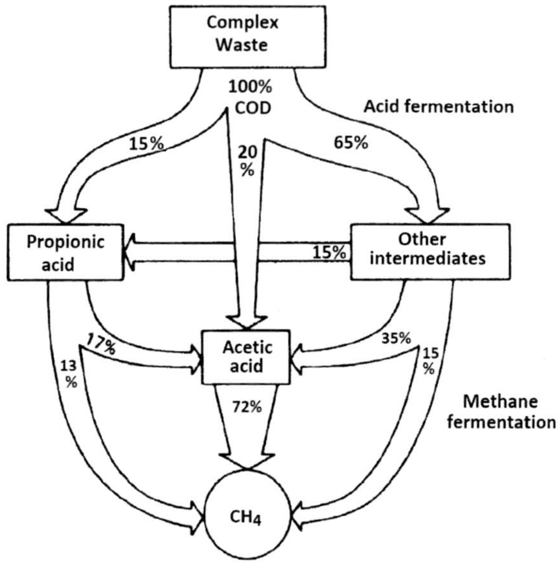

Fig. 15.5. Pathways in methane fermentation of organic wastes. Percentages represent the conversion of waste COD by various routes.

numbers between 10^6 and 10^8 mL^{-1} (Table 15.2). It is from acetic and, to a lesser extent, propionic acid that the greatest percentage of methane is derived during methane fermentation, whereas formic acid fermentation and methane fermentation associated with β-oxidation of long-chain fatty acids probably account for most of the small percentage of methane not derived from acetic and propionic acids (Fig. 15.5).

Methane formation from acetic acid is a single-step process carried out by one group of methanogenic bacteria (Eq. [15.3]), whereas methane fermentation of propionic acid is a two-step process involving two groups of methanogenic archaea, with acetic acid as the intermediate step (Eq. [15.2]). The overall reaction is shown in Eq. (15.12).

$$\underset{\text{(propionic acid)}}{CH_3CH_2COOH} + 0.5H_2O \rightarrow 1.25CO_2 + \underset{\text{(methane)}}{1.75CH_4}. \qquad (15.12)$$

The utilization of the organic fraction of wastewater is directly related to methane production and vice versa. McCarty (1964) estimated the theoretical methane production from the complete stabilization of 1 kg of COD as 0.348 m^3 at standard temperature and pressure. Where the exact chemical nature of the sludge is known, the quantity of methane produced can be estimated by the following equation (Buswell and Mueller, 1952):

$$C_cH_hO_o + \left(c - \frac{h}{4} - \frac{o}{2}\right)H_2O = \left(\frac{c}{2} - \frac{h}{8} + \frac{o}{4}\right)CO_2 + \left(\frac{c}{2} + \frac{h}{8} - \frac{o}{4}\right)CH_4. \quad (15.13)$$

This earlier equation has been superseded by Eq. (15.14), where the amount of ammonia and hydrogen sulfide can also be calculated:

$$\begin{aligned} C_cH_hO_oN_nS_s &+ 1/4(4c - h - 2o + 3n + 2s)H_2O \\ &= 1/8(4c - h + 2o + 3n + 2s)CO_2 \\ &+ 1/8(4c + h - 2o - 3n - 2s)CH_4 + nNH_3 + sH_2S, \quad (15.14) \end{aligned}$$

where c, h, o, n, and s are the number of atoms of carbon, hydrogen, oxygen, nitrogen, and sulfur, respectively. Actual gas yields and the composition of the biogas produced from basic substrates are carbohydrates (0.8 m^3 of biogas per kg dry matter at 50% CH_4 and 50% CO_2), proteins (0.7 $m^3\ kg^{-1}$ at 70% CH_4 and 30% CO_2), and lipids (1.2 $m^3\ kg^{-1}$ at 67% CH_4 and 33% CO_2) (Konstandt, 1976). The exact gas yield per kg of volatile solids removed depends on the composition of the waste but is approximately 1.0 m^3. Some food-processing industries produce a pure substrate waste with 80 to 95% of the organic (volatile) matter removed, whereas only 40 to 50% of the organic matter in sewage sludge and even slightly less in animal slurries will be utilized with a proportionately lower gas yield. In anaerobic lagoons, the biogas production rate is a function of organic acid concentration and the lagoon temperature. Safley and Westerman (1988) studied a number of covered lagoons and found a minimum organic loading rate of 0·06 kg VS m^{-3} was required for consistent biogas production, with rates varying between 0·02 and

0·50 $m^3m^{-2}d^{-1}$, which is equivalent to 0·03 to 0·23 $m^3m^{-3}d^{-1}$, although the carbon dioxide content was generally less than 35%. These researchers later determined a relationship between methane production and lagoon temperature (6.7–27.2°C) after a four-year study on an anaerobic lagoon treating dairy manure:

$$CH_4 = 89.514e^{-0.00381T}, \tag{15.15}$$

where CH_4 is the percentage methane concentration and T is the lagoon temperature (°C) (Safley and Westerman, 1992).

15.5.2 *Ammonia*

Ammonia can be found in high concentrations in anaerobic lagoons and can therefore be an important local source of atmospheric ammonia (NH_3). The form that ammonia takes is controlled by an equilibrium reaction, with NH_3-N increasingly formed as the temperature rises or the pH increases.

$$NH_3 + H_2O \leftrightarrow NH_4^+ + OH^-. \tag{15.16}$$

Aneja *et al.* (2001) studied a number of different anaerobic lagoons (i.e., primary, secondary and tertiary) treating swine wastewater with ammonia concentrations ranging from 37 to 909 mg N L^{-1}. The rate of ammonia volatilization was correlated with surface water temperature and the NH_x (i.e., NH_3 + NH_4+) concentration in the lagoon, with the highest mean emission rate of 120.3 μg N $m^{-2}min^{-1}$ (50–374 μg N $m^{-2}min^{-1}$) recorded in a primary lagoon and falling to the lowest rate of 40.7 μg N $m^{-2}min^{-1}$ (22–84 μg N $m^{-2}min^{-1}$) in a tertiary anaerobic lagoon over a temperature range of 10.3 to 23.3°C. They were able to model the emission rate from their anaerobic lagoons as:

$$\text{In}(NH_3\text{-N flux}) = 1.0788 + 0.0406\ T_L + 0.0015\ (NH_x), \tag{15.17}$$

where the NH_3-N flux is the loss of ammonia from the lagoon surface (μg N $m^{-2}min^{-1}$), T_L is the surface water temperature of the lagoon (°C), and NH_x is the total ammonia-nitrogen concentration (mg N L^{-1}). A similar study by Grant *et al.* (2016) expressed the annual mean ammonia emissions rate in terms of sow live weight, which ranged from 0.124 to 0.240 g N $d^{-1}kg^{-1}$ for breeding and weaning units and 0.072 to 0.208 g N $d^{-1}kg^{-1}$ for finishing units (i.e., growing on to market). They also produce several models: (i) incorporating wind and temperature that accounted for 74% of the variability of the emissions, and (ii) incorporating suspended organic concentration, pH, and nitrogen excretion rates, which accounted for 66% of the variability of the emissions. The solubility of NH_3 and dissociation of NH_4+ to NH_3 in the lagoon liquid is temperature-dependent, resulting in more NH_3 volatilizing from the surface at higher temperatures, creating a seasonal variation in emissions (Fig. 15.6). There is also a significant diurnal variation in the release of ammonia into the atmosphere from lagoons reaching a maximum during the warmest time of day (Aneja *et al.,* 2001) (Fig. 15.7). The temperature in the lagoon is directly related to air temperature, although the scum layer or cover, where used, insulates the wastewater to a certain extent (Fig. 15.8). Grant *et al.* (2016) calculated that for each 1°C rise in the mean daily temperature, there is a corresponding 9.5% increase in maximum ammonia emissions and a 6.9% increase in annual emissions. Ammonia emissions

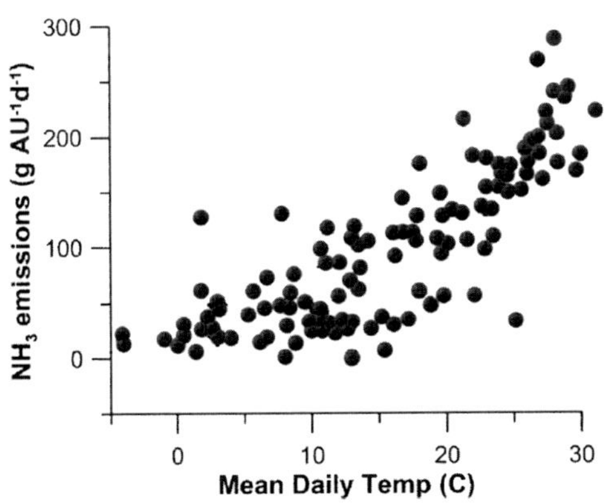

Fig. 15.6. Influence of air temperature on emissions from an anaerobic lagoon. The unit AU represents 500 kg of live pig weight (Grant *et al.,* 2016). Reproduced with permission from Elsevier Publishing.

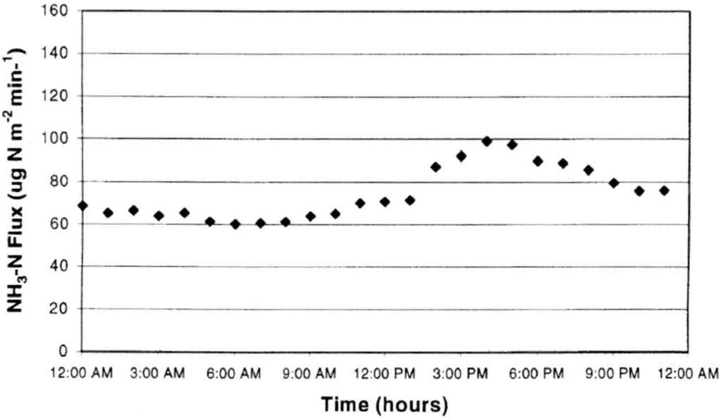

Fig. 15.7. The average diurnal variation in composite hourly ammonia-nitrogen flux concentrations for six anaerobic lagoons (Aneja *et al.*, 2001). Reproduced with permission from Elsevier Publishing.

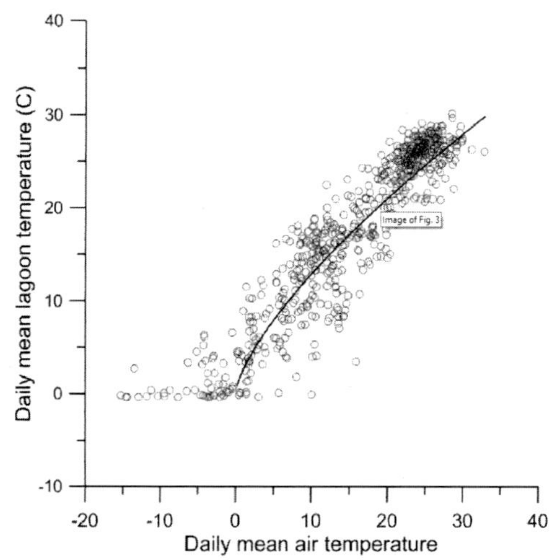

Fig. 15.8. Correlation between the daily mean air temperature, measured 2 m above the lagoon surface, with the daily mean wastewater temperature measured 0.3 m below the lagoon surface (R^2 0.87) (Grant *et al.*, 2016). Reproduced with permission from Elsevier Publishing.

from a dairy waste anaerobic lagoon were 55 kg NH_3 per cow per annum, which is 24% of the annual 180 kg N excreted per cow per annum (Rumburg *et al.*, 2008).

15.5.3 *Odors and hydrogen sulfide*

Many of the breakdown products of anaerobic fermentation in anaerobic lagoons are odorous. Odors are caused by the release of volatile or gaseous compounds into the atmosphere and can have very low odor thresholds, which is generally defined as the average concentration of the compound at which an odor is detected. The most common odor encountered is the smell of rotten eggs caused by hydrogen sulfide, which is produced by sulfate-reducing bacteria. The key organic acids produced during anaerobic breakdown are also odorous and so when the final methanogenic phase is inhibited, these can cause odor problems.

Some food processing wastewaters produce extremely strong odors, especially during treatment and storage (Gerick, 1984; Gray, 1988). For example, sugar beet wastewater undergoes partial anaerobic breakdown within the process and subsequently on storage in lagoons, with the production of a variety of odors. This is due to the processing season taking place from October to January, with the cold temperatures inhibiting the full completion of the fermentation process. The major odors come from the volatile acids that comprise most of the organic fraction of the effluent. The odor threshold concentrations for the volatile organic acids produced during the treatment of sugar beet wastewater and waste are 24.3 ppm for acetic acid, 20.0 ppm for propionic acid, 0.05 ppm for iso-butyric acid, 0.24 ppm for butyric acid, 0.7 ppm for iso-valeric acid, and 3.0 ppm for valeric acid. Therefore, by measuring the volatile acid concentration and dividing it by the appropriate odor threshold concentration, a measure of the odor production or concentration known as the odor number can be calculated:

$$\text{Odor number of compound} = \frac{\text{Compound concentration in wastewater}}{\text{Odor threshold concentration for compound}}. \tag{15.18}$$

Other odors associated with sugar beet processing include trimethylamine, which has a fishy odor, and organic sulfides, which produce a strong odor of rotting cabbage, as do the thiol compounds methyl mercaptan (CH_3SH) and ethyl mercaptan (C_2H_5SH), both of which have very low odor thresholds of 0.0011 ppm and 0.00019 ppm, respectively (Shore *et al.*, 1979). A detailed list of odor threshold values has been compiled by Fazzalari (1978), while a list of common odor compounds associated with anaerobic fermentation is given in Table 15.5.

Under normal operating conditions, no odors are released due to complete anaerobiosis and adequate scum cover. However, problems of odor because of sulfide formation at low temperatures or high sulfate concentrations in the wastewater can occur. Sulfate can also be a problem in some paper mill wastes that use alum as a sizing agent.

Hydrogen sulfide is produced by the reduction of sulfate by a wide range of sulfate-reducing bacteria that use the breakdown

Table 15.5. Some characteristic odors produced by compounds present in wastewaters. These degradation products can be categorized into two main groups: degradation products of nitrogenous or sulfurous compounds. There are other odorous compounds such as those associated with chlorine and phenolic wastes.

Compounds	General formulae	Odor produced
Nitrogenous		
Amines	CH_3NH_2, $(CH_3)_3N$	Fishy
Ammonia	NH_3	Ammoniacal, pungent
Diamines	$NH_2(CH_2)_4NH_2$, $NH_2(CH_2)_5NH_2$	Rotten flesh
Skatole	$C_8H_5NHCH_3$	Fecel, repulsive
Sulfurous		
Hydrogen sulfide	H_2S	Rotten eggs
Mercaptans	CH_3SH, $CH_3(CH_2)_3SH$	Strong decayed cabbage
Organic sulfides	$(CH_3)_2S$, CH_3SSCH_3	Rotten cabbage
Sulfur dioxide	SO_2	Pungent, acidic
Others		
Chlorine	Cl_2	Chlorine
Chlorophenol	$Cl.C_6H_4OH$	Medicinal, phenolic

products of organic matter, in particular, acetate, butyrate, ethanol, lactate, and propionate while using sulfate as a terminal electron acceptor. There are over 30 known genera of sulfate-reducing bacteria in anaerobic lagoons with Desulfobacterium, Desulfobulbus and Desulfovibrio all commonly recorded (Zeikus, 1980; Rabus, *et al.*, 2006; Cook *et al.*, 2008). Although present in relatively small numbers in wastewater, sulfate-reducing bacteria rapidly grow under the ideal conditions found in anaerobic lagoons. For example, *Desulfovibrio desulfuricans* is present in low numbers in sewage (60–600 mL^{-1}) but rapidly increases on storage up to concentrations over 100,000 mL^{-1} over 14 days. Ducey and Hunt (2013) recorded 26 genera of sulfate-reducing bacteria in the four lagoons they studied. The genera Desulfatirhabdium, Desulfobulbus, Desulfocurvus, Desulfomonile, Desulfovibrio, Desulfovirga, and Thermodesulfobium were recorded in all four separate lagoons; Desulfoglaeba, Desulfonispora, Desulfosporosinus, Desulfatiferula, and Desulforegula in three lagoons; Desulfobacca, Desulfofaba, Desulfofustis, Desulfococcus, and Desulfomicrobium in two lagoons; while Desulfarculus, Desulfonema, Desulforhabdus, Desulforhopalus, Desulfospira, Desulfovermiculus, Desulfurispora, Syntrophobacter, and Thermodesulforhabdus were recorded in just one lagoon. The redox is fairly critical for sulfate reduction, with an ORP of <–50 mV normally required for H_2S production, which normally occurs within the sludge layer. The mean concentration of H_2S released with biogas (CH_4 [80.4%]; CO_2 [14.7%]) during a four-year study on an anaerobic lagoon treating dairy manure was 272 ppm over a lagoon temperature range of 6.7 to 27.2°C (Safley and Westerman, 1992).

With better design, it is now not always critical to have a surface scum as long as sufficiently high organic loading can be maintained to ensure anaerobic conditions. Hydrogen sulfide is present in an aqueous solution as hydrogen sulfide gas (H_2S), the bisulfide ion (HS^-), or the sulfide ion (S_2^-). Thus, for any given total sulfide concentration, the amount present as odorous H_2S is pH-dependent (Fig. 15.9). Mara and Pearson (1998) explain that designers have been reluctant to use anaerobic lagoons at treatment plants in case of odor formation. They stress that odor can be controlled by not exceeding the

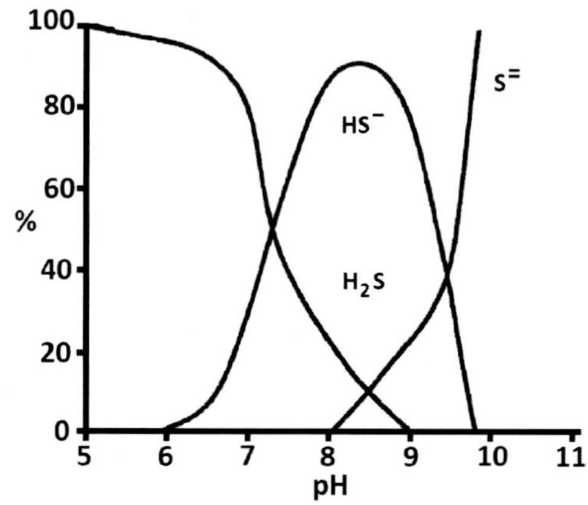

Fig. 15.9. The effect of pH on the equilibrium of hydrogen sulfide-bisulfide–sulfide.

design loadings (Table 15.3), and ensuring that influent wastewater sulfate concentrations are <500 mg SO_4 L^{-1}. Sulfide reacts with heavy metals to form insoluble metal sulfides, but if concentrations exceed <50 mg L^{-1}, there is a progressive inhibition of methanogenesis (Pfeffer, 1970). No odor should arise from a lagoon if the liquid temperature is above 20°C.

15.5.4 *Denitrification*

Jones *et al.* (2000) observed nitrogen gas emissions from anaerobic lagoons treating piggery wastewater. Under normal circumstances, denitrification requires the pre-step of nitrification, which is impossible under anaerobic conditions but may be possible under microaerobic conditions found at the air–water interface of open lagoons. A wide range of bacteria and archaea are able to denitrify under anoxic conditions, using NO_3^- and NO_2^- as terminal electron acceptors. The reduction of nitrate follows a sequence of steps, each mediated by a specific enzyme.

$$NO_3^- (aq) \rightarrow NO_2^- (aq) \rightarrow NO (g) \rightarrow N_2O (g) \rightarrow N_2 (g). \quad (15.19)$$

It is during incomplete denitrification that both NO and N_2O can be released into the atmosphere, causing damaging effects to the environment, both in terms of toxicity and global warming (Sec. 3.1).

Denitrification

$$NO_3^- + 1.25CH_2O \rightarrow 0.5N_2 + OH^- + 0.75H_2 + 1.25CO_2. \quad (15.20)$$

As can be seen from the equation above, denitrification releases OH^- ions into the effluent, increasing the alkalinity. However, under the strict anaerobic environment of the lagoons, nitrification would be very limited or at best incomplete. Jones and his colleagues suggested alternative microbial mediated denitrification processes such as anaerobic ammonium oxidation, better known as anammox. Ammonium is used as an energy source by the anammox bacteria and CO_2 as a carbon source, with NO_2^- and NO_3^-, if present, used as electron acceptors (Gray, 2021). Kartel *et al.* (2011) have described the process in detail where the enzyme nitrite oxidoreductase reduces nitrite to NO, which then reacts with ammonium to form hydrazine (N_2H_2), enabled by the enzyme hydrazine synthase, which is oxidized by hydrazine dehydrogenase to N_2. The bacteria are extremely slow-growing and strict anaerobes, unable to function if the dissolved oxygen concentration rises above 0.5 mg L^{-1}.

Anammox

$$NH_4^+ + NO_2^- \rightarrow N_2 + 2H_2O. \quad (15.21)$$

The autotrophic bacteria responsible for the oxidation of ammonia and the simultaneous reduction of nitrite belong to the widely distributed phylum *Planctomycetes.* Not all the bacteria in this phylum are able to oxidize ammonium, this being restricted to the Anammox-*Planctomycetes*, which have a characteristic membrane-bound cell compartment called the anammoxosome (Fuerst and Sagulenko, 2011).

15.6 Covered Lagoons for Biogas Collection

In the early 1990s, with increasing concern over the release of methane into the atmosphere, which has a high Global Warming Potential (GWP) (Table 3.1), operators began to cover anaerobic lagoons (Green *et al.,* 1995; Hodgson and Paspaliaris, 1996). This was initially done using a sheet of geotextile membrane floated over the surface of the lagoon, with the biogas passively recovered and flared off. Covering also means that scum formation or the use of plastic balls to act as a surface barrier is no longer required and that odor control can be managed, opening up the process to a wider range of wastewaters that can be treated, and also increasing its sustainability.

Although mesophilic digestion using standard digesters has been used for agriculture, the cost and complexity of operation have seen its limited application. In the past decade, the use of farm-based covered anaerobic lagoons has spread throughout Europe, the US, Australia, New Zealand, and elsewhere for the treatment of animal wastewaters, primarily cattle and pig manures. This is hugely advantageous for both farmers and the environment. For the farmers, it provides safe storage, commercial biogas production, also the on-site generation of electricity, as well as heat, allowing farmers to diversify into new production areas such as high-profit greenhouse fruit, herbs and vegetables (Fig. 15.10). Farm-based lagoons only need to be desludged every 3 to 5 years, depending on the depth, total capacity in relation to loading, and degree of anaerobic breakdown achieved (Fig. 15.11). A highly stabilized and largely odorless sludge is produced, which is generally used for land disposal (Owusu-Twum and

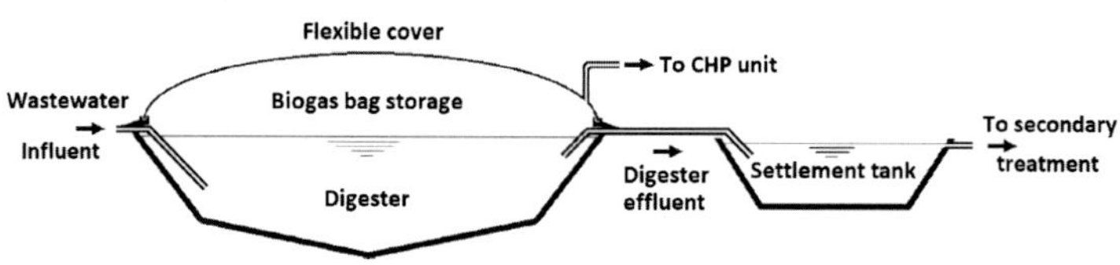

Fig. 15.10. Covered anaerobic lagoons can be constructed in any size and are a comparatively cheap option to recover bigas from small to large-sized agricultural, industrial and wastewater operations.

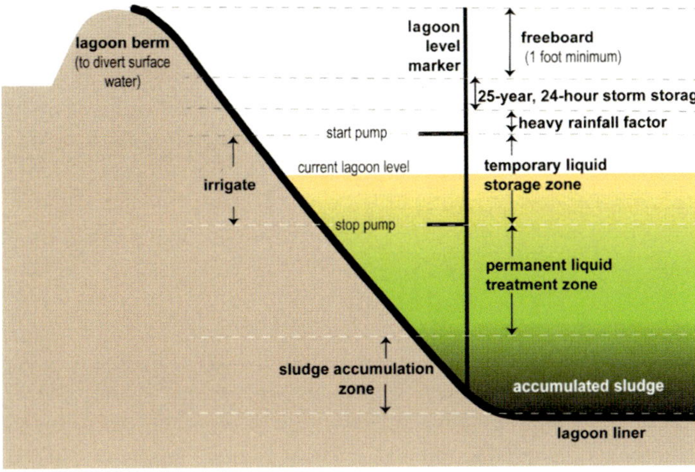

Fig. 15.11. Section through a single-stage anaerobic lagoon used for swine treatment (Owsusu-Twum and Sharara, 2020). Reproduced with permission from Elsevier Publishing.

Sharara, 2020). The effluent is rich in nutrients and can be used for surface irrigation of grassland for silage or hay.

Anaerobic lagoons are widely used in New Zealand for the treatment of predominantly dairy and also other animal wastewaters. Although originally open, lagoons are increasing being covered to collect the biogas to reduce greenhouse gas emissions as well as generate energy. Such ponds are operated below optimum mesophilic temperature and so have a reduced volatile solids (VS) loading of 0.02 to 0.46 kg VS $m^{-3}d^{-1}$. Park and Craggs (2007) compared two anaerobic lagoons — one treating piggery wastewater (influent BOD of 11,500 mg L^{-1}; the BOD loading rate of 0.30 kg BOD $m^{-3}d^{-1}$; VS of 5,450 mg L^{-1}) and the other dairy wastewater (influent BOD of 1,147 mg L^{-1}; BOD loading rate of 0.01 kg BOD $m^{-3}d^{-1}$; VS of 1,980 mg L^{-1}). The mean BOD and VS removal efficiencies were 60.5 and 73.8% for the piggery, and 81.1 and 78.5% for the dairy wastewater anaerobic lagoons, respectively. The mean biogas production of 0.78 and 0.03 $m^3m^{-2}d^{-1}$, contained 72.0 and 80.3% methane from piggery and dairy wastewater, respectively, which is equivalent to 0.81 and 0.54 m^3 CH_4 $kg^{-1}VS$.

Biogas production is closely linked to temperature, as is COD reduction and process stability. Optimal biogas production occurs in mesophilic systems at 38°C, which is rarely achieved under field conditions without the use of heat exchangers. In a study of covered anaerobic lagoons treating abattoir wastewater in Queensland, Australia, Schmidt *et al.* (2019) found that even the highest natural lagoon temperature of 31°C represented a reduction of potential biogas production of 30%. This increased to an 80% reduction at the lower winter temperature of 25.5°C. They concluded that the natural ambient temperatures found in covered anaerobic lagoons cannot ensure sustainable or profitable production of biogas and recommended the use of heat exchangers or waste heat where available, to maximize biogas production. Some farms have incorporated passive solar heating to optimize the temperature. The move from low-rate anaerobic lagoons operated at ambient temperatures to high-rate lagoons maintained at 38°C maximizes return on investment and overall confidence in the technology, as operating at higher temperatures also increases operational stability and improves both effluent and sludge quality.

Anaerobic lagoons are often used in series with a facultative pond followed by at least one maturation pond. The loading to the facultative pond after the anaerobic stage is calculated using Table 15.3, which gives the expected percentage BOD reduction at the medium operating temperature (Chap. 14).

References

Aneja, V.P., Bunton, B., Walkr, J.T. and Malik, B.P. (2001) Measurement and analysis of atmospheric ammonia emissions from anaerobic lagoons. *Atmospheric Environment*, **35**, 1949–1958.

Bapteste, E., Brochier, C. and Boucher, Y. (2005) Higher-level classification of the Archaea: Evolution of methanogenesis and methanogens. *Archaea*, **1**, 353–363.

Barnes, D., Forster, C.F. and Hrudey, S.E. (eds.) (1984). *Food and Allied Industries*. Pitman, London.

Brockett, O.D. (1976) Microbial Reactions in Facultative Ponds — 1. The Anaerobic Nature of Oxidation Pond Sediments. *Water Research*, **10**, 1, 45–49.

Buswell, A.M. and Mueller, H.F. (1952) Mechanisms of methane fermentation. *Industrial Engineering Chemistry*, **44**, 550–552.

Cook, K.L., Whitehead, T.R., Spence, C. and Cotta, M.A. (2008) Evaluation of the sulfate reducing bacterial population associated with stored swine slurry. *Anaerobe*, **14**, 172–180.

Cook, K.L., Rothrock, M.J., Lovanh, N., Sorrell, J.K. and Loughrin, J.H. (2010) Spatial and temporal changes in the microbial community in an anaerobic swine waste treatment lagoon. *Anaerobe*, **16**, 74–82. doi:10.1016/j.anaerobe.2009.06.003

Ducey, T.F. and Hunt, P.G. (2013) Microbial community analysis of swine wastewater anaerobic lagoons by next-generation DNA sequencing. *Anaerobe*, **21**, 50–57.

Fazzalari, F.A. (ed.) (1978) Compilation of odor and taste threshold values data. ASTM Data Series DS 48A. American Society for Testing and Materials, Philadelphia.

Fuerst, J.A. and Sagulenko, E. (2011) Beyond the bacterium: Planctomycetes challenge our concepts of microbial structure and function. *Nature Reviews Microbiology*, **9**, 403–413.

Gallert, C. and Winter, J. (2005) Bacterial metabolism in wastewater treatment systems. In *Environmental Biotechnology* (eds. H.J. Jordening and J. Winter). Wiley-VCH, Weinheim.

Gerick, J.A. (1984) Land application odour study. *Journal of Water Pollution Control Federation*, **56**, 287–291.

Gloyna, E.F. (1971) Waste stabilization ponds. WHO Monograph Series, 60. World Health Organization, Geneva.

Grant, R.H., Boehm, M.T. and Heber, A.J. (2016) Ammonia emissions from anaerobic waste lagoons at pork production operations: Influence of climate. *Agricultural and Forest Meteorology*, **228–229**, 73–84. http://dx.doi.org/10.1016/j.agrformet.2013.06.006

Gray, N.F. (1988) *Sugar Beet Processing in Ireland: A Case Study*. Water Technology Press, Dublin.

Gray, N.F. (2021) *Fixed-Film Reactors in Wastewater Treatment*. World Scientific Publishing, Singapore.

Green, F.B., Lundquist, T.J. and Oswald, W.J. (1995) Energetics of advanced integrated wastewater pond systems. *Water Science and Technology*, **31**(12), 9–20.

Hammer, M.J. (1977). *Water and Wastewater Technology.* John Wiley, New York.

Hawkes, H.A. (1983). Stabilization ponds. In *Ecological Aspects Of Used Water Treatment, Vol. 2. Biological Activities And Treatment Processes* (eds. C.R. Curds and H.A. Hawkes). Academic Press, London, UK, pp. 163–217.

Hodgson, B. and Paspaliaris, P. (1996) Melbourne water's wastewater treatment lagoons: Design modifications to reduce odours and enhance nutrient removal. *Water Science and Technology*, **37**(7), 157–164.

Hughes, D.E. (1980) What is aerobic digestion? An overview. In *Anaerobic Digestion* (eds. D. A. Stafford, B.I. Wheatley and D.E. Hughes). Applied Science Publishers, London, pp. 1–14.

Jones, M.L., Liehr, S.K., Classon, J.J. and Robarge, W. (2000) Mechanisms of dinitrogen gas formation in anaerobic lagoons. *Advances in Environmental Research*, **4**, 133–139.

Kartal, B., Maalcke, W.J., de Almeida, N.M., Cirpus, I., Gloerich, J., Geerts, W., den Camp, H.J.M.O., Harhangi, H.R., Janssen-Megens, E.M., Francoijs, K.-J., Stunnenberg, H.G., Keltjens J.T., Jetten, M.S.M. and Strousm, M. (2011). Molecular mechanism of anaerobic ammonium oxidation. *Nature*, **479**, 127–130.

Konstandt, H.G. (1976) *Microbial Energy Conversion* (eds. H.G. Schlegel and J. Barnea). Erich Goltze, Gotingen.

Liu, Y. (2010) Taxonomy of Methanogens. In *Handbook of Hydrocarbon and Lipid Microbiology* (ed. K.N. Timmis). Springer, Heidelberg, pp. 547–558. https://doi.org/10.1007/978-3-540-77587-4_42

Lovanh, N., Loughrin, J.H., Cook, K., Rothrock, M. and Sistani, K. (2009) The effect of stratification and seasonal variability on the profile of an anaerobic swine waste treatment lagoon. *Bioresource Technology*, **100**, 3706–3712. doi:10.1016/j.biortech.2008.09.024

Mara, D.D. and Pearson, H.W. (1998) Design manual for waste stabilization ponds in Mediterranean countries. Lagoon Technology International, Leeds, UK.

McCarty, P.L. (1964) Anaerobic waste treatment fundamentals. I. Chemistry and microbiology; II. Environmental requirements and control; III. Toxic materials and their control. *Public Works*, **95**, 91–94; 107–112; 123–126.

McInerney, M.L., Bryant, M.P. and Stafford, D.A. (1980) Metabolic stages and energetics of microbial anaerobic digestion. In *Anaerobic Digestion* (eds. D.A. Stafford, B.I. Wheatley and D.E. Hughes). Applied Science Publishers, London, pp. 91–98.

McLaughlin, M.R., Brooks, J.P. and Adeli, A. (2012) Temporal flux and spatial dynamics, fecal indicators, and zoonotic pathogens in anaerobic swine manure lagoon water. *Water Research*, **46**, 4949–4960. https://doi.org/10.1016/j.watres.2012.06.023

Nelson, N.O., Mikkelsen, R.L. and Hesterberg, D.L. (2003) Struvite precipitation in anaerobic swine lagoon liquid: Effect of pH and Mg:P ratio and determination of rate constant. *Bioresource Technology*, **89**, 229–236. doi:10.1016/S0960-8524(03)00076-2

Owusu-Twum, M.Y and Sharara, M.A. (2020) Sludge management in anaerobic swine lagoons: A review. *Journal of Environmental Management*, **271**, 110949. https://doi.org/10.1016/j.jenvman.2020.110949

Pandey, P.R., Cao, W., Vaddella, V., Castillo, A.R., Sousa, A. and del Rio, N.S. (2016) Simulating the effects of mesophilic anaerobic and aerobic digestions, lagoon system and composting on pathogen inactivation. *Ecological Engineering*, **97**, 633–641. http://dx.doi.org/10.1016/j.ecoleng.2016.10.047

Park, J.B.K. and Craggs, R.J. (2007) Biogas production from anaerobic waste stabilization ponds treating dairy and piggery wastewater in New Zealand. *Water Science and Technology*, **55**(11), 257–264.

Parker, C.D., Jones, H.L. and Taylor, W.S. (1950) Purification of sewage in lagoons. *Sewage and Industrial Waste*, **22**, 760–775.

Paterson, C. and Curtis, T. (2005) Physical and chemical environments. In *Pond Treatment Technology* (ed. A. Shilton). IWA Publishing, London, UK. https://doi.org/10.2166/9781780402499

Pearson, H. (2005) Microbiology of waste stabilisation ponds. In *Pond Treatment Technology* (ed. A. Shilton). IWA Publishing, London, UK. https://doi.org/10.2166/9781780402499

Pfeffer, J.T. (1970) Anaerobic lagoons: Theoretical considerations. In *Proceedings of the Second International Symposium on Water Treatment Lagoons* (ed. R.E. McKinney). University of Kansas, Laurence, KS, USA, pp. 310–320.

Pfeffer, J.T. (1980) Anaerobic digestion processes. In *Anaerobic Digestion* (eds. D.A. Stafford, B.I. Wheatley and D.E. Hughes). Applied Science Publishers, London, pp. 15–36.

Porges, R. and Brackney, D. H. (1962) Industrial waste disposal by stabilization ponds. *Engineering Experimental Station Bulletin*, **68**, 131–140, Louisiana State University.

Rabus, R., Hansen, T.A. and Widdel, F. (eds.) (2006) *Dissimilatory Sulfate- and Sulfur Reducing Prokaryotes, 3rd ed.* Springer Science, New York, NY.

Rumburg, B., Mount, G.H., Yonge, D., Lamb, B., Westberg, H., Neger, M., Filipy, J., Kinacid, R. and Johnson, K. (2008) Measurements and modelling of atmospheric flux of ammonia from an anaerobic dairy waste lagoon. *Atmospheric Environment*, **42**, 3380–3393. doi:10.1016/j.atmosenv.2007.02.046

Safley, L.M. and Westerman, P.W. (1988) Biogas production from anaerobic lagoons. *Biological Wastes*, **23**(3), 181–193. https://doi.org/10.1016/0269-7483(88)90033-X

Safley, L.M. and Westerman, P.W. (1992) Performance of a dairy manure anaerobic lagoon. *Bioresource Technology*, **42**, 43–52. https://doi.org/10.1016/0960-8524(92)90086-D

Schmidt, T., Harris, P., Lee, S. and McCabe, B.K. (2019) Investigating the impact of seasonal temperature variation on biogas production from covered anaerobic lagoons treating slaughterhouse wastewater using lab scale studies. *Journal of Environmental Chemical Engineering*, 7, 103077. https://doi.org/10.1016/j.jece.2019.103077

Sharpley, A., Beegle, D., Bolster, C., Good, L., Joern, B., Ketterings, Q., Lory, J., Mikkelsen, R., Osmond, D. and Vadas, P. (2011) Revision of the 590 nutrient management standard: SERA-17 recommendations. *Southern Cooperative Series Bulletin*, **412**. Virginia Technical University, Blacksburg, VA. https://sera17dotorg.files.wordpress.com/2015/02/590-sera-17-recommendations.pdf

Shore, M., Dutton, J.V., Broughton, N.W. and Bumstead, N. (1979) The effect of pH on the odours of transport and storage waters. British Sugar Corporation Research Laboratories Report 79/36. Annual Works Managers Meeting, British Sugar Corporation Ltd, Norwich.

USEPA. (1983) Design manual: Municipal wastewater stabilization ponds. EPA-625/1-83-015. Office of Water, US Environmental Protection Agency, Washington, DC, USA.

USEPA. (2011) Principles of design and operation of wastewater treatment pond systems for plant operators, engineers and managers. EPA/600/R-10/088. Office of Research and Development, US Environmental Protection Agency, Cincinnati, Ohio, USA.

Vuillot, M. and Boutin, C. (1987). Waste stabilization ponds in Europe: A state of the art review. *Water Science and Technology*, **19**(12), 1–6. https://doi.org/10.2166/wst.1987.0119

WHO. (1987) Wastewater stabilization ponds: Principles of planning and practice. Technical Publication, 10. Regional Office for the Eastern Mediterranean, World Health Organization, Alexandria.

WHO. (2006) Wastewater use in agriculture: WHO guidelines for the safe use of wastewater, excreta and greywater. World Health Organization, Geneva, Switzerland. https://www.who.int/water_sanitation_health/publications/gsuweg2/en/

Wu, B. and Chen, Z. (2011) An integrated physical and biological model for anaerobic lagoons. *Bioresource Technology*, **102**, 5032–5038. doi:10.1016/j.biortech.2011.01.076

Zeikus, J.G. (1980) Microbial populations in digesters. In *Anaerobic Digestion* (eds. D.A. Stafford, B.I. Wheatley and D.E. Hughes). Applied Science Publishers, London, pp. 61–89.

16
Oxidation Ponds

16.1 Introduction to Oxidation Ponds

The oxidation pond is designed to provide an aerobic breakdown of organic matter in wastewater, with a significant proportion of the dissolved oxygen provided by photosynthetic algae. Three types of stabilization ponds fall into this category: facultative ponds, high-rate algal ponds, and maturation ponds (Middlebrooks *et al.*, 1982). All three are different from each other and treated separately below. A fourth type of oxidation pond, the purification lake, is also considered in this chapter.

16.2 Facultative Ponds

16.2.1 *Basis of operation*

Facultative stabilization ponds are characterized by having an upper aerobic and lower anaerobic zone, with active purification occurring in both. They are the most common type of waste stabilization pond in use that can completely treat both crude and settled sewage, as well as a wide range of organic-based industrial wastewaters, normally after pretreatment by anaerobic or completely mixed aerated ponds (Chaps. 15 and 17). Facultative ponds are often categorized as either primary or secondary ponds, treating raw wastewaters (i.e., acting also as a primary settlement tank but after preliminary treatment to remove grit and rags) or settled wastewaters (i.e., a post anaerobic lagoon or septic tank), respectively. Many modern designs often use facultative ponds in series (i.e., primary, secondary and tertiary). When the term facultative pond is used on its own, then a secondary treatment stage is being referred to.

As organic matter enters the basin, the settleable and flocculated colloidal matter settle to the bottom to form a sludge layer where organic matter is decomposed anaerobically. The remainder of the organic matter, which is either soluble or suspended, passes into the body of the water where decomposition is mainly aerobic or facultative, although it is occasionally anaerobic. The biological basis of the treatment in facultative ponds is summarized in Fig. 16.1. Organic matter in solution or suspension is broken down principally by aerobic and facultative bacteria, which releases nitrogen and phosphorus compounds and carbon dioxide. Algae use these inorganic compounds for growth by using sunlight for energy and releasing oxygen into the solution. The dissolved oxygen is then utilized by the aerobic bacteria, thus completing the symbiotic cycle. Oxygen is also introduced by natural oxygen transfer, the rate of which is increased by turbulence caused by wind action and the oxygen deficit (i.e., the rate of oxygen transfer is directly proportional to the oxygen deficit).

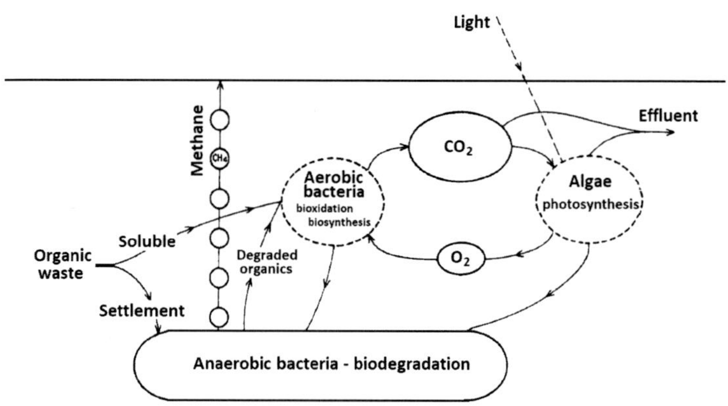

Fig. 16.1. Summary of biological interactions in a facultative pond (Hawkes, 1983a). Reproduced with permission from Elsevier Publishing.

In the sludge layer, the settled solids are anaerobically broken down, with methane, nitrogen and carbon dioxide being released along with a variety of soluble organic degradation products. The gases escape into the atmosphere. In a facultative pond, up to 30% of the biochemical oxygen demand (BOD) load can be dissipated as gas production. The soluble degradation products, such as ammonia, organic acids, and inorganic nutrients are also released and subsequently oxidized aerobically in the water layer. The hydrogen sulfide released from the deposited sludge is also oxidized in the water zone, thus preventing odors from being released. Anaerobic degradation is temperature-dependent and no significant activity occurs in the sludge layer below a water temperature of 17°C. A four-fold increase in activity occurs in the sludge with every 5°C rise in temperature, over the range of 4 to 22°C (Gloyna, 1971). The digestion rate of the sludge expressed as the rate of gas production (K_s), at a particular temperature (T) can be estimated by:

$$K_s = 0.002(1.35)^{-(20-T)}. \qquad (16.1)$$

16.2.2 *Ecology*

Aerobic and facultative bacteria are the primary decomposers in the pond system. Although fungi are also present in ponds (Cooke, 1963), they play a minor role and are thought to be restricted due to the high pH caused by the photosynthetic activity of the algae (Arceivala *et al.*, 1970). The dominant bacteria found are similar to those isolated from other aerobic treatment systems, and mainly belong to the genera *Pseudomonas*, *Achromobacter* and *Flavobacterium*. Apart from obligate aerobes, other types of bacteria such as coliform bacteria, methane bacteria, sulfate reducers, and purple sulfur bacteria such as *Thiocapsa floridana* and *Chromatium vinosum*, are all common (Holm and Vennes, 1970; Pike, 1975). Laboratory-based studies using experimental ponds have demonstrated that the degree of illumination has no effect on bacterial activity and that the BOD removal is related to bacterial density (Uhlmann, 1969). However, according to Uhlmann, the number of solids synthesized, which is largely algal biomass, is proportional to the degree of illumination, and an increase in algal biomass results in an increase in nitrogen and phosphorus incorporated into the sludge. In these experimental ponds, oxygen was never limiting, even in complete darkness, whereas oxygen is normally a limiting factor in full-size ponds. Therefore, the full importance of algae in the euphotic zone was not demonstrated. The oxygen for aerobic bacterial degradation is mostly supplied by the photosynthetic activity of the phototrophs.

Although there is a succession of dominant algal species during the year, generally, only one or two species will be dominant at any one time in a facultative pond. Porges and Mackenthun (1963) found a striking similarity between the algal species present in facultative ponds throughout the United States (US) and concluded that geographical location had little effect on speciation. The most recorded genera are: *Chlorella, Scenedesmus, Chlamydomonas, Micractinium, Euglena, Ankistrodesmus, Oscillatoria,* and *Microcystis* (Table 16.1). Motile genera such as *Chlamydomonas, Pyrobotrys,* and *Euglena* can alter their vertical position in the water column to take maximum advantage of light intensity and temperature, unlike non-motile genera such as *Chlorella*. The dominant algal species is determined by the organic loading, with those algae able to tolerate anaerobic conditions such as the *Chlamydomonas* spp. and *Euglena* spp. being recorded in ponds receiving heavy organic loads. However, the photosynthesis of many of the algae commonly found in ponds is inhibited by high ammonia conditions. The oxygenation capacity of different algae varies appreciably. For example, blue-green algae such as *Anabaena variabilis, A. cylindrica, Phormidium faveolarum,* and *Calothrix membranacea* are less efficient than green algae in ponds, with both the *Chlorella* spp. and *Scenedesmus* spp. being desirable because of superior oxygenating capability. There is also some evidence that certain algae are capable of growth by photo-assimilation of organic compounds. The utilization of organic matter as a supplementary source of carbon normally only occurs when light or carbon dioxide is limiting (Hawkes, 1983b). There are few studies on the community structure of algal species in waste stabilization ponds with average algal abundance between 10^4 and 10^5 cells per mL^{-1}. In a waste stabilization pond (WSP) system treating domestic wastewater in Ecuador, Pham *et al.* (2014) found species were associated with the BOD

Table 16.1. Presence and absence of key algal genera in different waste stabilization ponds (Mara and Pearson, 1998).

Algae	Facultative ponds	Maturation ponds
Euglenophyta		
Euglena	+	+
Phacus	+	+
Chlorophyta		
Chlamydomonas	+	+
Chlorogonium	+	+
Eudorina	+	+
Pandorina	+	+
Pyrobotrys	+	+
Ankistrodesmus	⊗	+
Chlorella	+	+
Micractinium	⊗	+
Scenedesmus	⊗	+
Selenastrum	⊗	+
Carteria	+	+
Coelastrum	⊗	+
Dictyosphaerium	⊗	+
Oocystis	⊗	+
Rhodomonas	⊗	+
Volvox	+	⊗
Chrysophyta		
Navicula	+	+
Cyclotella	⊗	+
Cyanophyta		
Oscillatoria	+	+
Anabaena	+	+

+ = present; ⊗ = absent

concentration, with species showing distinct saprobic preferences with the greatest diversity and abundance recorded in the maturation pond compared to the facultative pond (Table 16.2). Like other groups, algae have saprobic ratings allowing them to be used as

Table 16.2. The occurrence of algal species (+Present, –Absent) and abundance (where the volume of counting unit [VCU] is μm^3) in facultative and maturation ponds (Pham *et al.*, 2014). Reproduced with permission from Elsevier Publishing.

Species	Facultative pond	Maturation pond	VCU (μm^3)	Characteristics
Chlorophyta				
Actinastrum hantzschii	+	+	38	In colonies
Chlamydomonas sp.	+	+	2,351	Flagella, unicellular
Chlorella sp.	+	+	489	Unicellular
Closteriopsis sp.	–	+	713	Unicellular
Coelastrum microporum	+	+	87	In colonies
Desmodesmus costato granulatus	+	+	15	Unicellular or in colonies
Eudorina elegans	+	+	955	Flagella, in colonies
Korshikoviella sp.	+	+	204	Unicellular
Micractinium pusillum	–	+	58	Unicellular or in colonieswith setae
Monoraphidium sp.	+	+	99	Unicellular or in colonies
Oocystis sp.	+	+	361	Unicellular or in colonies
Pandorina morum	+	+	524	Flagella, in colonies
Planktosphaeria gelatinosa	+	+	3,444	Unicellular in gel matrix
Scenedesmus sp.	+	+	495	Unicellular or in colonieswith spines
Schroederia sp.	+	+	1,204	Unicellular
Unidentified species	+	+	110	Unidentified, similar to *Tetraedron* sp.
Unidentified species	+	+	795	Unidentified, similar to *Pandorina* sp.
Chrysophyta				
Cyclotella sp.	+	+	2,106	Diatom, unicellular or in colonies

(*Continued*)

Table 16.2. (*Continued*)

Species	Facultative pond	Maturation pond	VCU (μm^3)	Characteristics
Navicula sp.	+	+	538	Diatom, unicellular
Nitzschia sp.	–	+	82	Diatom, unicellular
Cryptophyta				
Chroomonas sp.	+	+	132	Flagella, unicellular
Cryptomonas sp.	+	+	1,345	Flagella, unicellular
Cyanobacteria				
Merismopedia minima	+	+	0.5	Mobile, in colonies
Oscillatoria sp.	+	+	94	Filamentous, mobile
Euglenophyta				
Colacium cyclopicola	+	+	621	Flagella, unicellular or attached in group
Euglena viridis and *E. acus*	+	+	8,468	Flagella, unicellular
Lepocinclis texta	+	+	4,312	Flagella, unicellular
Phacus pleuronectes and *P. pyrum*	+	+	1,900	Flagella, unicellular
Strombomonas acuminate	+	+	10,011	Flagella, unicellular
Trachelomonas volvocina and *T. armata*	+	+	1,651	Flagella, unicellular

indicators of organic loading and general pond health. Motile algal species such as the Euglena and Cryptophytes dominate at the inlet area and in the facultative pond, and their percentage abundance decreases as the BOD of the wastewater in the pond declines, being replaced by the Chlorophyta, which are the most common in the maturation pond (Table 16.3).

The main function of algae in facultative ponds is as phototrophs, producing oxygen to maintain and maximize aerobic conditions within the pond. A supplementary role, but a very important one, is the removal of plant nutrients such as nitrogen and phosphorus. Facultative ponds are the most effective of all the conventional biological treatment processes in removing these nutrients, and thus

Table 16.3. Percentage contribution (%) of algal phyla as wastewater passes through a two-pond system (facultative — maturation ponds) showing the effect of organic strength on abundance (Pham *et al.*, 2014). Reproduced with permission from Elsevier Publishing.

	Algal distribution			
Algal phylum	**Inlet facultative pond**	**Outlet facultative pond**	**Inlet maturation pond**	**Outlet maturation pond**
Chlorophyta	13.12 ± 1.72	28.00 ± 4.88	54.24 ± 3.16	47.61 ± 6.56
Chrysophyta	0.02 ± 0.03	0.05 ± 0.06	0.04 ± 0.02	0.14 ± 0.16
Cryptophyta	15.96 ± 2.93	9.38 ± 5.43	16.29 ± 2.66	28.20 ± 5.33
Cyanobacteria	0.67 ± 0.15	1.34 ± 0.15	0.33 ± 0.01	0.40 ± 0.05
Euglenophyta	70.22 ± 4.53	61.24 ± 0.33	29.10 ± 5.86	23.65 ± 1.34

reducing eutrophication in receiving waters, without dedicated redox zones or tertiary treatment. Apart from nutrient uptake to satisfy the normal nutrition requirements of algae, some species are able to take up nutrients, especially phosphorus, far in excess of their own requirements, which is known as "luxury uptake". Nutrients are also precipitated out of the solution as a consequence of the pH change brought about by photosynthesis, which reduces the concentration of carbon dioxide in the water. Above pH 8, phosphates are precipitated out as calcium phosphate and at higher pH values, nitrogen can be lost as ammonia. However, above pH 9, the conditions are no longer optimal for normal aerobic and facultative bacterial activity. Maximum dissolved oxygen concentrations reach a peak in mid-afternoon, falling to a minimum during the night and early morning as photosynthesis ceases but respiration continues. During periods of rapid photosynthesis, algal demand for carbon dioxide exceeds that produced by bacterial respiration. At this point, carbonate and bicarbonate ions dissociate to produce carbon dioxide, which is used by the algae, and hydroxyl ions, which accumulate, raising the pH even further to 10 and above.

$$2HCO_3^- \rightarrow CO_3^{2-} + H_2O + CO_2, \quad (16.2)$$

$$CO_3^{2-} + H_2O \rightarrow 2OH^- + CO_2. \quad (16.3)$$

Once photosynthesis declines, free carbon dioxide accumulates, reversing the above reactions and returning the pH to normal. This is a major mechanism for the destruction of fecal bacteria in ponds.

Due to the absence of suitable surfaces for the attachment of nitrifying bacteria, nitrification is largely absent from facultative ponds. Unless there is nitrate in the influent wastewater, denitrification is also negligible. However, removal rates of 80% for total nitrogen (TN) and 95% for ammonia are possible. Where the BOD removal is ≥90%, then removal efficiencies of the total phosphorus (TP) can be as much as 45% (Huang and Gloyna, 1984). Although it remains difficult to predict phosphorus removal, Pano and Middlebrooks (1982) and Reed (1985) have produced equations to predict the removal of ammoniacal nitrogen (Eqs. [16.4] and [16.5]) and TN (Eq. [16.6]), respectively, in both facultative and maturation ponds.

The removal of ammoniacal nitrogen at temperatures below 20°C:

$$C_e = C_i/\{1 + [(A/Q)(0.0038 + 0.000134T) \times \exp((1.041 + 0.044T)(\mathrm{pH} - 6.6))]\}. \quad (16.4)$$

The removal of ammoniacal nitrogen at temperatures above 20°C:

$$C_e = C_i/\{1 + [5.035 \times 10^{-3}(A/Q)] [\exp(1.540 \times (\mathrm{pH} - 6.6))]\}, \quad (16.5)$$

where C_e is the ammoniacal nitrogen concentration in pond effluent (mg N L^{-1}), C_i is the ammoniacal nitrogen concentration in the pond influent (mg N L^{-1}), A is the pond area (m^2), T is the temperature (°C), and Q is the influent flow rate (m^3d^{-1}).

The removal of total nitrogen:

$$C_e = C_i \exp\{-[0.0064(1.039)^{T-20}][\theta + 60.6(\mathrm{pH} - 6.6)]\}, \quad (16.6)$$

where C_e is the total nitrogen concentration in the pond effluent (mg N L^{-1}), C_i is the total nitrogen concentration in the pond influent (mg N L^{-1}), T is the temperature (°C) (range: 1–28°C), and θ is the hydraulic retention time (d^{-1}) (range: 5–231 days).

The pH in Eqs. (16.2) to (16.4) can be estimated using the equation below:

$$pH = 7.3 \exp(0.0005\mathbf{A}), \tag{16.7}$$

where **A** is the influent alkalinity (mg $CaCO_3$ L^{-1}).

The purple photosynthetic sulfur bacteria of the family *Thiorhodaceae* can occur in significant numbers in facultative ponds. They oxidize sulfides using carbon dioxide as a hydrogen acceptor and so no oxygen is liberated. Unlike other bacteria, they do not contribute to the breakdown of organic matter; they synthesize it, depositing sulfur granules within their cells. However, by reducing the sulfide concentration in the pond, they alleviate odors. As the group requires light, anaerobic conditions and the presence of sulfides, which are produced from protein degradation in the anaerobic sludge layer, the bacteria occur at the lowest depth in the pond where light can penetrate and as near the anaerobic sludge blanket as possible. The photosynthetic bacteria generally utilize longer light wavelengths than the green algae and so are found at depths below the algal zone. However, as the population density of the photosynthetic bacteria increases, they extend to the upper layers, excluding the algae due to the increased turbidity. It is at this stage that the pond will change from its normal green to brown color to pink, red or brown (Pike, 1975).

Unlike the other aerobic biological treatment processes, protozoans, except the phototrophic flagellates, do not appear to play a significant role in the process. Protozoa are unable to significantly reduce the number of algae in the final effluent and are out-competed for the dispersed bacteria by the filter-feeding Cladocera, which also feed on any protozoans that are present. Ponds have a range of zooplankton including Rotifera, Copepoda and Cladocera. However, Cladocera is by far the most abundant group found in facultative ponds although Rotifera and Copepoda dominate in maturation ponds.

Cladocera are found in ponds with a retention period in excess of 10 days, and under suitable conditions can develop large populations, which can significantly affect the operation of the pond, though not always beneficially (Tschörtner, 1967). The cladoceran population affects the pond through three activities: (i) feeding on the bacteria, (ii) feeding on the algae, and (iii) the formation of boluses. By feeding on bacteria and suspended particulate matter, Cladocera contribute directly to the stabilization process as well as clarifying the effluent by reducing the density of dispersed bacteria (Uhlmann, 1969). Although they may help to maintain the bacterial population in an actively dividing phase, thus enhancing the rate of carbon metabolism in the pond, Cladocera do not appear to significantly reduce bacteria numbers overall (Loedolff, 1965). The rejection of food in the form of settleable boluses is probably more important in the clarification process than direct predation, especially when turbidity is high. Most cladocerans in oxidation ponds are filter feeders. Water is forced, by the rhythmic beating of numerous flat limbs, to pass between barbs on the setae fringing the limbs, which act as a filtering apparatus (Dodson *et al.*, 2010). Particulate suspended matter, including bacteria and even colloidal solids, are filtered out of the water and collect in a food groove where it is concentrated before ingestion. If excess food is collected, which is normally the case when the pond is turbid, it is ejected from the food grooves as a bolus and rapidly settles out of suspension because of its density. Differences in the distance between the setae and barbs determine the size of particles collected by each species. For example, *Moina dubia*, which is a bacterial feeder, is more efficient at straining the bacteria *Escherichia coli* than *Daphnia magna*, an algal feeder, due to the difference in the coarseness of filtering setae (Loedolff, 1965).

Although some Cladocera readily utilize small unicellular algae such as *Chlorella*, it is doubtful whether any species can feed on the larger algae such as diatoms, blue-green algae, or the filamentoids to any great extent. Unlike the filter-feeding Cladocera, copepods seize and bite their food, being capable of utilizing even filamentoids. Effects of algal grazing can be beneficial by reducing the density of algal cells in the effluent and by reducing the overall algal biomass. Alternatively, it can be detrimental, as reduced algal density will reduce the rate of reaeration of water, especially at higher temperatures when bacterial oxygen demand is greatest and solubility of oxygen is at its lowest. In India, blooms of Daphnids feeding on *Chlorella* have resulted in low dissolved oxygen concentrations due to reduced photosynthesis (Lakshminarayana, 1965; Bopardikar, 1969). However, the daphnid population was controlled by the addition of lime. The role of Cladocera and the other zooplankton in controlling algae is unclear. Seasonal fluctuations in population and species dominance have been identified, and a general decrease in cladoceran numbers observed in the winter. It has been proposed that seasonal fluctuations in solar radiation and temperature, which affects algal activity, subsequently affects the cladocerans by altering the oxygen balance, pH and food supply. A more detailed summary of the microbial ecology and the major biochemical interactions that occur in facultative ponds is given in Fig. 16.2.

Suggestions have been made to utilize the zooplankton either by harvesting the cladocerans for sale as fish food (Dinges, 1982) or incorporating fish into the pond to feed directly on Cladocera, thus reducing the biomass and providing a more readily harvested end-product. The fish *Tilapia mossambica* is particularly successful in facultative ponds in warm climates as it feeds not only on the zooplankton

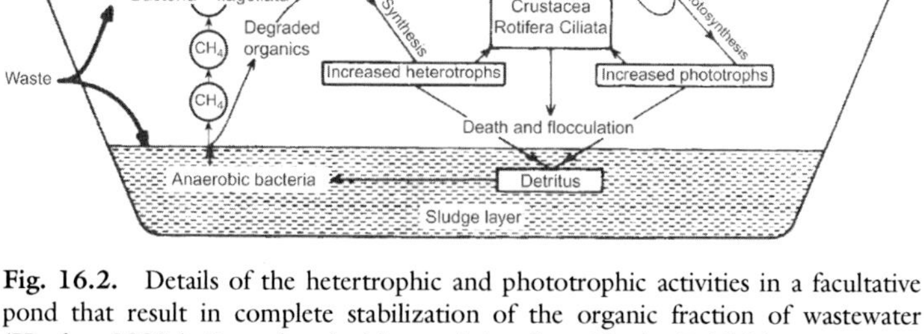

Fig. 16.2. Details of the hetertrophic and phototrophic activities in a facultative pond that result in complete stabilization of the organic fraction of wastewater (Hawkes, 1983a). Reproduced with permission from Elsevier Publishing.

and detritus, but also controls mosquitoes and other flies. However, it has been shown that the mortality constant for fecal coliforms decreases proportionally as the fish population increases in both facultative and maturation ponds. This is due to the reduction in the algal biomass by fish grazing and the potential elevation of the pH during the day (Rivas Hernandez, 1997) (Sec. 16.4).

Several groups of invertebrates are present in facultative ponds. In the sludge, benthic nematode worms, tubificid worms, and chironomid larvae are all common (Gloyna, 1971). The worms contribute to the stabilization of the organic sludge, and both the worms and midge larvae increase the rate of exchange of materials between the sludge and water at the solids–water interface. The dominance and abundance of species are related to the dissolved oxygen conditions of the sludge (Kimerle and Anderson, 1971) and the organic loading (Kimerle and Enns, 1968). It is important to prevent the development of surface mats of filamentous algae, or thin scum of encysted unicellular or blue-green algae (Hartley and Weiss, 1970). Such growths seriously reduce the light penetration and interfere with the mass transfer of oxygen into the pond from the atmosphere. Infestations of surface-dwelling macrophytes such as the *Lemna* spp. have the same adverse effects (Hawkes, 1983b). Surface material will eventually collect either on the banks or in the corners of ponds due to wind action, and unpleasant odors are produced as it decomposes.

16.2.3 *Factors affecting performance*

The hydrodynamics of facultative ponds are important in treatment efficiency and are influenced by the shape of the pond (i.e., length to width ratio), depth, the position of the inlet and outlet, and climate (i.e., temperature, solar radiation, and prevailing wind direction), which have all been reviewed by Li *et al.* (2018). Generally, these ponds behave as completely mixed reactors and are modeled as such (Torres *et al.*, 1999). While models have been developed for either pond hydrodynamics or biochemical processes, few have combined both (Ouarghi *et al.*, 2000; Dochain *et al.*, 2003; Shilton and

Table 16.4. The removal efficiency of fecal coliforms by ponds with different baffle arrangements. The influent fecal coliform concentration was 1 × 10^8 per 100 mL. (Li *et al.*, 2018). Reproduced with permission from Elsevier Publishing.

Baffle arrangement	No baffles	Traditional baffles	Modified long baffles	Stub baffles
	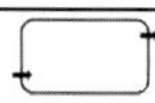	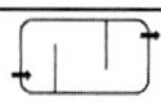	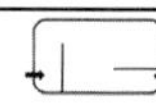	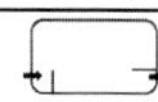
Outlet coliform concentration (f.c./100 mL)	6200 × 10^3	6.0 × 10^3	11 × 10^3	4.4 × 10^3

Mara, 2005). Sah *et al.* (2011) developed a new combined model to explore the effects of the inclusion of baffles and wind-induced mixing. Simulations using this model showed that while these factors did not increase chemical oxygen demand (COD) removal, they did improve the removal of *E. coli* (Table 16.4). This paper also lists useful mathematical equations for a range of key biological processes found in facultative ponds.

Apart from nutrient and carbon dioxide availability, temperature and solar radiation are the major factors affecting algal photosynthetic activity. Most algae grow over a wide temperature range of 4 to 40°C (Arceivala *et al.*, 1970), with optimum growth for the green algae (Chlorophyceae) at about 20°C but decreasing at 35°C. Above this temperature, other algae become dominant, such as the blue-green algae (Cyanophyceae) and euglenoids. Although different algae have different light intensity optima, high light intensity is not necessary for the successful operation of a facultative pond. For example, ponds operate successfully throughout the Indian sub-continent from latitudes 8°N to 34°N although different organic loading rates are required for ponds from different latitudes to allow for the variability in temperature and solar radiation, both of which affect the rate of microbial action (Arceivala *et al.*, 1970). Light penetration can be restricted by high densities of phytoplankton (> 1 × 10^6 mL^{-1}) in the surface layer of the pond. In many ponds, optimum light intensity for the algae occurs at <30 cm, although effective photosynthesis may not exceed depths of 45 cm under such conditions. Therefore, most of the pond is in darkness, and while heterotrophic activity is independent of light, mixing is required to ensure that the rest of the pond's depth is kept aerobic. Mixing, which can occur naturally in shallow ponds by wind action or in deeper ponds by currents induced by temperature differences, also ensures the transfer of materials between the photosynthetic and bio-oxidation phases, which ultimately governs the rate of stabilization. In dense algal layers, the water quickly becomes supersaturated during the day, with the excess dissolved oxygen being lost by transfer to the atmosphere unless mixed with the deeper unsaturated water. The circulation of algae in and out of the light ensures that a larger biomass of algae is supported and that more oxygen is made available to the bacteria. This is because algal cells fix a higher proportion of light energy under intermittent exposure to light compared to algae under constant exposure.

The aerobic bacterial breakdown is less sensitive to temperature than anaerobiosis and can proceed effectively over a range of 10 to 35°C. However, if the water temperature is only a few degrees above the freezing point, then the organic matter entering the pond will rapidly accumulate. Microbial activity will be reduced even further by ice and snow cover, which not only reduces light penetration but also effectively stops the phototrophic production of oxygen. Ice or snow cover will also prevent oxygen transfer from the atmosphere, normally helped by wind action, possibly leading to deoxygenation.

The major design function of facultative ponds is to maintain dissolved oxygen in the basin provided by the photosynthetic activity of the algae and therefore allow heterotrophic activity to proceed at an optimum rate. The site for such a pond should be at least 0.5 to 1.0 km from residential areas and, because of the production of odors during spring, it should be sited preferably downwind. It should be an exposed site to ensure maximum wind sweep over the surface of the pond, which is a function of wind speed and distance, to encourage mixing. Thus, planting trees or landscaping the site to shield the pond from view can severely reduce mixing. Leaf fall will also reduce light penetration. The shape of the pond is not important so long as the influent mixes rapidly with the pond supernatant (Pearson *et al.*, 1995).

Uniform flow throughout the pond is vital, therefore peninsulas and inlets should be avoided and corners of ponds should be rounded. If a rectangular pond is used, then the length should not be greater than four times its width. The pond should be located to ensure that its longest section lies in the direction of the prevailing wind to ensure maximum wind-induced mixing of the surface layer of the pond. If variable seasonally, then the wind direction in the hot season should be used to minimize thermal stratification.

Depth is always a critical factor. In temperate to tropical regions, ponds range from between 0.7 and 1.5 m in depth with a freeboard of 0.9 to 1.5 m above the high water level to accommodate wave action. A minimum depth of 0.7 m must be maintained to prevent the growth of rooted aquatic weeds, which not only damages the lining of the pond and affects circulation but also encourages mosquitoes and other flies. The recommended depth is between 1.0 and 1.5 m. Depths in excess of 1.8 m, which are used in colder regions (Sec. 14.1), may create odor problems due to excessive anaerobiosis in other areas. The sides should slope gently — a maximum slope of 1 in 3 is recommended to prevent earth movement and should be covered with paving slabs to prevent erosion by wave action and the growth of marginal vegetation. If the soil is not impervious, then it must be lined with bentonite clay or plastic to prevent groundwater pollution. The influent enters ponds horizontally at about 0.5 m above the bottom to induce a mixing current. The outlet is positioned to ensure adequate mixing and avoid short-circuiting. By ensuring that the wastewater flow is against the wind, short-circuiting is minimized. Simple baffles can also be used to enhance the plug flow nature of the pond and to minimize short-circuiting, although care must be taken in their precise location to avoid excessively high BOD loading in the inlet zone, resulting in odor production. Certain wind directions and high wind speeds can result in excessive mixing, which can impact treatment efficiency (Badrot-Nico and Brissaud, 2010), although this can be controlled by careful positioning of the inlet and outlet to ensure better flow control (Shilton and Harrison, 2003). The use of baffles reduces WSPs sensitivity to the wind (Badrot-Nico and Brissaud, 2010), and Shilton and Harrison (2003) recommend the use of short stub baffles in the place of longer traditional designs (Table 16.4). Shilton and Harrison also found that a maximum of two traditional baffles was sufficient and that any more only gave marginal improvements using an optimum length to width ratio of 4:1.

Grit and screenings are generally removed from the wastewater before entry into the pond, which ensures that the sludge accumulates slowly over the years. Accumulation rates range from 10 to 90 mm per year, and depending on the rate, ponds will only require desludging after long intervals. If ponds are operated in series, then the majority of sludge will accumulate in the primary pond, especially near the inlet (Nelson and Jiménez, 2000). As accumulation will reduce the volumetric capacity of the pond over the years, the rate of sludge accumulation must be taken into account when designing the pond depth and retention period. The use of sludge from waste stabilization ponds is reviewed by Keffala *et al.* (2013).

Photosynthetic activity is determined by the light entering the basin, which, in turn, is a function of the surface area of the pond. Organic loading is, therefore, expressed as kg BOD per hectare per day (kg BOD $ha^{-1}d^{-1}$) or g BOD per square meter per day (g BOD $m^{-2}d^{-1}$). The loading is limited by specific environmental factors such as insolation, temperature, and wind, which vary according to latitude. Therefore, different loading rates are recommended in temperate zones compared with tropical zones (Table 16.5). The maximum loading rate is limited by the need to maintain aerobic conditions in the pond during the most difficult season of the year when insolation and temperature are at their lowest and in temperate zones, ice formation occurs. Although ice formation reduces the radiation reaching the algae, snow accumulation will rapidly eliminate all light until photosynthesis is inhibited and the pond becomes anaerobic. Severe odors are produced as the ice breaks up due to anaerobic activity and although the odors become less severe, they continue to be given off until the algae are re-established and the pond becomes aerobic once more. This may take up to 4 weeks depending on temperatures, solar radiation, and the amount of accumulated organic matter (Porges and Mackenthun, 1963). The duration of odor production is a function of BOD loading and the duration of ice cover (Svore, 1968).

Table 16.5. Generalized loading and design criteria for facultative ponds constructed in different climatic zones (Gloyna, 1971). Reproduced with permission from the World Health Organization.

Surface loading (kg BOD $ha^{-1}d^{-1}$)	Population per hectare	Hydraulic retention time (d)	Environmental conditions
<10	<200	>200	**Frigid zones**: Seasonal ice cover, uniformly low water temperatures and variable cloud cover.
10–49	200–1,000	200–100	**Cold seasonal climate**: Seasonal ice cover and temperate summer temperatures for short season.
50–149	1,000–3,000	100–33	**Temperate to semi-tropical:** Occasional ice cover, no prolonged cloud cover.
150–350	3,000–7,000	33–17	**Tropical:** Uniformly distributed, sunshine and temperature, and no seasonal cloud cover.

To minimize this period, the organic loading can be reduced and the retention period increased, which, of course, results in the pond being underloaded for the rest of the year. This problem, as well as a reduction in the BOD removal as the temperature falls due to reduced bacterial activity, can be partially overcome by slowly reducing the pond level in the autumn and early winter when there is adequate dilution flow in the receiving waters. This allows the discharge from the pond to be minimal or stopped completely during winter so that the influent is stored until spring when discharge from the pond can be recommenced, as the aerobic activity is able to reduce the BOD of the effluent to an acceptable level. Little anaerobic activity occurs in the sludge during winter, so no soluble degradation products are released to enhance the overall BOD of the stored wastewater. However, the sludge gradually accumulates as sewage is continued to be discharged to the pond, and it is not until the summer at the enhanced temperatures that there is a net reduction in sludge volume due to the increased rate of degradation. The increase in soluble degradation products released from the sludge layer results in a rise in the BOD of the supernatant, although no fluctuation in effluent BOD is recorded due to the enhanced heterotrophic activity at the warmer temperatures, which utilizes the increase in organic load. The BOD of the stored wastewater in the pond remains almost constant throughout the year, whereas the actual oxygen demand on the pond varies considerably as it is a product of both the BOD and rate of degradation. Therefore, at higher temperatures, the oxygen demand can increase several-fold due to the higher degradation rate, although the BOD remains constant. This provides a problem to the operator, because at temperatures in excess of 20°C, the oxygen requirement of the pond cannot be determined by the standard BOD_5, nor can BOD_5 be used as an indicator of the onset of anaerobiosis (Hawkes, 1983a). The use of stabilization ponds in cold climates has been the subject of a workshop organized by the Environment Protection Directorate of Environment Canada (Townshend and Knoll, 1987). The effects of seasonality on performance are summarized in Fig. 16.3.

The depth of the aerobic zone fluctuates according to the photosynthetic activity and degree of mixing, therefore, the depth of the aerobic zone will always be less during the hours of darkness. As the organic loading increases, the depth of the aerobic zone tends to decrease, with a greater portion of the organic matter being degraded anaerobically (Fig. 16.4). Apart from currents induced by the influent wastewater entering the pond, circulation is also caused by wind action on the surface or currents caused by temperature differences within the pond. In shallow ponds, wind action will normally be sufficient to induce mixing as long as the sides of the pond are not too steep or trees have been planted, which may interfere with wind sweep. In deeper ponds, and under calm hot conditions, ponds can become stratified. This is due to the surface water being warmed, which reduces its density so that it forms a distinct layer (the epilimnion), which is isolated from the lower mass of cooler denser water (the hyperlimnion) by a thin layer of water called the thermoclime through which the temperature gradient falls rapidly. Once stratification has occurred, then mixing becomes restricted to the epilimnion only. As the density change per degree is greater at higher temperatures, the resistance to mixing will increase as the temperature rises.

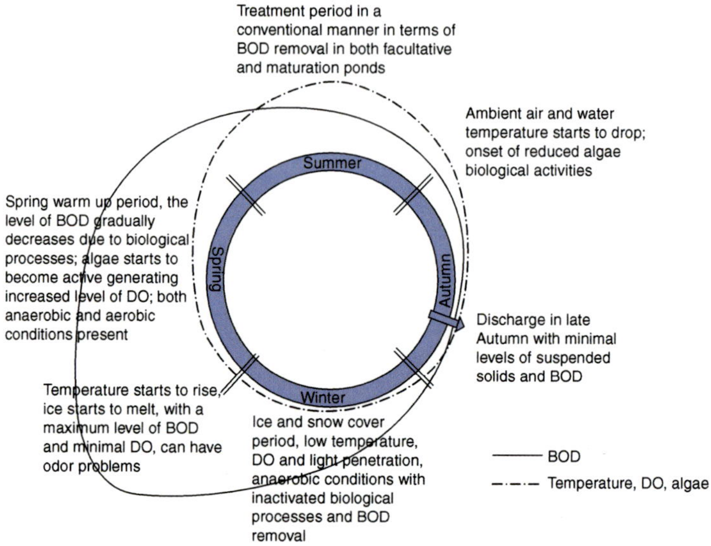

Fig. 16.3. Summary of the seasonal variation in the functional mechanisms of facultative and maturation ponds operated in a cold climate. The solid line (BOD) and the dotted line (temperature, dissolved oxygen and algae) show the variation in strength represented by the distance from the inner circle (Li *et al.,* 2018). Reproduced with permission from Elsevier Publishing.

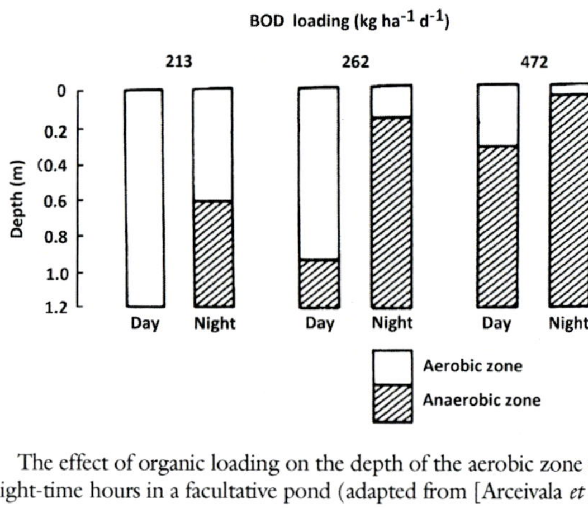

Fig. 16.4. The effect of organic loading on the depth of the aerobic zone during daylight and night-time hours in a facultative pond (adapted from [Arceivala *et al.,* 1970]).

The hyperlimnion rapidly becomes anaerobic, even though the epilimnion may be supersaturated with oxygen. Although the non-motile algae tend to sink in the less dense epilimnion and subsequently die because of a lack of light in the hyperlimnion, motile forms, such as *Euglena* are able to maintain a position in the pond where the light conditions are most favorable and they eventually become dominant, which is characteristic of stratified ponds. Stratification can lead to the short-circuiting of the wastewater through the epilimnion with a resultant reduction in both the BOD and pathogen removal. Under certain climatic conditions where hot days are followed by cold nights, daytime stratification will be followed by night-time mixing. However, in most cases, mechanical mixing is required using compressed air to break up the stratification, although the use of barriers to induce an under and over pattern of flow has been successful (Hartley and Weiss, 1970). Mechanical mixing should not be continuous as the increase in suspended solids from the sludge layer will reduce the light penetration and reduce photosynthesis. Therefore, mixing should be confined to short periods at night.

16.2.4 *Design*

In the northern states of the US, where the climatic conditions are similar to those of Western Europe, the maximum loading permissible is 2.2 g BOD $m^{-2}d^{-1}$ if an odor nuisance is to be avoided in spring. Where ice coverage of the pond does not occur, this loading can be increased, and in the southern and south-western states of the US, loadings of up to 5.6 g BOD $m^{-2}d^{-1}$ are used. However, this is only equivalent to a volumetric loading of 0.0015 to 0.0037 g $m^{-3}d^{-1}$ for a 1.5 m deep pond, which is a much lower organic loading compared to either the activated sludge or percolating filtration process. Retention time depends on the load, depth of the pond, evaporation rate, and loss by seepage, but periods of 3 to 6 days are common for ponds in non-tropical locations. Ponds normally achieve a 90% reduction in the BOD, which is less efficient compared with other biological treatment processes. Larger ponds are more effective because of the greater wind sweep over the surface, which enhances mixing and hence reaeration. Generally, ponds are operated in series to prevent short-circuiting, to provide a greater reduction in pathogens and an increased BOD reduction. The first pond in series has a higher organic load and sludge accumulation compared with other ponds, which may lead to excessive odor production. If ponds are operated in parallel, odors are prevented as the BOD load is distributed more evenly, although such an arrangement allows short-circuiting. Facultative ponds rarely achieve a final effluent suspended solids concentration of ≤30 mg L^{-1}, even when in series. The suspended material is predominantly algae and zooplankton, which may be present at concentrations of between 50 and 70 mg L^{-1} in the final effluent. Tertiary treatment is required to remove the plankton biomass and any other colloidal or particulate matter if the receiving water is unable to assimilate the material. The three most popular options are chemical flocculation followed by a gravity filter, upflow filter, or by spray irrigation onto land.

In practice, facultative ponds are designed on the basis of surface BOD loading (A_s) (kg $ha^{-1}d^{-1}$):

$$A_s = 10L_iQ/A_f, \tag{16.8}$$

where L_i is the influent BOD (mg L^{-1}), Q is the flow (m^3d^{-1}), and A_f is the area (m^2) of the facultative pond. The temperature (T) and A_s are related, allowing the maximum permissible surface BOD loading rate to be calculated using Eq. (16.9), developed by McGarry and Pescod (1970):

$$A_s = 60(1.099)T. \tag{16.9}$$

If this figure is exceeded, then the entire pond becomes anaerobic and fails. Mara (1987) has suggested a more conservative design equation:

$$A_s = 350(1.107 - 0.002T)^{T-25}. \tag{16.10}$$

Table 16.6 gives permissible surface BOD loadings based on Eq. (16.10). It is important that the temperature selected for design

Table 16.6. The permissible surface loading rates (A_s) to facultative ponds at different temperatures.

T(°C)	A_s (kg ha^{-1} d^{-1})	T(°C)	A_s (kg ha^{-1}d^{-1})
≤8	80	17	199
9	89	18	217
10	100	19	235
11	112	20	253
12	124	21	272
13	137	22	291
14	152	23	311
15	167	24	331
16	183	25	350

purposes should be the lowest mean monthly temperature. Once A_s has been selected, the pond area (A_f) is calculated using Eq. (16.8). The hydraulic retention time (HRT) (θ_f) is calculated as:

$$\theta_f = A_f \, D/Q_m \text{ d}, \tag{16.11}$$

where D is the depth (m) and Q_m is the mean flow (m^{3}d^{-1}). As there is often considerable evaporation and infiltration from ponds, the mean flow is calculated by taking the mean of the influent and effluent flows. At temperatures below 20°C, θ_f should be a minimum of 5 days, whereas 4 days is sufficient for temperatures above 20°C. Sufficient HRT is important not only to prevent hydraulic short-circuiting but to allow algae sufficient time to multiply to counterbalance the loss of algae via the final effluent. Further details on design and upgrading ponds can be found in Middlebrooks *et al.* (1982).

16.2.5 *Application*

Facultative ponds are best suited to isolated communities where land is cheap and no industrial expansion is planned. In Europe, waste stabilization ponds are increasingly used as a sewage treatment option for small rural communities — they are robust, low maintenance, and have low energy systems. Capital costs are about 30% of that of a conventional system and 50% of a reed bed system. When treating sewage, ponds are capable of producing odor-free effluents throughout the year, except if ice formation is a problem in the spring. Industrial or strong organic wastes normally require pretreatment before discharge to a facultative pond to reduce the organic content, preventing overloading and permanent odors. The use of an anaerobic pond, initially to maximize the BOD removal followed by the facultative pond, may be adequate. Incremental feeding, where some of the influent to the first pond in series is transferred to the second pond, or recirculation from the second pond to the primary pond may also improve efficiency and prevent odors. Facultative ponds are most popular in areas where there is sufficient solar radiation throughout the year. Where the climatic conditions are suitable, oxidation ponds are the cheapest form of sewage treatment in constructional and operational terms. The advantage of the system is the ease of operation — no skilled labor, mechanical parts, or power is required. However, among the disadvantages are a number of operational problems, which include poor assimilative capacity for industrial wastes, production of odors, and failure to meet the minimum standards for discharge.

New ponds are easily commissioned once construction is complete and the basin cleared of any loose debris, soil and vegetation. The pond cannot be operated at its full design load until a balanced community of bacteria and algae has been established. This is most effectively achieved by allowing raw wastewater into the center of the pond to a depth of about 200 mm. Where the area of the basin is large, then a smaller area should be partitioned off by constructing an embankment at one corner (or in the center) and filled with wastewater. Bacteria become established first, followed more slowly by algae. Colonization occurs naturally, although the speed of colonization can be increased by the addition of a few bucketloads of algal rich water from a pond or tank. Ensure that green algae is used and not blue-green algae, which may quickly dominate the pond. Additional wastewater must be added at intervals to maintain the level in the pond, due to evaporation. Once an algal bloom has become established, the pond can be gradually filled with wastewater, trying to maintain the rich algal bloom until the pond is full. The pond should be allowed to stand for a further 3 to 4 days before full continuous-flow operation is commenced. In temperate climates, this process will take up to 6 weeks even in the summer months. Care must be taken to prevent weed growth developing during the start-up process, and to this end, it may be necessary to increase the initial depth to 1 m if it becomes a problem. Full details of commissioning all types of stabilization ponds are given by Arceivala *et al.* (1970) and Gloyna (1971).

The biologist plays a major role in the successful operation of facultative ponds by conducting regular microscopic examinations of the algae to give a clear quantitative assessment of the community structure within the pond (Arceivala *et al.,* 1970). The overall photosynthetic activity of the pond can be determined either by measuring the biomass of the algae or the chlorophyll *a* concentration (American Public Health Association *et al.,* 1985). Changes in the color of the pond are also a useful guide to changes in condition. Healthy ponds are green or brownish due to the presence of green algae. However, a change to bluish-green indicates a change in algal dominance from green to blue-green algae and as the latter are less efficient than the green algae, this indicates a deterioration in conditions. Changes in color to pink, red or reddy-brown in the summer or autumn are caused by blooms of photosynthetic sulfur bacteria because of an increase in sulfate or sulfide concentration resulting from increased anaerobiosis. This is generally associated with overloading, thermal stratification, or another operational problem. The most common forms associated with sewage are *Chromatium* and *Thiospirillum*, whereas *Thiopeda rosea* and smaller spiral and rod forms are associated with industrial wastes (Gloyna, 1971). A blue-green algae *Merismopedia tenuissina*, which is also associated with anaerobic conditions where sulfides are abundant, also imparts a pink color to the pond. When the pond has broken down and gone completely anaerobic, then it takes on a gray or black color. This is normally accompanied by rising sludge due to excessive gas production in the sludge layer. Such a condition is normally due to a rapid change in temperature, severe overloading, change in wastewater characteristics, or an input of toxic waste. Waste stabilization ponds are considered further in Pearson *et al.* (2000).

16.3 Maturation Ponds

Maturation ponds are widely used throughout the world as a tertiary treatment process for improving the effluent quality from secondary biological processes, including facultative ponds. Effluent quality is improved by removing suspended solids, reducing ammonia, nitrate and phosphate concentration, and by reducing the number of enteric microorganisms. The primary function of such ponds in most parts of the world is the removal of pathogenic organisms (Maynard *et al.*, 1999). Their design, size and number, in series, is determined by the required bacteriological quality of the final effluent (Marais, 1974; Mara and Pearson, 1998). They are of the same depth as secondary facultative ponds (1.0–1.5 m), although shallower ponds are used where insolation is high to achieve maximum light penetration. Unlike high-rate aeration ponds, maturation or tertiary lagoons are specially designed for maximum purification and not algal production (Hawkes, 1983a). The retention period is normally 10 to 15 days, although shorter periods can be used for suspended solids (4 days) or phosphate removal (7–10 days) (Gloyna, 1971; Toms *et al.*, 1975). The biochemical oxygen demand removal efficiency remains low due to the initial low organic loading (Toms *et al.*, 1975). By acting as a buffer between the secondary biological phase and the receiving water, maturation ponds can reduce the effects of toxic loads and fluctuations in the performance of the secondary biological phase by dilution, similar to an equalization pond. Baffles improve the hydraulic efficiency of large maturation ponds by up to 74%, although three channel configurations are even more efficient (69–92%) (Aldana *et al.*, 2005).

Pathogenic bacterial die-off is associated with flow patterns in maturation ponds and inversely proportional to the dispersion number. The dispersion number (d) can be increased by including baffles in ponds, hence increasing the length to width (L:W) ratio. For example, the removal of fecal coliforms in an open lagoon with a dispersion

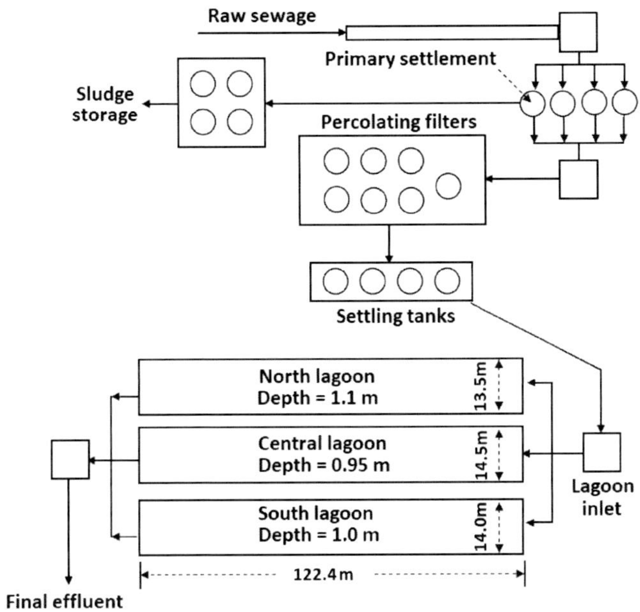

Fig. 16.5. Layout of Lidsey Wastewater Treatment Plant, which uses three maturation ponds in parallel for polishing the effluent from percolating (trickling) filters (Bracho *et al.*, 2006). Reproduced with permission from Elsevier Publishing.

number (d) of 0.79 was 90%, but in a channeled lagoon (d = 0.40), this rose to 98% removal. This was explored in a comparative study at the Lidsey Wastewater Treatment Plant in West Sussex, United Kingdom (UK), where the effluent from percolating filters is treated by three equally loaded maturation ponds in parallel (Fig. 16.5) (Bracho *et al.*, 2006). One pond was converted into three channels using baffles, increasing the L:W ratio from 9:1 to 79:1, and reducing d from 0.37 to 0.074, thus increasing the HRT of 4.02 days by 5 hours overall and reducing any short-circuiting, significantly increasing exposure times to UV radiation and other bactericidal actions. This led to a significant increase in fecal coliform die-off from 90.2% to 99.84% during the summer.

Maturation ponds differ from facultative ponds in showing less vertical and physicochemical stratification and remaining fully oxygenated during the day. They have a more diverse algal population than facultative ponds (Tables 16.1 and 16.2), with the diversity increasing with each pond in series. Non-motile genera tend to dominate. Pathogenic bacteria are removed by a number of mechanisms including settlement, natural die-off, which is a function of time and temperature, elevated pH (>9), UV radiation, and predation (Sec. 16.6).

Due to the development of algae, which is constantly washed out of the pond with the final effluent, the unfiltered BOD can be much higher than the influent with up to 70 to 90% of the final effluent BOD attributed to its algal content (Mara *et al.,* 1992). As algal cells are predated by zooplankton, or eaten by fish such as carp, there is normally an overall reduction in the final effluent BOD. The filtered BOD of the final effluent can be estimated by assuming a 90% cumulative removal in the anaerobic and facultative ponds, and a further 25% reduction for each maturation pond in series (Mara and Pearson, 1998). The BOD removal that occurs in maturation ponds is due to oxidation by the common genera of the heterotrophic bacteria found in other secondary biological treatment processes, in particular, the *Pseudomonas* spp., *Flavobacterium* spp., *Archromobacter* spp., and *Alcaligenes* spp. In Europe, the EU Urban Wastewater Treatment Directive permits higher concentrations of the BOD and suspended solids in the effluent from lagoons than other systems (European Community, 1991). Therefore, the BOD and suspended solids analysis are carried out on filtered samples, whereas all other analyses must be unfiltered. Unfiltered suspended solids concentrations of up to 150 mg L^{-1} are permitted (European Community, 1991). The algal biomass is a function of the nutrient concentration. Therefore, if the availability of nutrients declines, so does the density of algal in the final effluent. Different planktonic organisms vary in their efficiency in removing nitrogen and phosphorus. For example, the algae *Scenedesmus obliquus* is particularly efficient in utilizing organic and ammoniacal nitrogen, whereas *S. quadricauda* var. *alterans* is more efficient at using nitrate (Kalisz and Suchecka, 1966). Cladocera such as *Daphnia pulex* reduce turbidity and enhances nitrogen removal by ingesting particulate matter and algae.

The removal of nitrate in maturation ponds is not linked with the algal biomass and although small amounts are utilized by the algae, the majority is lost by denitrification at the sludge–water interface (Toms *et al.,* 1975). Optimum nitrate removal is achieved by ensuring the development of a sludge layer and by using shallow ponds to give the maximum sludge–water surface contact. In this way, up to 0.8 g N $m^{-2}d^{-1}$ can be removed. It has been reported that nitrate levels tend to be much higher in the effluent than in the feed and that nitrate concentrations can build up during autumn and winter (Van der Post and Englebrecht, 1973). However, the volatilization of ammonia and sedimentation of organic nitrogen via biological uptake are thought to be the principle mechanisms of nitrogen removal (Maynard *et al.,* 1999). However, studies carried out on the University of Dar es Salaam pond system treating domestic wastewater (840 m^3d^{-1}), which comprises a primary facultative pond discharging into parallel lines of two secondary facultative ponds and a maturation pond each, examined the physicochemical and biological nitrogen transformation mechanisms and losses through the system (Fig. 16.6) (Mayo and Abbas, 2014). The retention time and volume of the primary facultative and a single maturation pond were 11.5 days and 6,080 m^3, and 7.6 days and 1,824 m^3, respectively. Nitrogen transformation and losses are summarized in Table 16.7 and showed that the volatilization of ammonia was a minor route for nitrogen loss in ponds at 1.4% overall. In the primary facultative pond, nitrogen losses (13.2%) were primarily due to the settlement of organic nitrogen (9.8%) and denitrification (3.4%). Losses in the maturation pond were slightly higher (15.2% of the influent TN), mainly by denitrification (13.6%). Mineralization and the ammonia uptake were the main transformation processes in the primary pond while the ammonia uptake by microorganisms was the most important transformation process in the maturation pond. A nitrogen balance model was produced by the authors.

Work on the maturation ponds at Rye Meads sewage treatment works, which discharges to the River Lee in the UK, has shown that

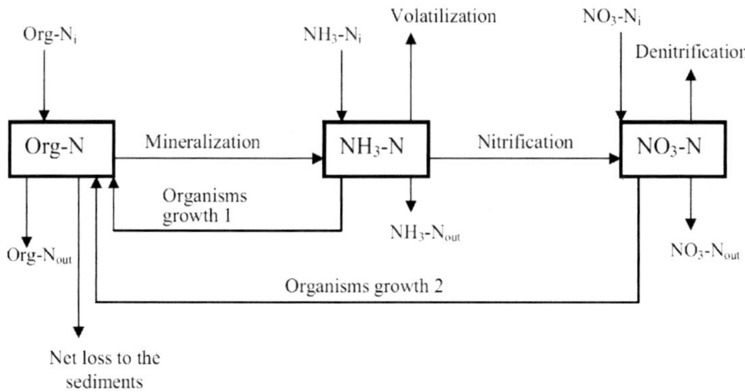

Fig. 16.6. Flow diagram showing nitrogen transformation and removal in waste stabilization ponds (Mayo and Abbas, 2014). Reproduced with permission from Elsevier Publishing.

Table 16.7. Comparison of nitrogen transformation and losses between a primary facultative and maturation ponds at the University of Dar es Salaam (UDSM) pond system (Mayo and Abbas, 2014). Reproduced with permission from Elsevier Publishing.

		Primary pond		**Maturation pond**	
S/no.	**Transformation mechanism**	**kg d^{-1}**	**Percent**	**kg d^{-1}**	**Percent**
1	Mineralization of organic nitrogen to NH_3	3.515	39.1	1.149	34.7
2	Uptake of NH_3 by microorganisms	3.177	35.4	1.188	35.9
3	Organic nitrogen lost to sediments	1.413	15.7	0.020	0.6
4	Nitrification of NH_3 to NO_3^-	0.372	4.1	0.380	11.5
5	Volatilization of NH_3 to gas	0.010	0.1	0.043	1.3
6	Denitrification of NO_3^- to N_2 gas	0.495	5.6	0.527	15.9
7	Total	8.982	100.0	3.307	100.0

phosphate removal is directly correlated with the algal biomass, reaching a maximum removal of 73% in May and a minimum of 2% in January (Toms *et al.,* 1975). However, only a small proportion (<20%) of the phosphate removal can be attributed to nutritional uptake by the phytoplankton, as the majority is removed by precipitation at enhanced pH values (>pH 8.2), which is due to photosynthetic activity. For every unit increase in pH, the concentration of phosphate remaining decreased by a factor of 10. In the temperate climate of the UK, it is only possible to maintain sufficient algal biomass to remove a reasonable proportion of the phosphate from the pond during the spring and autumn by having a long retention time. In winter, this is not feasible although the higher dilution factor, reduced temperature, and insolation all significantly reduce the risk of eutrophication in receiving waters. It has been reported that phosphorus can be regenerated during the night or the winter at lower pH values (Hemens and Mason, 1968). However, this was not observed at Rye Meads by Toms and his colleagues. In practice, phosphorus removal in maturation ponds is poor (Maynard *et al.*, 1999). The design of maturation ponds is explored by Martínez *et al.* (2016), who gives full sets of equations for both design, performance and pathogen removal.

In conclusion, although algae are beneficial in terms of phosphate removal and to a lesser extent nitrogen removal, their presence in the final effluent from a maturation pond increases the suspended solids and turbidity. Where strict effluent standards have to be met, simple algal removal systems can be added. Rock filters are most widely used for this purpose and consist of submerged beds, 1.5 to 2.0 m deep, of crushed rock, 75 to 100 mm in diameter (Middlebrooks, 1988). The algae are retained within the filter and decompose, releasing nutrients that are utilized by attached bacteria. The effluent is added below the surface of the medium to prevent odor production from cyanobacterial films that can develop over any exposed wet surfaces. Significant nitrification occurs as nitrifying bacteria colonize the filter. It has also been suggested that algae can reduce the mortality of fecal bacteria (Potten *et al.,* 1971: Potten, 1972).

16.4 Fishponds

Bellamy (1975) reviewed the increasing use of biological treatment systems incorporating many different trophic levels. Nutrients released during the degradation of organic matter by heterotrophs are utilized by algae which, in turn, are consumed by zooplankton, and so on up the food chain to the top trophic level that is usually occupied by a species of fish. The fish can then be harvested for human consumption, with coarse fish being especially popular in Central Europe. This form of treatment has been popular in China for centuries and is increasing in popularity in both developing and developed countries around the world (Edwards and Pullen, 1990). However, the system requires careful management to ensure that the rate of organic loading does not disturb the ecological balance. European wastewater maturation ponds can provide ideal conditions for fish culture, especially roach, carp, and, to a lesser extent, chub and perch. Experiments on carp culture have shown that a high standing crop can be maintained in Western European maturation ponds, with good growth and production rates. However, problems with diseases were encountered at higher stocking densities, and harvesting fish proved very difficult because of the layout of the ponds and the nature of the banks. The inclusion of fish in maturation ponds has a significant effect on the ecological balance. In their original model for the control of algae by *Cladocera*, Green and Watts (1973) did not include predation of the grazers. Unfortunately, the fish, especially the young fingerlings which give such high productivity, feed on the zooplankton which, in turn, feed on the phytoplankton. Therefore, as the fish population increases so does the density of algae in the pond, with a corresponding increase of algae in the final effluent (Williams *et al.*, 1973) (Fig. 16.7). This may not be a problem in countries where the quality of the final effluent is less important in relation to the production of fish, or if the effluent is to be used for irrigation. The effluent from stabilization ponds in India, which are rich in both phyto- and zooplankton, the natural food of so many fish species, has been profitably exploited for rearing fish (pisciculture). As fish are extremely sensitive to oxygen depletion, even for short periods, and the dissolved oxygen concentration in stabilization ponds fluctuates diurnally, often reaching zero at night, fish are rarely stocked in the main stabilization pond itself but normally in secondary ponds (Fig. 16.8). The major species of fish cultured in lagoons and ponds are all carp

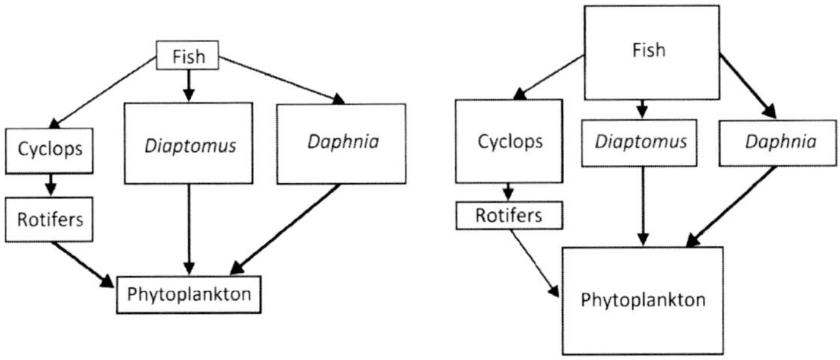

Fig. 16.7. The effect of overstocking with fish on the food web in a fishpond system (Hawkes, 1983b). Reproduced with permission from Elsevier Publishing.

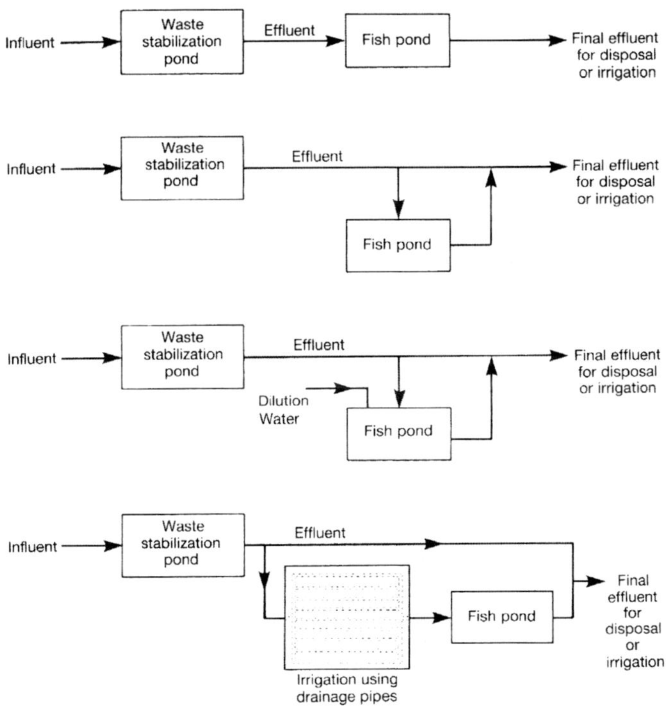

Fig. 16.8. Possible incorporation of fishponds with waste stabilization pond systems.

species, with *Catla catla*, *Labeo rohita*, *L. calbasu*, and *Cirrhinus mrigala* all grown successfully in both the stabilization and the subsidiary lagoon. In Bhilai, 2,000 fingerlings of *L. rohita* and *C. catla*, with mean weights of 18.7 g and 5.5 g, respectively, were stocked in a stabilization pond and after 12 months their weight had increased to 140 g and 990 g, respectively (Chatterjee *et al.*, 1967). Fish fry grown in a pond using sewage effluent in Bhopal has been shown to have a higher rate of growth than regularly grown nursery pond fry. The main species of carp grow best in the subsidiary rather than the main stabilization ponds, and the conditions can be more carefully controlled for pisciculture in the former, and dissolved oxygen depletion is prevented. However, *Ophiocephalus* is a popular fish in some regions of India because of its high iron content, and reports have indicated that it may be particularly suited to intensive rearing in stabilization ponds as it can utilize both dissolved and atmospheric oxygen; hence its local name of "the surface-breathing fish".

Normal fish yields from freshwater ponds have also been reported as doubling (i.e., 7–8 t $ha^{-1}yr^{-1}$) when effluents from stabilization ponds are added in the ratio of 1 to 2 volumes of effluent to each volume of pond water (Arceivala *et al.*, 1970). Optimum yields are obtained using small ponds of up to 1 ha in areas that are stocked with fingerlings, fertilized with facultative pond effluent, and then harvested four months after stocking. The growth of fish within this period is rapid, rising from 20 g to 250 g.

There are serious microbial risks associated with the reuse of effluents and Mara *et al.* (1992) have proposed a design procedure to minimize these risks. The design ensures the protection of the fish from free ammonia within the pond, maximum treatment of wastewater using anaerobic and facultative ponds in series, and the maximum production of microbially safe fish. The fishpond receiving the effluent from the facultative pond is designed using a surface loading rate of 4 kg total N $ha^{-1}d^{-1}$. The loading rate is critical as too little nitrogen results in low algal biomass giving small fish yields, while too much nitrogen results in a high concentration of algae and consequent oxygen depletion at night. The area of a fishpond (A_{fp}) is estimated by a modified version of Eq. (16.8):

$$A_{fp} = 10\ C_i Q/\lambda_s^{TN}, \tag{16.12}$$

where C_i is the total nitrogen concentration in facultative pond effluent (mg N L^{-1}) (use Eq. [16.6]), Q is the flow rate (m^3d^{-1}), and λ_s^{TN} is the surface loading of total nitrogen (kg $ha^{-1}d^{-1}$).

The retention time (ϕ_{fp}) is calculated as:

$$\phi_{fp} = A_{fp}\ D/Q, \tag{16.13}$$

where the depth (D) is normally 1 m.

The concentration of fecal coliforms in the fishpond (N_{fp}) must be ≤1,000 per 100 ml. This is achieved by increasing the retention time or incorporating a maturation pond after the facultative pond and before the fishpond. N_{fp} is calculated as:

$$N_{fp} = N_i/(1 + k_T\phi_a)(1 + k_T\phi_f)(1 + k_T\phi_{fp}), \tag{16.14}$$

where N_i is the number of fecal coliforms per 100 mL in the influent wastewater, k_T is the first-order rate constant for fecal coliform removal (d^{-1}) (Gray, 2004), ϕ_a is the retention time of the anaerobic pond, ϕ_f is the retention time of the facultative pond, and ϕ_{fp} is the retention time of the fishpond. To protect the fish from ammonia toxicity, the free ammonia (NH_3) should not exceed 0.5 mg N L^{-1}. The percentage of free ammonia (p) in solution is a function of the temperature and pH (Table 16.8) and can be calculated as:

$$p = 1 = [10(pk_a^{-pH}) + 1], \tag{16.15}$$

where pk_a = 0:09018 + (2729.92/T) and T is given in degrees Kelvin (°K = °C + 273.15). The pH in fishponds varies between 6.5 and 7.5, and so when designing such ponds, a pH of 7.5 should be assumed. The total ammonia concentration is calculated using Eq. (16.4) or (16.5) in Sec. 16.2.2.

16.5 River Purification Lakes

The development of large impounded lakes to remove and degrade the residual pollution from a river is becoming increasingly popular.

Table 16.8. Concentrations of total ammonia (mg L^{-1}) in freshwater that contains an unionized ammonia concentration of 0.02 mg NH_3 L^{-1}. The threshold concentration of 0.02 mg NH_3 L^{-1} is the long-term toxic effect level for salmonoid fish while the lethal concentration is about ten times greater.

Temp. (°C)	pH							
	6.0	6.5	7.0	7.5	8.0	8.5	9.0[a]	9.5[a]
5	160	51	16	5.1	1.6	0.53	0.18	0.071
10	110	34	11	3.4	1.1	0.36	0.13	0.054
15	73	23	7.3	2.3	0.75	0.25	0.09	0.043
20	50	16	5.1	1.6	0.52	0.18	0.07	0.036
25	35	11	3.5	1.1	0.37	0.13	0.055	0.031
30	25	7.9	2.5	0.8	0.27	0.10	0.045	0.028

[a]Criteria may be unduly low if there is low free carbon dioxide in the water.

Such lakes are very similar to maturation ponds. Five lakes have been constructed in the Ruhr area (Germany) with Lake Baldeney, south of Essen, the largest with a retention period of 60 hours at low flows (Imhoff, 1982; 1984) (Fig. 16.9; Table 16.9). Together, they provide an area of 4,488 km^2, with Lake Baldeney alone providing a purification capacity equivalent to the BOD produced by a population of 100,000 per day. Lake Baldeney is also able to remove dissolved phosphorus from the River Ruhr (Fig. 16.10). The phosphorus is removed by algal growth and subsequent algal settlement. Up to 170 tons of phosphorus per annum, together with heavy metals, are removed by sludge settlement in the lake (Table 16.10). It was estimated that by 1984, Lake Hengstey, constructed in 1929, had accumulated some 400,000 m^3 of sludge, whereas in Lake Harkort, which was impounded in 1930, some 350,000 m^3. In Lake Baldeney, some 60,000 m^3 of sludge accumulates each year, but at flows 10 times greater than the average flow rate, the sludge starts to become mobilized. The sludge is subsequently flushed out of the lake when the river is in flood and flows are in excess of 15 times the average flow rate (Imhoff, 1982). Because of the reduction in the depth of these lakes, desludging begun in 1983. Some 1,100,000 m^3 of sludge is to

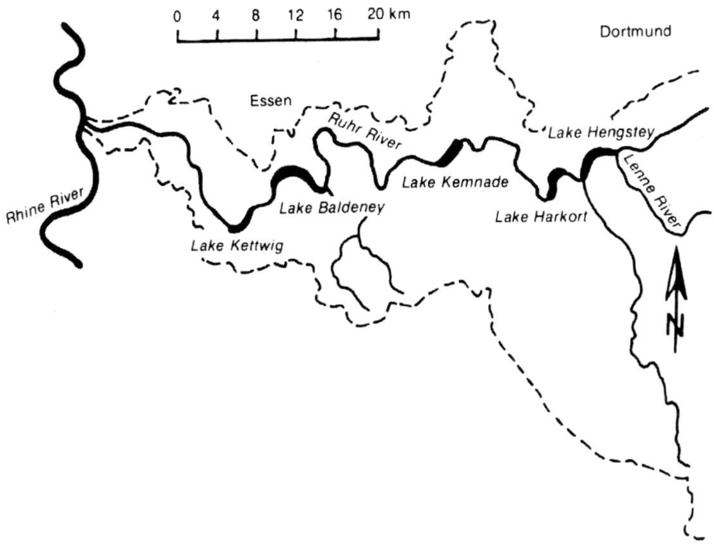

Fig. 16.9. Position of impounded lakes on the River Ruhr, which operated as purification lakes (adapted from [Imhoff, 1984]).

Table 16.9. Details of the five river purification lakes built to improve water quality and generate electricity in the Ruhr area of Germany.

	Lake Hengstey	Lake Harkort	Lake Kemnade	Lake Baldeney	Lake Kettwig
Year of impoundment	1929	1930	1979	1932	1949
Original lake storage capacity (m^3)	3.3×10^6	3.1×10^6	3.0×10^6	8.3×10^6	1.4×10^6
Lake surface area (km^2)	1.36	1.37	1.25	2.64	0.55
Length (km)	4.2	3.2	3.0	7.8	5.2
Average width (m)	296	335	420	355	130
Average depth (m)	1.94	2.21	2.40	3.14	2.54

be removed from Lake Baldeney alone by hydraulic dredging. Lake Harkort, originally with an average depth of 2.2 m, had become so silted after 70 years of operation that parts were <50 cm in depth. From 1999 to 2003, some 450,000 m^3 of sludge was removed,

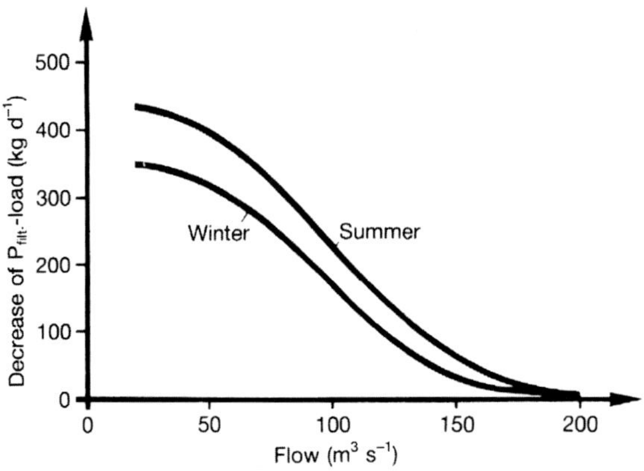

Fig. 16.10. The effect of flow rate ($m^3 s^{-1}$) and season on the removal of phosphorus in the purification lake, Lake Baldeney (Imhoff, 1984). Reproduced with permission from the Chartered Institution of Water and Environmental Management.

Table 16.10. Comparison between sediments collected from purification lakes in the UK and the Ruhr catchment (Woods *et al.*, 1984). Reproduced with permission from the Chartered Institution of Water and Environmental Management.

				Metal concentrations ($mg\ kg^{-1}$ dry solids)					
Lake Site	**Period**	**Dry solids (%)**	**Volatile solids (%)**	**Ni**	**Cu**	**Cr**	**Zn**	**Cd**	**Pb**
Lea Marston (sed. tank)	1969–1970	10.5	40.5	554	1,743	1,275	8,000	44	617
Lea Marston (exp. lake 1)	1972–1974	11.4	35.1	1,000	2,350	1,340	5,160	70	950
Lea Marston (exp. lake 2)	1972–1974	7.5	32.7	970	2,230	1,650	5,400	80	950
Lea Marston (dredged sludge)	1981–1982	4.5	24.9	410	1,335	510	3,570	39	780
Hengstey	1981–1982	39.4	12.3	261	1,240	440	4,030	18	450
Harkort	1981–1982	45.8	11.0	250	830	280	3,120	29	480
Baldeney	1981–1982	45.6	14.0	312	730	400	3,540	36	525

restoring it to almost its original depth. Even with the expensive cost of desludging, it has been estimated that treatment by purification lakes is 30% cheaper than conventional treatment. The purification lakes form an important part of the water quality management strategy for the 4,488 km^2 Ruhr catchment that includes 92 wastewater treatment plants and 473 stormwater treatment plants. The total phosphorus concentration in the Ruhr at Essen has shown a steady decline over the past three decades from 0.8 mg L^{-1} in the 1970s to 0.15 mg L^{-1} in 2000 (Bode and Klopp, 2001). These lakes have been improved by dredging to improve their ecology, with the inclusion of fish passes where necessary. This has led to the creation of a series of important bird and wildlife reserves. Each lake was also constructed to allow for hydroelectric power generation. The five purification (impounded) lakes have a joint generation capacity of 28.1 MW, equivalent to 92.5 million kWh annually.

On the River Tame, a polluted tributary of the River Trent in the West Midlands, UK, a purification lake has also been built to provide *in-situ* treatment. The River Trent received vast quantities of sewage and industrial effluent reaching in excess of 80% of the total volume of the river at low flows. The Lea Marston purification lake was built at a disused gravel works after the last major conurbation and sewage input. It was opened in 1980 and removes settleable solids from the river, which are especially a problem during storms. The solids are removed from the main body of water by settlement in the lake and subsequently by dredging. In contrast to the Ruhr lakes, the Lea Marston lake is continuously desludged although it takes about 12 months to cover the whole flow area of the lake. The lake also reduces the need for tertiary treatment at Coleshill and Minworth sewage treatment plants and acts as a buffer in case of accidental spillages. The results have been very encouraging, with the water quality in the Rivers Tame and Trent significantly improved. A coarse fishery has been restored and fish are now being caught in the River Tame for the first time in 100 years. The river enters the system and passes through large booms that remove floating debris (Fig. 16.11). Grit traps and grit channels remove >2,000 tons of dry matter per annum from the river, with the lake itself removing a further 15,000 tons of

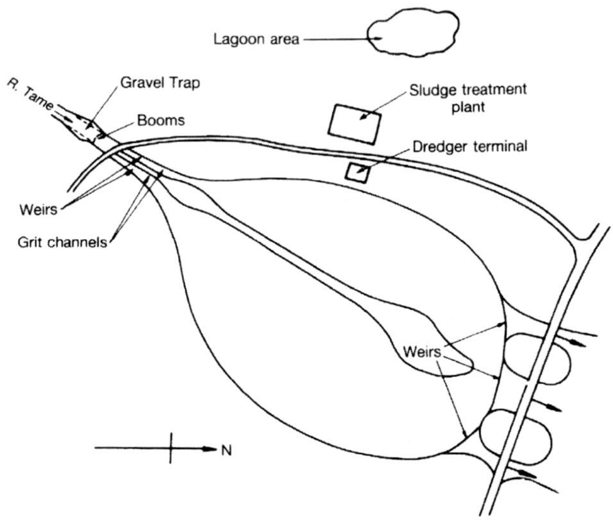

Fig. 16.11. Schematic plan of the Lea Marston purification lake (Woods *et al.*, 1984). Reproduced with permission from the Chartered Institution of Water and Environmental Management.

sludge each year. The system was expanded by the addition of two further gravel pits downstream of the main lake and has been shown to be a cost-effective method for improving water quality in the river (Crabtree *et al.*, 1999). Forty years on, the three purification lakes now form part of the Tame Valley Wetlands, a large nature reserve. With improvements at the upstream wastewater treatment plants, including tertiary treatment, coupled with the decline in heavy and polluting industries in the area, the lakes no longer need to be desludged and have been allowed to naturally transform into rich lake habitats, although they still play a role in naturally improving water quality. http://www.tamevalleywetlands.co.uk/lea-marston-lakes/

16.6 Pathogen Removal in Waste Stabilization Ponds

Waste stabilization ponds are extremely efficient at removing helminth and protozoan eggs and cysts by sedimentation, often achieving 100%

removal rates. However, the removal of enteric bacteria and viruses is less efficient. Removal is by a combination of factors including ultraviolet radiation, high dissolved oxygen (DO), settlement, pond temperature, nutrient level, predation, an increase in pH due to daytime assimilation of carbon dioxide by algae (which can reach in excess of pH 9), and anti-bacterial toxins produced by algae (Pike, 1975; von Sperling, 2007).

It is most appropriate to consider ponds as completely mixed reactors, so the survival of pathogens (x_t/x_0) can be calculated as:

$$(x_t/x_0) = 1/[1 - (\mu/D)]. \tag{16.16}$$

Ponds are normally in series, and if they have similar dilution rates, the survival of pathogens can be calculated as:

$$(x_n/x_0) = 1/[1 - (\mu/D)^n], \tag{16.17}$$

where n is the number of ponds. A problem arises if the number of ponds in series is large (i.e., >5) because the system then behaves more like a plug flow reactor. In this case, the relationship becomes:

$$(x_n/x_0) = \exp - (\mu/D_n), \tag{16.18}$$

where D_n is the dilution rate of the complete system.

Pond design is often based on fecal coliform or helminth egg reduction, especially where effluents are used for irrigation. In tropical and sub-tropical countries, the removal of pathogens is as important as BOD removal is in temperate zones, and in many countries, more so. Both facultative and maturation ponds are very efficient in the removal of pathogens (Arceivala *et al.*, 1970; Gloyna, 1971; Mara and Pearson, 1998), with removal rates of coliform and streptococci bacteria being >99%, and viruses inactivated by light and removed by absorption onto settleable solids and settlement so that >90% removals are achieved. The greater efficiency of facultative ponds compared with other systems is because of the much longer retention times (Coetzee and Fourie, 1965), with ponds in series achieving greater removals than single ponds (Marais, 1974). Major removal mechanisms are the high pH values created by photosynthesis and the higher zooplankton predation rate. However, the function of maturation ponds is primarily to improve the bacterial quality of the effluent from facultative ponds. The size and number of ponds are governed by the required removal of both fecal coliforms and helminth eggs. Maturation ponds only remove small quantities of BOD, but due to a diverse population of algae, nutrient removal can be significant. Their shallow design (1.0 to 1.5 m in depth) and long retention times ensure maximum pathogen immobilization.

A study of the coliform decay rate in 186 different WSPs determined the median removal rates (von Sperling, 2005), which were 98% for primary facultative ponds (1.8 $\log_{10}$ units), 90% for secondary facultative ponds (1.0 $\log_{10}$ units), and 94% for each maturation pond in series (1.2 $\log_{10}$ units) (Table 16.11). Two relationships were identified for calculating the coliform dieoff coefficient (K_b) in d^{-1} using detention time (t) in °C and pond depth (D) in m:

$$K_b = 0.682\ D^{-1.286} t^{-0.103}\ (R^2 = 0.874), \qquad (16.19)$$

$$K_b = 0.549\ D^{-1.456}\ (R^2 = 0.845). \qquad (16.20)$$

Table 16.11. Summary of coliform removal efficiencies expressed as log units removed for the different pond types evaluated. Log units removed = $-\log_{10}$ (1–Efficiency/100) where Efficiency (%) = $100(N_0 - N)/N_0 = 100(1 - 10^{-\text{log units removed}})$. N is the effluent coliform concentration (MPN [most probable number] per 100 mL) and N_0 is the influent coliform concentration (MPN per 100 mL) (von Sperling, 2005). Reproduced with permission from Elsevier Publishing.

	Facultative ponds		Maturation ponds		
Statistics	**Primary pond**	**Secondary pond**	**Pond 1**	**Pond 2**	**Ponds 3–5**
Number of data	99	51	48	34	29
Mean	1.83	1.16	1.38	1.19	1.09
Standard deviation	0.65	0.74	0.70	0.46	0.51
25% percentile	1.40	0.65	0.83	0.80	0.72
50% percentile (median)	1.76	0.95	1.25	1.16	1.13
75% percentile	2.30	1.46	1.65	1.55	1.35
Minimum	0.46	0.33	0.46	0.21	0.20
Maximum	3.70	3.30	3.31	2.22	2.08

Coliform removal efficiencies increased at higher temperatures, longer detention times, an increase in length to breadth ratio, and a decrease in depth. Von Sperling produced tables giving estimates of removal efficiencies at both 20°C and 25°C, the latter is shown in Table 16.12.

Table 16.12. The expected mean coliform removal efficiencies (expressed in terms of logarithmic units removed) from a facultative or maturation pond for different values of the hydraulic detention time *t*, depth *D* and length to breadth (*L/B*) ratio (dispersed flow) at 25°C (von Sperling, 2005). Reproduced with permission from Elsevier Publishing.

		Log units removed *L/B* ratio								
t(d)	*D*(m)	1	2	3	4	6	8	10	12	16
3	1.0	0.61	0.67	0.71	0.75	0.79	0.83	0.85	0.87	0.89
	1.5	0.40	0.43	0.45	0.46	0.48	0.49	0.50	0.51	0.52
	2.0	0.29	0.30	0.31	0.32	0.33	0.34	0.34	0.34	0.35
5	1.0	0.85	0.97	1.05	1.11	1.20	1.26	1.31	1.35	1.40
	1.5	0.58	0.63	0.67	0.70	0.74	0.77	0.79	0.81	0.83
	2.0	0.43	0.46	0.48	0.50	0.52	0.53	0.55	0.55	0.56
10	1.0	1.29	1.52	1.68	1.81	2.00	2.15	2.26	2.35	2.49
	1.5	0.91	1.03	1.12	1.19	1.30	1.37	1.42	1.47	1.53
	2.0	0.70	0.77	0.83	0.87	0.93	0.97	1.00	1.03	1.06
20	1.0	1.90	2.30	2.60	2.84	3.21	3.50	3.73	3.92	4.23
	1.5	1.37	1.62	1.80	1.94	2.16	2.32	2.44	2.54	2.70
	2.0	1.08	1.24	1.36	1.45	1.59	1.69	1.77	1.83	1.93
30	1.0	2.34	2.89	3.30	3.64	4.16	4.57	4.90	5.19	5.65
	1.5	1.72	2.06	2.32	2.53	2.84	3.08	3.28	3.44	3.69
	2.0	1.36	1.60	1.78	1.92	2.13	2.29	2.42	2.52	2.67
40	1.0	2.72	3.39	3.90	4.31	4.96	5.48	5.91	6.27	6.87
	1.5	2.00	2.44	2.76	3.02	3.43	3.75	4.00	4.22	4.56
	2.0	1.60	1.91	2.14	2.32	2.60	2.82	2.98	3.12	3.34

Temperature = 25°C.
K_b (dispersed flow) = 0.549 $D^{-1.456}$ $d = 1/(L/B)$.
Log units removed = $-\log_{10}$ (1-Efficiency, 100).
Efficiency (%) = 100 $(N_0 - N)/N_0$ = 100 $(1-10^{-\text{log units removed}})$.
Log units removed in a system with ponds in series = sum of the log units removed in each individual pond in the series.

In temperate climates, only maturation ponds are commonly used in the treatment of domestic sewage, which can be effective in the removal of bacteria and viruses (Windle Taylor, 1966; Adams *et al.*, 1972; Toms *et al.*, 1975). The removal efficiencies of *Streptococci* spp. and *Salmonella* spp. are equal to or better than the removal rate for *E. coli*, although *Clostridium perfringens*, both as active cells and spores, are less effectively removed (Toms *et al.*, 1975). Removal efficiency in maturation ponds is related to retention time, with removal rates falling rapidly as the retention time is reduced. For example, high removals of *E. coli* have been reported in Britain at retention times of 8.9 days (99.9%) and 17.0 day (99.5%) (Windle Taylor, 1966; Adams *et al.*, 1972), whereas at 3.5, 3.0 and 2.3 days retention, the removal rates were 94.0, 92.4 and 81.0%, respectively (Fish, 1966; Windle Taylor, 1966). Retention time is critical and any short-circuiting within ponds will reduce removal efficiency.

Solar radiation is a major removal mechanism in maturation ponds. The sun promotes the die-off of fecal bacteria in waste stabilization ponds by increasing the pond temperature and providing the energy for algal photosynthesis that both increases the pH above 9 and promotes photo-oxidative damage or photoinactivation (Mara and Pearson, 1998) (Fig. 16.12). Photoinactivation by UV radiated from the sun is caused by photons being absorbed by cellular molecules damaging genetic material (Maraccini *et al.*, 2016). UV radiation does not kill pathogens; rather, it inactivates them so that they are unable to reproduce and cause infection, but remain alive. UV light can pass across the cell wall of microorganisms and be absorbed by the protein and nucleotides that comprise the cell's components. While UV over a wide range (200–300 nm) can enter the cell, proteins only absorb UV at wavelengths <230 nm, although water and suspended solids tend to absorb most of the light at these wavelengths, limiting their inactivation capability, while the nucleotides absorb wavelengths >230 nm. Most of the protein is found in the cell wall, so high UV exposure is required to inactivate the cell compared to that required to damage the nucleotides that make up the DNA and RNA.

The cell nucleus in bacterial and protozoan waterborne pathogens contains DNA, which is comprised of strands of nucleotides

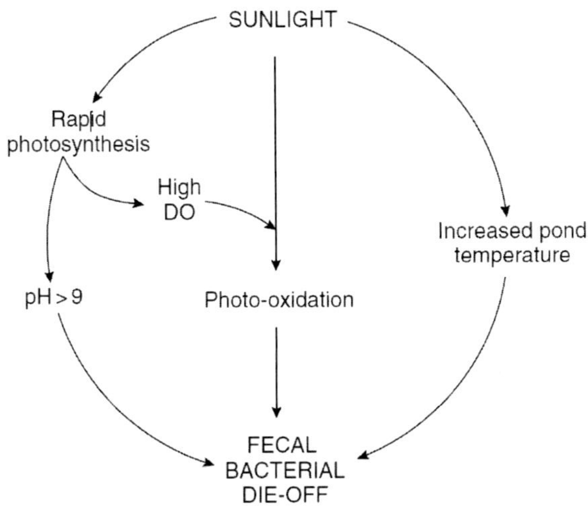

Fig. 16.12. Summary of removal mechanisms and the fate of organic micro-pollutants in waste stabilization ponds (Garcia-Rodríguez, *et al.*, 2014). Reproduced with permission from Springer Nature.

that join to form pairs held together by hydrogen bonds. These form the characteristic double stranded helix. There are four nucleotides that form pairs — cytosine with guanine and adenine with thymine. Viruses do not have a nucleus but can contain either DNA or RNA within a protein shell. RNA also has four types of nucleotides like DNA, but thymine is replaced by uracil. Microbial inactivation by UV radiation occurs when two thymine molecules are adjacent in the DNA strand and one of them is excited by a UV photon (Fig. 16.13). This leads to the two thymine molecules bonding, creating a new molecule called a thymine dimer. When a critical number of thymine dimmers are formed, estimated to be about 100 dimers per chain, then the DNA is inactivated and unable to replicate (Oguma *et al.*, 2002). RNA is inactivated when two adjacent uracil molecules bond and form a uracil dimer. Bacteria and viruses are both able to reactivate damage done to their DNA and RNA by UV light employing repair mechanisms that are classified either as

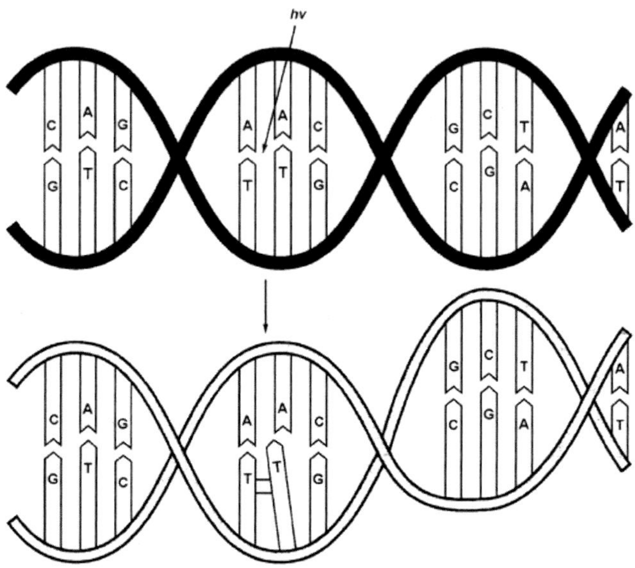

Fig. 16.13. The formation of thymine dimers caused by disruption to double stranded DNA by UV radiation. A = adenine, G = guanine, T = thymine, and C = cytosine. Reproduced from (Bolton and Cotton, 2008). Reproduced with permission from the American Water Works Association.

dark or light reactivation. Bacteria rely mainly on reactivation under dark conditions. This involves either the replacement of the thymine dimer and the adjacent section with a resynthesized section of nucleotides or whole strands of DNA being resynthesized and replaced. Light reactions, known as photoreactivation, are caused by UV-A, which causes the light-activated enzyme photolyase to be produced in both bacteria and viruses, which is able to reverse the formation of thymine dimers, thus restoring the DNA to its original structure (Harm, 1980). The former reactivation is dependent on the degree of damage done to the bacteria. Light reactivation is far more problematic for waste stabilization ponds if they are discharged to surface waters. Studies on drinking water disinfection have shown that *Cryptosporidium* spp. and *Giardia* spp. have not been recorded as being able to reactivate and so it is unlikely that they will become infective again after UV treatment (Belosevic *et al.,* 2001). Waterborne pathogens have different sensitivities to UV exposure at a particular wavelength. UV sensitivity is the greatest in bacteria and protozoa > viruses > bacterial spores > adenovirus (an exception as it is the only virus with low sensitivity to UV) > algae, which are least sensitive.

In waste stabilization ponds, the effectiveness of UV on pathogen destruction is dependent on the total UV light exposure, both in terms of time and intensity. As sunlight is quickly absorbed in turbid water, the vertical position of pathogens in the pond is critical with die-off rates declining with depth, making less turbid, shallower ponds most effective (Brissaud *et al.,* 2003). Thermal stratification can trap a portion of the pathogens near the surface at higher temperatures, which can increase die-off, while little die-off will occur in the dark cooler water in the lower stratified layer (Yang *et al.*, 2015). Shallow ponds may also temporarily thermally stratify during daylight with natural convection reinstating mixing as the surface layer cools during the evening. However, mixing in shallow ponds ensure that all pathogens are exposed to maximum exposure to sunlight, ensuring optimal die-off, which is especially effective in maturation ponds where water turbidity is lowest. Further information on pathogen inactivation in maturation ponds is given by Dahl *et al.* (2017).

The die-off rate is dependent on the presence of oxygen, being significantly enhanced at high pH. In South Africa, the mean removal efficiency of *E. coli* during the summer was 99.6% compared with 96.9% in winter (Drews, 1966). This is more clearly illustrated in Britain where the difference between solar radiation in the summer and winter is more pronounced. At a retention time of 3.5 days, the removal rate of *E. coli* was >90% in the summer but fell to 40% in the winter (60% for fecal streptococci) because of the seasonal difference in light intensity (Toms *et al.,* 1975). In the British Isles, the algal density is never sufficient to significantly shift the pH so that the pathogens are killed. The mortality of bacteria and viruses can be reduced by algae because of the reduction of light intensity by shading. The depth is also an important factor with removal rates being reduced with depth. Under ideal conditions, pond algae can consume carbon dioxide faster than it can be replaced by bacterial respiration. This results in carbonate and bicarbonate ions dissociating to release carbon dioxide (Eqs. [16.2] and [16.3]). The carbon dioxide is fixed by the algae and the hydroxyl ions accumulate, raising the pH as high as 10. At pH >9, fecal bacteria die rapidly, except for *Vibrio cholerae* (Pearson *et al.*, 1987).

Helminth eggs and protozoan cysts are primarily removed by settlement. The settling velocities of eggs are quite high (e.g. *Ascaris lumbricoides*, 3.4×10^{-4} m s^{-1}), thus pond systems appear extremely effective in the removal of parasites because long retention times allow maximum settlement. The average number of helminth eggs in wastewater in developing countries can be as high as 500 eggs per liter in poor, recently sewered, areas. Although this high incidence of eggs gradually declines as the opportunities for re-infection gradually falls due to the introduction of sewerage. The removal of eggs from pond effluents is extremely important if they are to be used for irrigation or fishpond fertilization. Cysts of the protozoans *Entamoeba histolytica* and *Giardia lamblia* are almost completely removed by well-designed pond systems, and the helminth parasites, such as *Schistosoma, Ascaris, Enterobius, Ancylostoma*, and *Trichuris* are also effectively removed (Gloyna, 1971). Maximum removal occurs in the first pond, and as the nematode eggs, in particular, are highly resistant to oxidative treatment and desiccation, extreme care must be taken with the disposal of the raw sludge from the pond if soil contamination is to be prevented. Removal efficiencies for three ponds in India have been given by Veerannan (1977). For helminth ova, these were 89 to 93% for *Ascaris lumbricoides*, >92% for hookworms, >68% for *Trichuris*, >95% for *Enterobius*, and >83% for Hymenolepis. For protozoan cysts removals, these were 39 to 77% for *Entamoeba coli*, >62% for *E. histolytica*, and >98% for *Giardia lamblia*. The efficiency of a pond system in removing bacteria and viruses, comprising an anaerobic pond (P1) with a mean retention time of 1 day, a facultative pond (P2) with a retention time of 5 days, followed by three maturation ponds (P3–5), each with a retention time of 5 days, is shown in Table 16.13.

Table 16.13. The geometric mean of bacterial and viral numbers[a] and percentage removals in raw wastewater (RW) and the effluents from five waste stabilization ponds in series (P1 to P5[b]) at a mean depth pond temperature of 26°C (Oragui *et al.*, 1987). Reproduced with permission from the International Water Association.

Organism	RW	P1	P2	P3	P4	P5	Percentage removal
Fecal coliforms	2×10^7	4×10^6	8×10^5	2×10^5	3×10^4	7×10^3	99.97
Fecal streptococci	3×10^6	9×10^5	1×10^5	1×10^4	2×10^3	300	99.99
Clostridium perfringens	5×10^4	2×10^4	6×10^3	2×10^3	1×10^3	300	99.40
Campylobacters	70	20	0.2	0	0	0	100.00
Salmonellae	20	8	0.1	0.02	0.01	0	100.00
Enteroviruses	1×10^4	6×10^3	1×10^3	400	50	9	99.91
Rotaviruses	800	200	70	30	10	3	99.63
BOD (mg L^{-1})	215	36	41	21	21	18	92

[a]Bacterial numbers per 100 mL, viral numbers per 10 liters.
[b]P1 was an anaerobic pond with a mean hydraulic retention time of 1 day: P2 and P3–P5 were secondary facultative and maturation ponds respectively, each with a retention time of 5 days.

Pathogens accumulate in the sludge of facultative ponds. In a study of pathogen survival in Australia where WSPs are widely used to treat domestic and municipal wastewater, Schwarz *et al.* (2019) studied decay rates of the number of fecal microorganisms. Background concentrations ranged from 10^4 to 10^6 colony-forming units (CFU) mL^{-1} for *Escherichia coli* 104, 10^1 to 10^3 for enterococci, 10^1 to 10^3 for F-specific bacteriophage (MS2), 10^1 to 10^3 for somatic coliphage, and 10^1 to 10^2 for human adenovirus gene copies (mL^{-1}), which were all studied in the primary facultative pond. They discovered that the pond environment creates conditions where pathogens can survive for long periods within the sludge layer, especially bacteria, although these decayed faster than viruses, which had very slow decay rates during winter. The time for a one $\log_{10}$ reduction (T_{90}) in the upper and lower (>1 m) sludge levels, respectively, are 15 and 54 days for *E. coli*, 13 and 14 days for the *Enterococcus* spp., 69 and 122 days for F-specific bacteriophage MS2, 41 and 41 days for somatic coliphage, and >180 and 156 days for human adenovirus. Decay rates were influenced by a number of factors such as sludge depth, pond temperature, pH, conductivity, season, and rainfall. Due to regrowth, they concluded that *E. coli* was not a reliable indicator of pathogen content in WSP sludge and that sludge from facultative ponds needs to be further treated before use on agricultural land due to the risk of pathogen transfer. Pathogens in the upper sludge layers can also be readily resuspended during times of high-induce mixing or high-flow rates, especially common in smaller plants due to high rainfall events resulting in surface water contamination (Keffala *et al.*, 2013).

16.7 Contaminant Removal in Waste Stabilization Ponds

There have been many studies on the removal efficiency of nutrients and pathogens in WSPs, but few on the removal of organic pollutants (Qiang *et al.*, 2013; Rivera-Utrilla *et al.*, 2013). Gruchlik

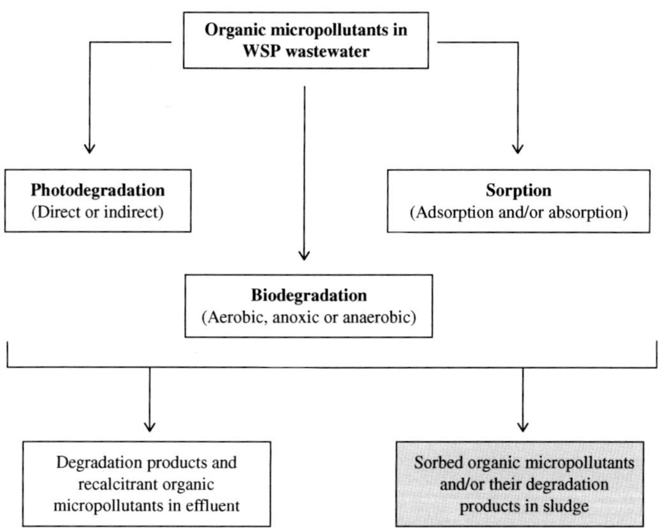

Fig. 16.14. Removal mechanisms for organic micropollutants in waste stabilization ponds (adapted from [Garcia-Rodríguez *et al.*, 2014]).

et al. (2018) have reviewed the efficiency and mechanisms for the removal of micropollutants such as personal care products, pharmaceuticals, and endocrine-disrupting compounds by WSPs. They identified that removal primarily took place in the anaerobic lagoon or facultative pond stages by a combination of processes such as biodegradation, photodegradation and sorption (Fig. 16.14).

Table 16.14. The reported average removal rates of common micropollutants by waste stabilization ponds adapted from (Gruchlik *et al.*, 2018). Reproduced with permission from Elsevier Publishing.

Compound	Average removal efficiency (%)	Compound	Average removal efficiency (%)	Compound	Average removal efficiency (%)
Pharmaceuticals		***Personal care product chemicals***		***Endocrine disrupting compounds***	
Carbamazepine	6–29	Cashmeran	82	17α-Ethinyl estradiol (EE2)	35–99
Diclofenac	65–90	Furosemide	99	17β-Estradiol (E2)	42–62
Diphenhydramine	85	Galaxolide	75–96	17β-Trenbolone	100
Gemfibrozil	15–20	Methyl dihydrojasmonate	80–99	4-Nonylphenol (NP)	75
Ibuprofen	75–100	Methylparaben	81	Androstendione	93–100
Ketoprofen	87–100	Oxybenzone	99	Androsterone	100
Naproxen	75–100	Tonalide	76–96	Bisphenol A (BPA)	24
Paracetamol	99	Triclocarban	85	Estriol (E3)	26–77
Propranolol	10	Triclosan	77–100	Estrone (E1)	31–100
Salicylic acid 9	5–100			Eticholanolone	100
Sulfamethoxazole	82–100			Octylphenol	71
Trimethoprim	44–100	***Other compounds***			
		5-Methyl benzotriazole	76		
		Benzothiazole	81		
		Benzotriazole	50		
		Caffeine	70–100		
		OH-Benzothiazole	86		
		Tributyl phosphate	59		
		Triphenyl phosphate	76		
		Tris(2-chloroethyl) phosphate	21		

Light (both as a driver of photosynthesis and UV [200–400 nm] photodegradation), temperature, pH, and the presence of algae have all been identified as key factors in their removal (Li *et al.*, 2013; Heaven *et al.*, 2005; Matamoros *et al.*, 2015). The removal rates have been given by Ying *et al.* (2009), Coleman *et al.* (2010), Camacho-Muñoz *et al.* (2012), Li *et al.* (2013), and Matamoros *et al.* (2016) and summarized in Table 16.14.

References

Adams, I.M., Ayres, P.A. and Wood, P.C. (1972) Bacterial reduction during tertiary treatment of sewage effluents. *Institute of Public Health Engineers Journal*, **71**, 108–125.

Aldana, G.J., Lloyd, B.J., Guganesharajah, K. and Bracho, N. (2005) The development and calibration of a physical model to assist in optimising the hydraulic performance and design of maturation ponds. *Water Science and Technology*, **51**(12), 173–181.

American Public Health Association, American Water Works Association and Water Pollution Control Federation. (1985) *Standard Methods for the Examination of Water and Wastewater, 16th ed.* American Public Health Association Inc., New York.

Arceivala, S.J., Lakshminarayana, J.S.S., Alagarsamy, S.R. and Sastry, C.A. (1970) *Waste Stabilization Ponds: Design, Construction and Operation in India.* Central Public Health Engineering Research Institute, Nagpur.

Badrot-Nico, F., Guinot, V. and Brissaud, F. (2010) Taking wind into account in the design of waste stabilisation ponds. *Water Science and Technology*, **61**(4), 937–944.

Bellamy, W.D. (1975) Conversion of insoluble agricultural wastes to SCP by thermophylic micro-organisms. In *Single Cell Protein II* (eds. S.R. Tannenbaum and D.I.C. Wang). MIT Press, Mass, pp. 2163–2172.

Belosevic, M., Craik, S.A., Stafford, J.L., Neumann, N.F., Kruithof, J. and Smith, D.W. (2001) Studies on the resistance/reactivation of *Giardia muris* cysts and *Cryposporidium parvum* oocysts exposed to medium pressure ultraviolet radiation. *FEMS Microbiology Letters*, **204**, 197–203.

Bode, H. and Klopp, R. (2001) Nutrient removal in the river Ruhr — A German case study. *Water Science and Technology*, **44**(1), 15–24.

Bolton, J.R. and Cotton, C.A. (2008) *The Water Disinfection Handbook.* American Water Works Association, Denver, CO.

Bopardikar, M.V. (1969) Microbiology of a waste stabilization pond. In *Advances in Water Pollution Research* (ed. S.H. Jenkins). Pergamon, Oxford, pp. 595–601.

Bracho, N. Lloyd, B. and Aldana, G. (2006) Optimisation of hydraulic performance to maximise faecal coliform removal in maturation ponds. *Water Research*, **40**, 1677–1685.

Brissaud, F., Tournoud, M.G., Drakides, C. and Lazarova, V. (2003) Mixing and its impact on faecal coliform removal in a stabilisation pond. *Water Science Technology*, **48**(2), 75–80.

Camacho-Muñoz, D., Martin, J., Santos, J.L., Aparicio, I. and Alonso, E. (2012) Effectiveness of conventional and low-cost wastewater treatments in the removal of pharmaceutically active compounds. *Water Air and Soil Pollution*, **223**, 2611–2621.

Chatterjee, S. N., Arora, B.K. and Gupta, D.R. (1967) Some observations on the utilization of sewage for fish culture in oxidation ponds. *Environmental Health (India)*, **9**, 156–161.

Coetzee, O.J. and Fourie, N.A. (1965) The efficiency of conventional sewage purification works, stabilization ponds and maturation ponds with respect to the survival of pathogenic bacteria and indicator organisms. *Journal and Proceedings of the Institute Sewage Purification* **3**, 210–215.

Coleman, H.M., Le-Minh, N., Khan, S.J., Short, M.D., Chernicharo, C. and Stuetz, R.M. (2010) Fate and levels of steroid oestrogens and androgens in waste stabilisation ponds: Quantification by liquid chromatography-tandem mass spectrometry. *Water Science and Technology*, **61**(3), 677–684.

Cooke, W.B. (1963) A laboratory guide to fungi in polluted waters, sewage and sewage treatment works. Public Health Service Publication No. 999-WP-1. US Department of Health, Education and Welfare, Washington, DC.

Crabtree B., Hickman, M. and Martin, D. (1999) Integrated water quality and environmental cost-benefit modelling for the management of the River Tame. *Water Science and Technology*, **39**(4), 213–220.

Dahl, N.W., Woodfield, P.L., Lemckert, C.J., Statton, H. and Roiko, A. (2017) A practical model for sunlight disinfection of a subtropical maturation pond. *Water Research*, **108**, 151–159.

Dinges, R. (1982). *Natural Systems for Water Pollution Control.* Van Nostrand Reinhold, New York.

Dochain, D., Gregoire, S., Pauss, A. and Schaegger, M. (2003) Dynamic modelling of a waste stabilization pond. *Bioprocess and Biosystems Engineering*, **26**, 19–26.

Dodson, S.L., Cáceres, C.E. and Rogers, D.C. (2010) Cladocera and other Brachiopoda. In *Ecology and Classification of North American Freshwater Invertebrates, 3rd ed.* (eds. J.H. Thorp and A.P. Covich). Academic Press, London, pp. 773–827.

Drews, R.J.L.C. (1966) Field studies of large scale maturation ponds with respect to their purification efficiency. *Journal and Proceedings of the Institute of Sewage Purification*, 280–294.

Edwards, P. and Pullen, R.S.V. (1990) Wastewater fed aquaculture. Report: 684. Environmental Sanitation Information Centre, Asian Institute of Technology, Bangkok, Thailand.

European Community. (1991) Council Directive concerning urban waste water treatment. *Official Journal of the European Community*, **L135**, 21.5.1991, 40–52.

Fish, H. (1966) Some investigations of tertiary methods of treatment. *Institute of Public Health Engineers Journal*, **65**, 33–47.

Garcia-Rodríguez, A., Matamoros, V., Fontàs, C. and Salvadó, V. (2014) The ability of biologically based wastewater treatment systems to remove emerging organic contaminants — A review. *Environmental Science and Pollution Research*, **21**, 11708–11728.

Gloyna, E.F. (1971) Waste stabilization ponds. WHO Monograph Series, 60. World Health Organization, Geneva.

Gray, N.F. (2004) *Biology of Wastewater Treatment, 2nd ed.* Imperial College Press, London.

Green, J. and Watts, E. (1973) Sewage effluent lagoons: Zooplankton in theory and practice. In *Symposium on Sewage Effluent as a Water Resource.* Institution of Public Health Engineers, London, pp. 36–44.

Gruchlik, Y., Linge, K. and Joll, C. (2018) Removal of organic micropollutants in waste stabilization ponds: A review. *Journal of Environmental Management*, **206**, 202–214.

Harm, W. (1980) *Biological Effects of Ultraviolet Radiation.* Cambridge University Press, Cambridge.

Hartley, W.R. and Weiss, C.M. (1970) Light intensity and vertical distribution of algae in tertiary oxidation ponds. *Water Research*, **4**, 751–763.

Hawkes, H.A. (1983a) Activated sludge. In *Ecological Aspects of Used Water Treatment, Vol. 2. Biological Activities and Treatment Processes* (eds. C.R. Curds and H.A. Hawkes). Academic Press, London, pp. 77–162.

Hawkes, H.A. (1983b) Stabilization ponds. In *Ecological Aspects of Used Water Treatment, Vol. 2. Biological Activities and Treatment Processes* (eds. C.R. Curds and H.A. Hawkes). Academic Press, London, pp. 163–217.

Heaven, S., Banks, C.J. and Zotova, E.A. (2005) Light attenuation parameters for waste stabilisation ponds. *Water Science and Technology*, **51**(12), 143–152.

Hemens, J. and Mason, M.H. (1968) Sewage nutrient removal by a shallow algal stream. *Water Research*, **2**, 277–287.

Holm, H.W. and Vennes, J.W. (1970) Occurrence of purple sulphur bacteria in sewage treatment lagoons. *Applied Microbiology*, **19**, 988–996.

Huang, H.J.S. and Gloyna, E.F. (1984) Phosphorus models for waste stabilization ponds. *Journal of the Environmental Engineering Division, American Society of Civil Engineers*, **110**, 550–561.

Imhoff, K.R. (1982) Experiences with purification lakes and polishing lagoons. *Water Science and Technology*, **14**, 189–203.

Imhoff, K.R. (1984) The design and operation of the purification lakes in the Ruhr valley. *Water Pollution Control*, **83**, 243–253.

Kalisz, L. and Suchecka, T. (1966) L'importance de certains organismes de phytq-et zooplancton dans le processus d'épuration des eaux d'é'gouts dans les étangs biologiques. *International vereinigung fur Theoretische und Angewandte Limnologie Verhandlungen*, **16**, 948–951.

Keffala, C., Harerimana, C. and Vasel, J. (2013) A review of the sustainable value and disposal techniques, wastewater stabilisation ponds sludge characteristics and accumulation. *Environmental Monitoring and Assessment*, **185**(1), 45–58.

Kimerle, R.A. and Anderson, N.H. (1971) Production and bioenergetic role of the midge *Glyptotendipes barbipes* (Staeger) in a waste stabilization lagoon. *Limnology and Oceanography*, **16**, 646–659.

Kimerle, R.A. and Enns, W.R. (1968) Aquatic insects associated with midwestern stabilization lagoons. *Journal of the Water Pollution Control Federation*, **40**(2), R31–R41.

Lakshminarayana, J.S.S. (1965) Discussion of paper presented by Loedolff. In *Advances in Water Pollution Research, Vol. 1* (ed. O. Jaag). Pergamon, Oxford, pp. 320–324.

Li, M., Zhang, H., Lemckert, C., Roiko, A. and Sratton, H. (2018) On the hydrodynamics and treatment efficiency of waste stabilization ponds: From a literature review to a strategic evaluation framework. *Journal of Cleaner Production*, **183**, 495–514.

Li, X., Zheng, W. and Kelly, W.R. (2013) Occurrence and removal of pharmaceutical and hormone contaminants in rural wastewater treatment lagoons. *Science of the Total Environment,* **445–446**, 22–28.

Loedolff, C.J. (1965). The function of cladorea in oxidation ponds. In *Advances in Water Pollution Research, Vol. 1* (ed. O. Jaag). Pergamon, Oxford, pp. 307–317.

Mara, D.D. (1987) Waste stabilization ponds: Problems and controversies. *Water Quality International,* (1), 20–22.

Mara, D.D. and Pearson, H.W. (1998) Design manual for waste stabilization ponds in Mediterranean countries. Lagoon Technology International, Leeds, UK.

Mara, D.D., Alabaster, G.P., Pearson, H.W. and Mills, S.W. (1992) Waste stabilization ponds: A design manual for Eastern Africa. Lagoon Technology International, Leeds, UK.

Maraccini, P.A., Mattioli, M.C., Sassoubre, L.M., Cao, Y., Griffith, J.F., Ervin, J.S. and van de Werfhorst, L.C. (2016) Solar inactivation of enterococci and *Escherichia coli* in natural waters: Effects of water absorbance and depth. *Environmental Science and Technology,* **50**(10), 5068–5076.

Marais, G.v.R. (1974) Faecal bacterial kinetics in waste stabilization ponds. *Journal of the Environmental Engineering Division, American Society of Civil Engineers,* **100** (EE1), 119–139.

Martínez, F.C., Cansino, A.T., López, A.S. Barrios, J.L.G. and de la Cruz Acosta, F.J. (2016) Mathematical modeling and optimization in the design of a maturation pond. *Journal of Applied Research and Technology,* **14**, 93–100.

Matamoros, V., Gutierrez, R., Ferrer, I., García, J. and Bayona, J.M. (2015) Capability of microalgae-based wastewater treatment systems to remove emerging contaminants: A pilot-scale study. *Journal of Hazardous Materials,* **288**, 34–42.

Matamoros, V., Rodríguez, Y. and Albaiges, J. (2016) A comparative assessment of intensive and extensive wastewater treatment technologies for removing emerging contaminants in small communities. *Water Research,* **88**, 777–785.

Maynard, H.E., Ouki, S.K. and Williams, S.C. (1999) Tertiary lagoons: A review of removal mechanisms and performance. *Water Research,* **33**(1), 1–13.

Mayo, A.W. and Abbas, M. (2014) Removal mechanisms of nitrogen in waste stabilization ponds. *Physics and Chemistry of the Earth,* **72–75**, 77–82.

McGarry, M.G. and Pescod, M.B. (1970) Stabilization pond design criteria for tropical Asia. In *Proceedings of the Second International Symposium on Waste Treatment Lagoons* (ed. R.E. McKinney). University of Kansas, Laurence, KS, USA, pp. 114–132.

Middlebrooks, E.J. (1988) Review of rock filters for the upgrade of lagoon effluents. *Journal of the Water Pollution Control Federation*, **60**(9), 1657–1662.

Middlebrooks, E.J., Middlebrooks, C.H., Reynolds, J.H., Walters, G.Z., Reed, S.C. and George, D.B. (1982) *Wastewater Stabilization Lagoon Design, Performance and Upgrading*. Macmillan Publishing, New York, USA.

Nelson, K.L. and Jiménez, B.C. (2000) Sludge accumulation, properties and degradation in a waste stabilization pond in Mexico. *Water Science and Technology*, **42**(10–11), 231–236.

Oguma, K., Katayama, H. and Ohgaki, S. (2002) Photoreactivation of *Escherichia coli* after low or medium pressure UV disinfection determined by endonuclease sensitive site assay. *Applied Environmental Microbiology*, **68**, 6029–6035.

Oragui, J.I., Curtis, T.P., Silva, S.A. and Mara, D.D. (1987) The removal of excreted bacteria and viruses in deep waste stabilization ponds in Northeast Brazil. *Water Science and Technology*, **19**(3–4), 569–573.

Ouarghi, H.E., Boumansour, B.E., Dufayt, O., El Hamouri, B. and Vasel, J.L. (2000) Hydrodynamics and oxygen balance in a high-rate algal pond. *Water Science and Technology*, **42**(10–11), 349–356.

Pano, A. and Middlebrooks, E.J. (1982) Ammonia nitrogen removal in facultative wastewater stabilization ponds. *Journal of the Water Pollution Control Federation*, **54**(4), 344–351.

Pearson, H.W., Mara, D.D. and Arridge, H.A. (1995) The influence of pond geometry and configuration on facultative and maturation waste stabilization pond performance and efficiency. *Water Science and Technology*, **31**(12), 129–139.

Pearson, H.W., Mara, D.D. and Azov, Y. (eds.) (2000) Waste stabilization ponds: Technology and the environment. *Water Science and Technology*, **42**(10–11), 1–408.

Pearson, H.W., Mara, D.D., Smallman, D.J. and Mills, S. (1987) Physicochemical parameters influencing faecal coliform survival in waste stabilization ponds. *Water Science and Technology*, **19**(12), 145–152.

Pham, D.T., Everaert, G., Janssens, N., Alvarado, A., Nopens, I. and Goethals, P.L.M. (2014) Algal community analysis in a waste stabilization pond. *Ecological Engineering*, **73**, 302–306.

Pike, E.B. (1975). Aerobic bacteria. In *Ecological Aspects of Used Water Treatment, Vol. 1. The Organisms and Their Ecology* (eds. C.R. Curds and H.A. Hawkes). Academic Press, London, pp. 1–64.

Porges, R. and Mackenthun, K.M. (1963) Waste stabilization ponds: Use, function and biota. *Biotechnology and Bioengineering*, **5**, 255–273.

Potten, A.H. (1972) Maturation ponds — Experiences in their operation in the United Kingdom as a tertiary treatment process for a high quality sewage effluent. *Water Research*, **6**, 375–391.

Potten A.H., Toms R.G., Hinge D.C., White C.M. and Ridley J.E. (1971) Rye Meads Sewage Purification Works: Part I. Tertiary treatment by lagoons of the nitrified effluent. Part II. Biological and chemical observations of the effect upon the river Lee. Part III. The waterworks aspect with special reference to algal productivity. *Journal of the Institute of Public Health Engineers*, **70**, 9–60.

Qiang, Z., Dong, H., Zhu, B., Qu, J. and Nie, Y. (2013) A comparison of various rural wastewater treatment processes for the removal of endocrine-disrupting chemicals (EDCs). *Chemosphere*, **92**, 986–992.

Reed, S.C. (1985) Nitrogen removal in wastewater stabilization ponds. *Journal of the Water Pollution Control Federation*, **57**(1), 39–45.

Rivas Hernandez, A. (1997) Effect of fish management on faecal coliform removal in facultative ponds. *Ingenieria Hidraulica en Mexico*, **12**(3), 27–33.

Rivera-Utrilla, J., Sanchez-Polo, M., Ferro-Garcia, Prados-Joya, G. and Ocampo-Perez, R. (2013) Pharmaceuticals as emerging contaminants and their removal from water. A review. *Chemosphere*, **93**, 1268–1287.

Sah, L., Rousseau, D.P.L., Hooijmans, C.M. and Lens, P.N.L. (2011) 3-D model for a secondary facultative pond. *Ecological Modelling*, **222**, 1592–1603.

Schwarz, K.R., Sidhu, J.P.S., Toze, S., Li, Y., Lee, E., Gruchlik, Y. and Prichard, D.L. (2019) Decay rates of *Escherichia coli*, *Enterococcus* spp., F-specific bacteriophage MS2, somatic coliphage and human adenovirus in facultative pond sludge. *Water Research*, **154**, 62–71.

Shilton A.N. and Harrison, J. (2003) Guidelines for the hydraulic design of waste stabilization ponds. *Water Science and Technology*, **48**(2), 173–180.

Shilton, A.N. and Mara, D.D. (2005) CFD (computational fluid dynamics) modelling of baffles for optimizing tropical waste stabilization pond system. *Water Science and Technology*, **51**(12), 103–106.

Svore, J.H. (1968) Waste stabilization pond practices in the United States. In *Advances in Water Quality Improvement* (eds. E.F. Gloyna and W.W. Eckenfelder). University of Texas, Austin, Texas, pp. 427–434.

Toms, I.P., Owens, M., Hall, J.A. and Mindenhall, M.J. (1975) Observations on the performance of polishing lagoons at a large regional works. *Water Pollution Control*, **74**, 383–401.

Torres, J.J., Soler, A., Salez, J., Leal, L.M. and Aguilar, M.I. (1999) Study of the internal hydrodynamics in three facultative ponds of two municipal WSPs in Spain. *Water Research*, **33**(5), 1133–1140.

Townshend, A.R. and Knoll, H. (1987) Cold climate sewage lagoons. Report EPS 3/NR/1. Environment Canada, Ottawa.

Tschörtner, U.S. (1967) The determination of chlorophyll *a* in algae and its application in South African oxidation ponds — I. *Water Research*, **1**, 785–793.

Uhlmann, D. (1969). Discussion on paper presented by Bopardikar, M.V. In *Advances in Water Pollution Research* (ed. S. H. Jenkins). Pergamon, Oxford, pp. 607–610.

Van Der Post D.C. and Engelbrecht R.J. (1973) The maturation process in the reclamation of potable water from sewage effluent. *Water Pollution Control*, **72**, 457–465.

Veerannan, K.M. (1977) Effect of sewage treatment by the stabilization pond method on the survival of intestinal parasites. *Indian Journal of Environmental Health*, **19**, 100–106.

von Sperling, M.V. (2005) Modelling of coliform removal in 186 facultative and maturation ponds around the world. *Water Research*, **39**, 5261–5273.

von Sperling, M.V. (2007) *Waste Stabilisation Ponds*. IWA Publishing, London.

Williams, W.P., White, R.W.G. and Noble, R.P. (1973) In *Symposium on Sewage Effluent as a Water Resource*. Institution of Public Health Engineers, London, pp. 45–53.

Windle Taylor, E. (1966). 41st Report on the bacteriological, chemical and biological examination of the London Waters for the years 1963–64. Metropolitan Water Board, London.

Woods, D.R., Green, M.B. and Parish, R.C. (1984) Lea Marston purification lake: Operation and river quality aspects. *Water Pollution Control*, **83**, 226–242.

Yang, P., Xing, Z., Fong, D.A., Monismith, S.G., Tan, K.M. and Lo, E.Y.M. (2015) Observations of vertical eddy diffusivities in a shallow tropical reservoir. *Journal of Hydrology and Environment Research*, **9**(3), 441–451.

Ying, G.G., Kookana, R.S. and Kolpin, D.W. (2009) Occurrence and removal of pharmaceutically active compounds in sewage treatment plants with different technologies. *Journal of Environmental Monitoring*, **11**, 1498–1505.

17
Partially and Completely Mixed Aerated Ponds

17.1 Introduction

Aeration lagoons, also referred to as aerated ponds, were originally designed for the first-stage treatment of sewage or the pretreatment of strong industrial wastewaters such as sugar beet waste before secondary treatment, usually by either facultative ponds or activated sludge. The basins are deeper than other types of stabilization ponds by 3.0 to 6.0 m, with oxygen being provided mechanically and not solely by the photosynthetic activity of algae. Oxygen is normally provided by mechanical pier-mounted or floating surface aerators, which ensures that the microbial biomass is kept in suspension and that sufficient dissolved oxygen is provided for maximum aerobic activity. Diffuse bubble aeration, provided by compressed air pumped through plastic tubing laid across the bottom or suspended above the base of the pond, is also used. Compared to facultative ponds, aerated ponds require less land area but have significantly higher capital and operational (energy) costs. Originally designed to rely completely on algal photosynthesis for aeration, facultative ponds are very susceptible to organic overloading, leading to operational failure due to deoxygenation. This is especially a problem when used in colder climates or where the population is variable such as in holiday resorts. Under these conditions, it becomes necessary to use supplementary oxygenation, and from this, two distinct designs of aerated ponds have developed: ***partially mixed aerated ponds*** (also known as aerated facultative ponds) and ***completely mixed aerated ponds*** (i.e., aeration lagoons).

17.2 Partially Mixed Aerated Ponds

Aerated facultative ponds or partially mixed aerated ponds (PMAPs) are defined as using only sufficient aeration to maintain oxygenation but not to keep any biomass and as little solids as possible in suspension, allowing these to settle and form a sludge layer where anaerobic breakdown can occur. Oxygen is still supplied by algae with supplementary aeration supplied by surface aerators as required. Aeration may be continuous or only at times when the algae are unable to provide sufficient oxygen (e.g. at night-time, during winter or on cloudy days). Partially mixed systems require a depth of 3 to 6 m and a hydraulic retention time (HRT) of 3 to 10 days, rising to 20 to 40 days in colder climates. A length to width ratio of 3–4:1 is used to create a plug flow movement, which is increased by using baffles or cells in series (Fig. 17.1). However, like complete mixed aeration ponds, PMAPs are modeled using completely mixed first-order models (Eqs. [17.1] and [17.9]). Partially mixed systems are primarily used for small communities and are simple to operate, with aeration automatically controlled by oxygen probes within the pond. Algae no longer play a role in oxygenation where aeration is continuous; however, the aeration should not achieve complete mixing, thereby preserving separate anaerobic and aerobic layers. These designs are

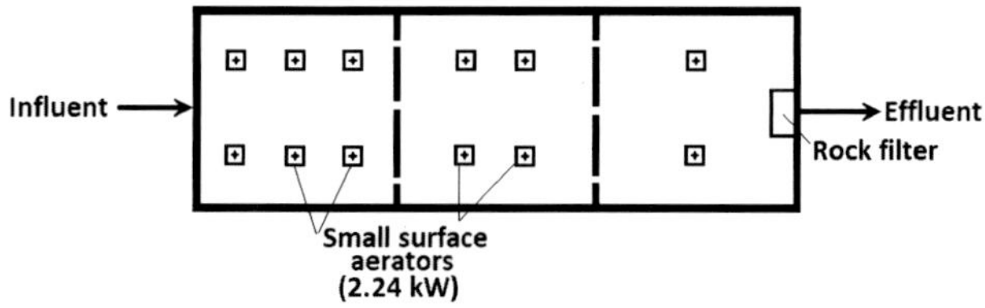

Fig. 17.1. The layout of a partially mixed aerated pond with an integrated rock filter to remove any algae present in the final effluent.

known as facultative aerated lagoons and are fully described by von Sperling (2007). These can be loaded at 20 to 30 g BOD $m^{-3}d^{-1}$.

Ponds are usually made up of a number of cells with aeration levels adjusted in each, as the oxygen demand decreases as the wastewater moves through the cells. Partially mixed aerated ponds use 10 times less energy than completely mixed aerated ponds. Surface aerators are normally used for PMAPs, providing between 1.5 and 2.1 kg O_2 kWh^{-1} as compared to 3.7 to 4 kg O_2 kWh^{-1} with piped diffused aeration, thus making the latter more cost-effective although the initial capital costs are higher.

Biochemical oxygen demand (BOD) and suspended solids (SS) concentrations of <30 mg L^{-1} are possible after settlement to remove algae. Nitrification is generally limited to summer months when there is sufficient dissolved oxygen (DO) to satisfy both the heterotrophs and nitrifiers, although sufficient HRT is required for nitrifiers to develop. Phosphorus removal is poor at <25%. Removal of coliforms is a function of retention time and temperature, with disinfection used depending on receiving water standards.

The design and operation of PMAPs are discussed in detail in *Principles of Design and Operations of Wastewater Treatment Pond Systems for Plant Operators, Engineers and Managers* (USEPA, 2011).

17.3 Completely Mixed Aerated Ponds

Aeration lagoons or completely mixed aerated ponds (CMAPs) are used as a first treatment stage for very strong wastewaters or a second stage after an anaerobic lagoon, in series with facultative and maturation ponds. In these ponds, the oxygen is supplied only by aeration, with algal development generally limited. These ponds have the smallest footprint of any other waste stabilization pond. A predominantly bacterial microbial biomass develops and as there is no provision for settlement or sludge return, the process relies on sufficient mixed liquor developing within the basin, which carries out oxidation of the incoming wastewater. Retention times vary between 2 and 5 days, depending on the degree of treatment required, temperature and strength of the influent wastewater. A retention time of 5 days at

20°C provides an 85% reduction in the BOD of domestic sewage, although a fall in temperature by 10°C results in a reduction in BOD removal to 65% (Hammer, 1977). However, only partial treatment is achieved in CMAPs, so further treatment is required after settlement.

Table 17.1 summarizes the loading and BOD removals of an experimental CMAP fed on raw sewage. The pond, 4.6 m wide and 25.2 m long was split into six cells, each 3.7 m deep, aerated and mixed using diffuse air fed from the base via a perforated pipework. The overall percentage removals were 92% for the BOD, 83% for the chemical oxygen demand (COD), and 91% for SS (USEPA, 1979). Design examples and case studies are presented in (USEPA, 1979), with further information available in (USEPA, 2011).

Bubble aeration is most successful in keeping ponds aerobic and completely mixed in locations where pond surfaces may become frozen in the winter. Under these conditions, surface aerators cannot operate due to ice formation. In this type of climate, unaerated facultative ponds are generally unsuccessful and tend to become anaerobic. As these are completely mixed reactors, individual cells are often square in plan; alternatively, cells can be separated by baffles within a single pond with a length to width ratio of 3–4:1 and a depth of 2.5 to 4.0 m is recommended. Surface aerators must be carefully spaced to ensure that there are no dead areas where settlement could occur, and for that reason, it is better to use smaller aerators than one or two larger ones per cell (Fig. 17.2).

As a unit process, aeration lagoons fall between the activated sludge process and facultative oxidation ponds. If the degree of agitation is sufficient, then all the solids may be kept in suspension under aerobic conditions corresponding to the activated sludge process. In systems with less efficient agitation, the solids may tend to settle and form an anaerobic sludge layer similar to PMAPs. Where the aeration is inadequate, then enhanced deposition of particulate solids and the reduced dissolved oxygen concentration in the basin will lead to anaerobic conditions, a loss of efficiency, and the production of odors. Therefore, sufficient agitation and mixing coupled with a strict loading regime are required for the successful operation of aeration

Table 17.1. Example of a performance of a completely mixed aerated pond comprising 6 cells and operated over 18 months in a northern climate showing the BOD and suspended solids reductions of 66 and 45% after an HRT of 5 days, and 81 and 65% after 8 days of HRT, respectively (USEPA, 1979). Reproduced with permission from the US Environmental Protection Agency.

	Pond HRT (d)		Organic loading (g BOD $m^{-3}d^{-1}$)		BOD				Suspended solids			
Cell	Per cell	Accum.	Per cell	Accum.	Mean (mg L^{-1})	SD	Cell removal (%)	Accum. removal (%)	Mean (mg L^{-1})	SD	Cell removal (%)	Accum. removal (%)
Influent	—	—	—	—	319	109	—	—	261	128	—	—
1	1.8	1.8	177.2	177.2	151	71	53	53	138	46	47	47
2	2.9	4.7	51.5	67.8	108	47	28	66	143	86	0	45
3	3.6	8.3	30.0	38.4	60	34	44	81	91	67	36	65
4	5.4	13.7	11.1	23.2	40	25	33	87	67	69	27	75
5	7.3	21.0	5.5	15.2	26	11	35	92	39	29	42	85
6	9.0	30.0	2.9	10.6	26	10	3	92	23	12	42	91

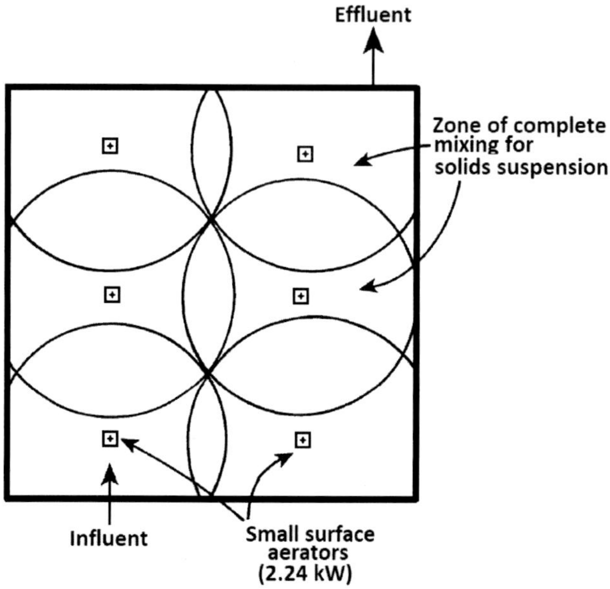

Fig. 17.2. Position of surface aerators to ensure complete mixing and aeration within a cell of a completely mixed aerated pond (USEPA, 1979). Reproduced with permission from the US Environmental Protection Agency.

lagoons. Between 30 and 50% of the effluent BOD is from the solids fraction, so a settlement basin must be incorporated in the design (Fig. 17.3). Settlement ponds may need a long HRT to ensure adequate settlement with regular desludging, although CMAPs are usually followed by a facultative pond and smaller ones with a sand filter. Some ponds have the facility to recycle the effluent to control loading, but where the solids are returned, this makes the pond more of a modified activated sludge system rather than an aeration pond. There are a number of modifications including high-performance aerated pond systems and the BIOLAC® process (USEPA, 2011).

Unlike activated sludge systems, aerated lagoons develop diverse and abundant populations of zooplankton that feed mainly on bacteria (Cauchie *et al.*, 2000). The active biomass comprises a mixture of

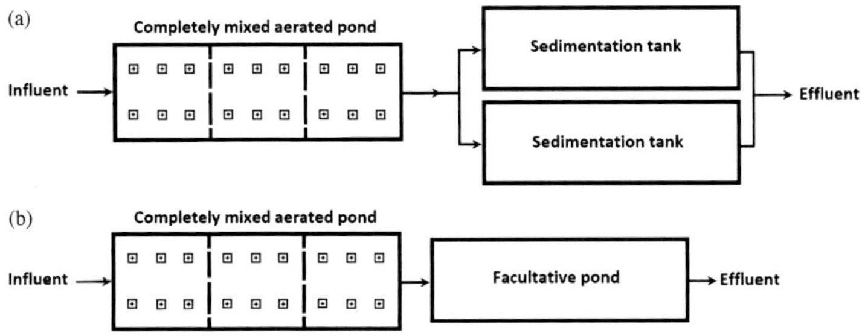

Fig. 17.3. The common design of completely mixed aerated pond systems: (a) with sedimentation tanks, and (b) followed by a facultative pond. Both systems are subsequently followed by maturation ponds.

wastewater solids as well as small bacterial flocs with high densities of freely dispersed bacteria. The mixed liquor suspended solids (MLSS) or biomass concentration in a CMAP is 20 to 30 times less than those in activated sludge, where the biomass is returned. The sludge's age is equal to the HRT, so unlike the activated sludge process, it is the HRT that is the key design parameter. The biomass is discharged with effluent from the lagoon and enters a sedimentation tank where as much of the solids are removed before the effluent goes onto the second stage of treatment.

Aeration lagoons are susceptible to large inputs of biodegradable or toxic waste that can severely reduce efficiency. High infiltration and stormwater result in reduced retention times and flush the microbial biomass out of the basin, thus reducing both the MLSS and the sludge age. Where this is a potential problem, provision should be made to divert a portion of the enhanced flow directly to the next stage, thereby maintaining optimum hydraulic loading to the aeration lagoon.

Middlebrooks *et al.* (1982) have proposed using the Marais–Shaw design model for CMAPs using first-order kinetics **assuming n cells of equal dimensions in series** (Marais and Shaw, 1961):

$$\frac{C_n}{C_o} = \frac{1}{\left[1 + \left(Kt/n\right)\right]^n}, \tag{17.1}$$

where C_n is the effluent BOD_5 (mg L^{-1}) in cell n, C_o is the influent BOD_5 (mg L^{-1}), K is the BOD removal coefficient (d^{-1}), t is the total HRT (d) of the pond system, and n is the number of cells. A default value for K when treating raw or settled sewage of 0.30 d^{-1} at 20°C can be used for all cells in series. The value K is a function of substrate concentration and temperature. In oxidation ponds, the substrate concentration is largely dependent on the HRT; the longer the retention, the greater the dilution of the incoming wastewater, and so the lower the reaction rate. This means that K will decrease in each following cell in series. Also, as the fraction of the readily assimilable BOD is quickly oxidized, the remaining more complex organic substrate takes longer to break down, which also slows the rate. The design values for K for PMAPs are the same as for facultative ponds at 0.25 to 0.40, as compared to 1.0 to 1.5 for CMAPs (von Sperling, 2007). The removal coefficient can be roughly estimated using the equations below that were developed from CMAP data collected in northern climates (USEPA, 1979):

For BOD loadings > 10 g BOD $m^{-3}d^{-1}$

$$K = 0.031\ L^{0.57}. \tag{17.2}$$

For BOD loadings <10 g BOD $m^{-3}d^{-1}$

$$K = 0.022\ L^{0.69}, \tag{17.3}$$

where L is the BOD loading rate in g BOD $m^{-3}d^{-1}$. The BOD removal coefficients can be measured directly using pilot trials and are discussed by Ramalho (1977) and summarized by Gray (2017).

As removal coefficients are a function of temperature that must be taken into account when using the model, this can be adjusted using Eq. (17.4):

$$K_T = K_{20}\theta^{T_w - 20}, \tag{17.4}$$

where K_T is the reaction rate (d^{-1}) at temperature T (°C), K_{20} is the reaction rate at 20°C (d^{-1}), θ is the temperature coefficient 1.036, and T_w is the temperature of the cell water (°C).

To estimate the pond water temperature (*Tw*), the equation developed by Mancini and Barnhart (1976) can be used:

$$T_w = \frac{AfT + QT}{Af + Q}, \tag{17.5}$$

where T and T_w are the ambient air and pond water temperature (°C), respectively, A is the pond surface (m^2), f is the proportionality factor (0.5), and Q is the wastewater flow rate (m^3 d^{-1}).

While cells of equal size are most efficient in terms of performance, this is not always possible to achieve due to landscape factors resulting in **cells of unequal volume**, thus requiring an alternative design model:

$$\frac{C_n}{C_o} = \left(\frac{1}{1 + Kt_1}\right)\left(\frac{1}{1 + K_2 t_2}\right)\cdots\left(\frac{1}{1 + K_n t_{1n}}\right), \tag{17.6}$$

where K_1, K_2, …, K_n are the reaction rates (d^{-1}) in cells 1, 2, …, n, while t_1, t_2 and t_n are the HRTs of each cell, respectively (d).

Partially mixed aerated and completely mixed aerated ponds are transitional treatment systems between facultative waste stabilization ponds and activated sludge, but offer a more sustainable and flexible treatment option than conventional biological treatment systems. They are also useful in extending the range that facultative ponds are normally used for in domestic wastewaters and allow for variable organic loadings to be accommodated.

References

Cauchie, H.M., Hoffmann, L. and Thome, J.P. (2000) Metazooplankton dynamics and secondary production of *Daphnia magna* (Crustacea) in an aerated waste stabilization pond. *Journal of Plankton Research*, **22**(12), 2263–2287.

Gray, N.F. (2017) *Water Science and Technology: An Introduction, 4th ed.* CRC Press, London.

Hammer, M.J. (1977) *Water and Wastewater Technology*. John Wiley & Sons, New York.

Mancini, J.L. and Barnhart, E.L. (1976) Industrial waste treatment in aerated lagoons. In *Ponds as a Wastewater Alternative* (eds. E.F. Gloyna, J.F. Malina and E.M. Davis). University of Texas Press, Austin, Texas.

Marais, C.V.R. and Shaw, V.A. (1961) A rational theory for the design of sewage stabilization ponds in Central and South Africa. *Transactions, South Africa Institute of Civil Engineers,* **3**, 205–214.

Middlebrooks, E.J., Middlebrooks, C.H., Reynolds, J.H., Watters, G.Z., Reed, S.C. and George, D.B. (1982) *Wastewater Stabilization Lagoon Design, Performance, and Upgrading*. Macmillan Publishing Inc., New York.

Ramalho, R.S. (1977) *Introduction to Wastewater Treatment Processes.* Academic Press, New York.

USEPA. (1979) Performance of aerated lagoons in northern climates. EPA:600/3-79-003. US Environmental Protection Agency, Corvallis, OR.

USEPA. (2011) Principles of design and operations of wastewater treatment pond systems for plant operators, engineers and managers. EPA:600/R-10/088. US Environmental Protection Agency, Cincinnati, OH. https://www.epa.gov/sites/production/files/2014-09/documents/lagoon-pond-treatment-2011.pdf

von Sperling, M. (2007) *Waste Stabilization Ponds.* IWA Publishing, London.

Part V
The Future

18

Sustainable Wastewater Treatment Systems

18.1 Assessing Sustainable Options

Wastewater treatment has changed as well as saved countless millions of lives by transforming the health of communities, protecting them from a wide range of waterborne diseases, enhancing their quality of life and wellbeing, and protecting the environment from a wide range of contaminants including organic pollution, nutrient enrichment leading to eutrophication, metal pollution and contamination from a wide variety of organic contaminants from industrial and pharmaceutical compounds to personal care products, and especially estrogen-mimicking compounds. Although 95% of wastewaters in developed countries are collected and treated, many smaller communities (e.g. <2,000 population equivalent [PE] in the European Union [EU]) and isolated housing are not subject to the same level of treatment, if any (EU Directive 91/271/EEC). However, with surface waters globally under threat from increased pollution due to reduced flows and higher temperatures, and because of a combination of over-abstraction and climate change, many existing treatment plants will need to be improved to meet stricter limits, especially in relation to nutrients. In the EU, this is being tackled through the Water Framework Directive (EU 2000/60/EU), which provides a catchment-based co-ordinated approach to the management of the water cycle, which requires wastewater to be fully treated in order to preserve good ecological status as defined in the legislation.

While wastewater treatment is rapidly installing circular economy measures into its new designs and current operation (Chap. 2), many barriers must be overcome. These include regulation, pricing and public perception. This is particularly evident within the water pathway (Fig. 2.4) and Voulvoulis (2018) stresses that extreme care must be exercised with regulating and monitoring levels of contaminants in recycled water, with new standards required for specific reuse options. The selection of appropriate technologies for these new wastewater treatment plants and upgrading existing plants not only requires that they be efficient in terms of final effluent quality, but they meet economic, environmental and social criteria as outlined in Table 1.7 in Sec. 1.3 (Flores-Alsina *et al.*, 2010).

The use of sustainability indicators is an easy way to compare individual wastewater treatment plants and technologies (Sec. 1.3). Molinos-Sanante *et al.* (2014) have given an example of using weighted sustainability indicators for assessing wastewater treatment. Using 17 indicators (Table 18.1) classified under just three sustainability dimensions, each dimension and indicator was given a ranking by a group of experts, each adding up to 100%, with each dimension being: environmental (47.1%), economic (30.8%), and social (22.1%) (Table 18.2). This reflects the relative importance given by these experts under local or national conditions. The method is very similar to water quality indices, which have been used successfully for over fifty years (Gray, 1996). The calculation method is given elsewhere with some indicators raising the value (positive) with others having a negative effect on scores; the higher the score, the more sustainable the plant or technology is based on the indicators and weighting selected (Molinos-Sanante *et al.*, 2014). Using this sustainability index, the authors compared two natural wastewater treatment systems — constructed wetlands (CWs) and pond systems (PSs) — with conventional systems like extended aeration (EA), membrane bioreactors (MBRs), rotating biological contactors (RBCs), trickling filters (TFs) and sequencing batch reactors (SBRs) (Table 18.3).

Table 18.1. Details of the sustainability indicators used by Molinos-Sanante *et al.* (2014). To assess the sustainability of different wastewater technologies, information on each indicator came from a combination of literature review, real data, and expert knowledge. Reproduced with permission from Elsevier Publishing.

Dimension	Indicator	Acronym	Direction	Formula	Source of information*	Unit
Economic	Investment cost	IECS1	Negative	$IECS1 = \frac{\text{Investment cost}}{\text{p.e.}}$	LR	€/p.e
	Operation and maintenance costs	IECS2	Negative	$IECS2 = \frac{\text{Operation costs}}{m^3 \text{ treated water}}$	RD	€/m^3
Environmental	Organic matter efficiency removal	IENS1	Positive	$IENS1 = \frac{[COD]_{inf} - [COD]_{eff}}{[COD]_{inf}}$	LR	—
	Suspended solids efficiency removal	IENS2	Positive	$IENS2 = \frac{[SS]_{inf} - [SS]_{eff}}{[SS]_{inf}}$	LR	—
	Nitrogen efficiency removal	IENS3	Positive	$IENS3 = \frac{[N]_{inf} - [N]_{eff}}{[N]_{inf}}$	LR	—
	Phosphorus efficiency removal	IENS4	Positive	$IENS4 = \frac{[P]_{inf} - [P]_{eff}}{[P]_{inf}}$	LR	—
	Energy consumption	IENS5	Negative	$IENS5 = \frac{\text{KhW}}{m^3 \text{ treated water}}$	LR and RD	KhW/m^3

(*Continued*)

Table 18.1. (*Continued*)

Dimension	Indicator	Acronym	Direction	Formula	Source of information*	Unit
	Land area required	IENS6	Negative	$\text{IENS6} = \frac{\text{m}^2 \text{ land}}{\text{p.e.}}$	LR	m^2/p.e.
	Sewage sludge production	IENS7	Negative	$\text{IENS7} = \frac{\text{Kg sludge}}{\text{m}^3 \text{ treated water}}$	LR and RD	Kg/m^3
	Potential for water reuse	IENS8	Positive	Qualitative	LR and EK	—
	Potential to recover products	IENS9	Positive	Qualitative	LR and EK	—
	Reliability	IENS10	Negative	Qualitative	LR	—
Social	Odors	ISS1	Negative	Qualitative	LR and EK	—
	Noise	ISS2	Negative	Qualitative	LR and EK	—
	Visual impact	ISS3	Negative	Qualitative	LR and EK	—
	Public acceptance	ISS4	Positive	Qualitative	LR and EK	—
	Complexity	ISS5	Negative	Qualitative	LR and EK	—

* LR: literature review, RD: real data, EK: expert knowledge.

Table 18.2. The weighting allocated to each sustainability indicator within each sustainability dimension (a total of 100%) and for each of the three dimensions: economic, environmental and social (Molinos-Sanante *et al.*, 2014). Reproduced with permission from Elsevier Publishing.

Dimension	Indicator	Weight of indicators (%)	Weight of dimensions (%)
Economic	Investment cost (IECS1)	33.6	30.8
	Operating and maintenance costs (IECS2)	66.4	
Environmental	Organic matter efficiency removal (IENS1)	14.6	47.1
	Suspended solids efficiency removal (IENS2)	12.4	
	Nitrogen efficiency removal (IENS3)	10.7	
	Phosphorus efficiency removal (IENS4)	12.4	
	Energy consumption (IENS5)	8.8	
	Land area required (IENS6)	7.8	
	Sewage sludge production (IENS7)	9.7	
	Potential for water reuse (IENS8)	9.4	
	Potential to recover products (IENS9)	6.5	
	Reliability (IENS10)	7.8	
Social	Odors (ISS1)	37.2	22.1
	Noise (ISS2)	18.2	
	Visual impact (ISS3)	9.2	
	Public acceptance (ISS4)	23.0	
	Complexity (ISS5)	12.4	

Table 18.3. Comparison of the scores for each dimension and the overall index of sustainability using Tables 18.1 and 18.2, to compare seven different wastewater treatment technologies (Molinos-Sanante *et al.*, 2014). The abbreviations for the treatment types are given in the text. Reproduced with permission from Elsevier Publishing.

Sustainability	CW	PS	EA	MBR	RBC	TF	SBR
Economic (ECS)	0.964	0.783	0.629	0.063	0.533	0.520	0.427
Environmental (ENS)	0.325	0.395	0.617	0.856	0.485	0.614	0.684
Social (SS)	0.628	0.628	0.551	0.510	0.456	0.318	0.616
Index of sustainability	0.589	0.566	0.606	0.535	0.493	0.519	0.590
Overall rank	3	4	1	5	7	6	2

The results generated by the index are extremely useful as they indicate for each plant or technology where improvements can be made, although interpretation can be confusing. For example, Molinos-Sanante *et al.* (2014) only allocated a weighting of 8.8% within the environmental dimension, representing just 4.1% of the overall weighting of the index of sustainability, which appears very low, given the importance of global warming in sustainable development. Problems with the weighting allocation are also particularly reflected in both TLs and RBCs, which are both considered to be reasonably energy efficient (Gray, 2021). CWs have a high economic and social score but low environment dimension scores. This is again due to the low weighting of energy use with both systems scoring 0.06 kWh m^{-3} compared to 0.30 kWh m^{-3} for TLs and RBCs, and 0.50 and 0.70 kWh m^{-3} for EAs and MBRs, respectively.

So, the percentage weighting given to each dimension significantly affects the overall sustainability score, for example, within built-up areas, the social dimension may need to be as high as 70 to 80%, while for a plant with high hydraulic loadings in a European country where land is limited (e.g. $<10^6$ PE), then the economic dimension may represent 60 to 80% of the overall weighting. The increasing urgency to combat global warming, preserve the aquatic ecosystems, and protect public health, requires that the environmental dimension should be given as much priority in terms of weighting as possible, regardless of cost and to some extent public acceptance. This is reflected in the indicators selected by Cossio *et al.,* (2020), which included greenhouse gas emissions as one of their 17 indicators using a fourth sustainability dimension, *Technical* (Table 18.4). They also included local indicators under a fifth sustainability dimension, *Institutional* (Table 1.9). So, the selection of harmonized national or international sustainability indicators for wastewater treatment is urgently required to ensure investment, and the correct planning is aimed at where it is most needed. Sustainability indicators also have an important role to play in achieving the UN Sustainable Development Goals (Malik *et al.,* 2015).

Table 18.4. The 17 standard sustainability indicators identified by a literature review by Cossio *et al.* (2020). Reproduced with permission from Elsevier Publishing.

Dimension	Indicator	Description	Suggested units
Environmental	Energy use	Energy used per volume unit of treated wastewater or inhabitant.	kW/p.e. kWh/m^3
	Global warming potential	Emissions of greenhouse gases (GHGs) into the atmosphere.	kg CO_2eq./p.e.-year kg CO_2eq./m^3
	Removal of BOD, TSS, TN, TP and FC	Contaminant removal efficiency to mitigate health and environmental risks.	%
	Land area	Land area required for the wastewater treatment facility.	m^2/p.e.
	Quality of effluent and sludge	Pollutants discharged into water and toxic compounds discharged into soil.	mg/L kg/p.e.-year
	Potential recycling	Reuse of treated wastewater nutrients (N, P) and energy.	g/p.e.-year %
Social	Public acceptance	Opinion of the local population affected by the plant.	Qualitative
	Aesthetics	Measured level of nuisance deriving from, e.g. odor, noise, visual impact, insects and other pests.	Qualitative
Economic	Investment costs	Cost of construction and installation of the WWTP.	\$/m^3 \$/p.e.-year
	Operating and maintenance costs	Operating costs per volume unit of wastewater treated.	\$/m^3
Technical	Reliability	Infrastructure or mechanical reliability, resilience; security, ability to endure shock loads and/or seasonal effects; potential for overflow.	Qualitative
	Complexity of construction and O&M	Ease of construction, complexity of plant construction and system installation; complexity of operation and maintenance; professional skills required for operation and maintenance.	Qualitative

p.e. = population equivalent that expresses the ratio between the sum of the pollution load produced during a 24-h period by institutions (i.e., schools or health centers) and the individual load produced by one person in a household.

18.2 Emerging Technologies

Emerging technologies in wastewater treatment are almost exclusively associated with the development of conventional biological and/or physicochemical processes (USEPA, 2013), although algal technologies have been widely accepted as low-cost, sustainable and efficient alternatives for the treatment of contaminants of emerging concern (Shah *et al.*, 2020). Natural treatment systems have not been fully exploited or sufficient research and investment made in them, and yet they hold the key to sustainable "net zero" wastewater treatment.

A key element in the circular economy is the recovery and reuse of wastewater components. The current processes being used and developed further include:

Recovery of energy: Biogas as a biofuel, on-site wind and solar generation, combined heat and power (CHP), incineration of biosolids, heat pumps, microalgae, effluent hydropower, and bioelectrical systems.

Reuse of nutrients: Biosolids application to farmland, wastewater irrigation, struvite production, urine separation, and microalgae.

Reuse of water: Irrigation, indirect potable reuse, direct potable reuse, industrial reuse, and household reuse.

For example, fertilizers are largely produced by the fixation of nitrogen from the atmosphere by the Haber–Bosch process, which means that this extra nitrogen in the form of ammonia or oxidized nitrogen entering the hydrological cycle from agriculture drives eutrophication. Wastewater treatment plays an important role in the transfer of nitrogen back into surface waters. Originally, the emphasis was on oxidizing ammonia to prevent toxicity to fish, converting it to nitrate that could safely be discharged into surface waters with sufficient dilution. But as the population has grown, the need to remove nitrate and phosphorous to prevent eutrophication has become imperative, resulting in often energy-hungry solutions. In a remarkable accounting study at the Amsterdam-West wastewater treatment plant in the Netherlands, van der Hoek and his colleagues looked at alternative

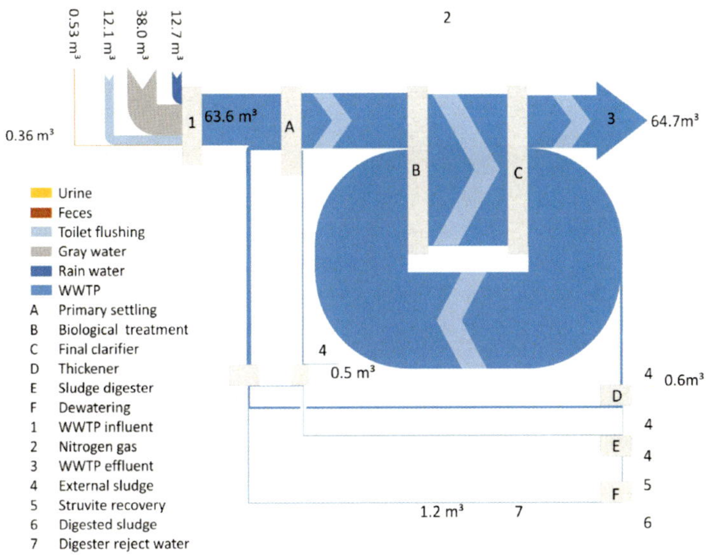

Fig. 18.1. The water balance at the Amsterdam-West wastewater treatment plant serving 1,044,584 inhabitants, as measured by van der Hoek *et al.* (2018) in 10^6 m^3 yr^{-1}. In terms of liters per capita per day, this constituted 31.7 L from toilet flushing, 99.6 L of gray water, 0.94 L of urine, and 1.4 L of feces. Rainwater as runoff contributed 20% of the flow. The authors noted an imbalance of 1.8% over the entire treatment process, which was added to the final effluent flow, due to evaporation.

ways of removing nitrogen alongside normal treatment and the recovery of other useful resources including biogas, phosphorus and cellulose (van der Hoek *et al.*, 2018). To do this, they carried out a detailed analysis of water and nitrogen flows through the treatment plant (Figs. 18.1 and 18.2), which allowed suitable adaption pathways to be explored (Fig. 18.3). Those methods selected for further analysis and the percentage removal of total nitrogen (TN) at the plant of each were for: struvite precipitation (1.1%), air stripping of the digester reject water (24%), vacuum membrane filtration (75%), hydrophobic membrane filtration (75%), and the treatment of air

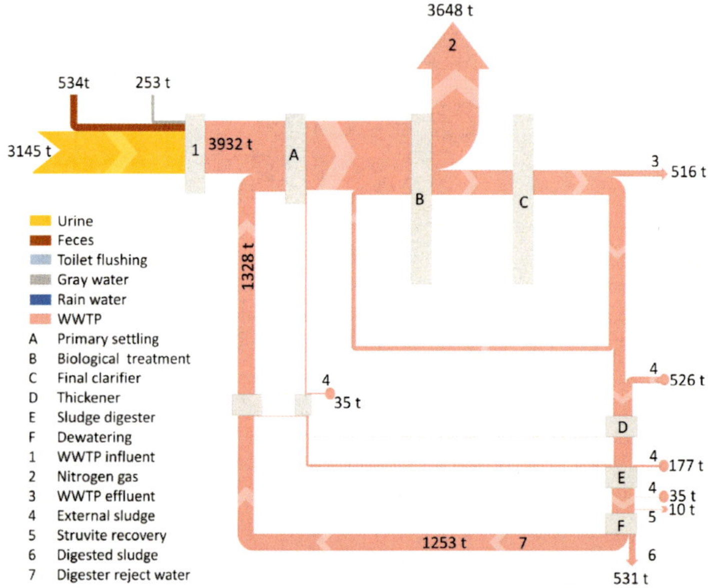

Fig. 18.2. The nitrogen balance at the Amsterdam-West wastewater treatment plant as measured by van der Hoek *et al.* (2018) in tons per year (t yr^{-1}). The incoming nitrogen came from urine, feces, graywater and rainwater, with urine contributing 80% of the incoming mass. Reproduced under the Creative Commons Attribution License CC BY (http: //creativecommons.org/licenses/by/4.0/).

from thermal sludge drying (2.1%). However, urine collection and it being reused as a fertilizer could reduce the TN in wastewater by 60%. The adaption pathway map (Fig. 18.3) can be used to explore the feasibility and effectiveness of a series of different scenarios, usually used in parallel, to reach set goals. The authors have used this approach to simulate four different scenarios, each increasing the amount of nitrogen removed but each step involving increasing levels of change to the operation. This approach can also be used with natural treatment systems, where options such as urine separation may be more feasible. However, nitrification–denitrification and plant uptake remain useful and underused natural solutions to the problem of nitrogen control.

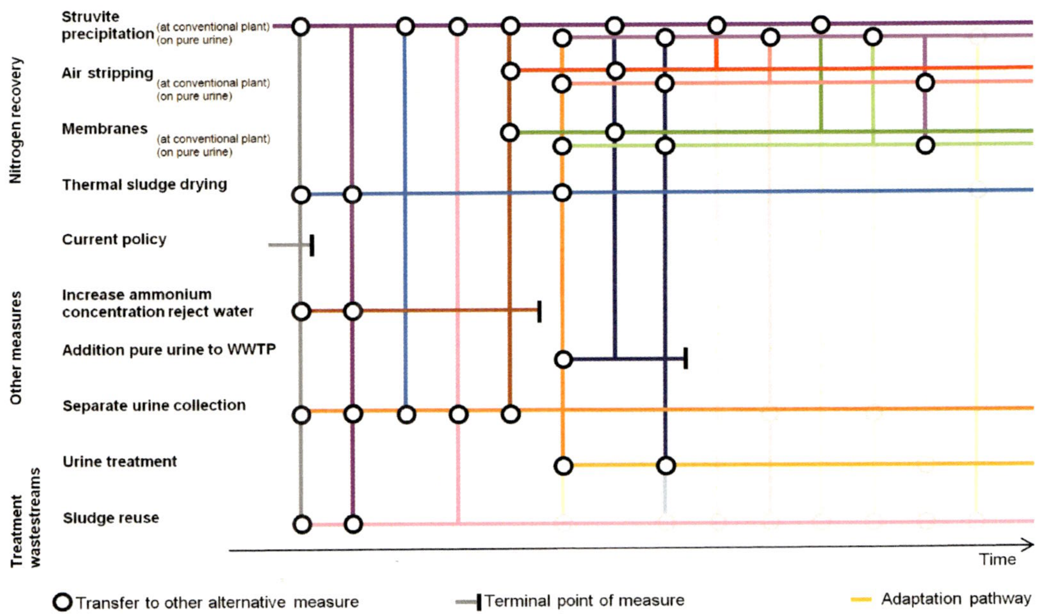

Fig. 18.3. An adaptive pathway approach to exploring the best treatment options for nitrogen removal at the Amsterdam-West waste water treatment plant (Hoek *et al.*, 2018). Each alternative is represented by a colored horizontal line, with vertical lines showing alternatives at different stages. Terminal points indicate where the option can no longer be considered useful after this point. Reproduced under the Creative Commons Attribution License CC BY (http://creativecommons.org/licenses/by/4.0/).

Anaerobic digestion is a key element in wastewater sustainability and the circular economy.

In order to make digestion more efficient and robust, pretreatment is required, in particular, removal of contaminants that either inhibit the process reducing gas production or by reducing the value or limiting the use of the biomass. Using new operational practices and technologies such as macerating the sludge before digestion or using thermal hydrolysis to condition the sludge, are all key to the success and reliability of the process (i.e., Advanced Anaerobic Digestion [AAD]). The inclusion of other waste materials, such as food waste, helps to enhance the rates of reaction. Natural treatment systems are able to generate significant biomass, which can be used in digestion. Biogas collection from anaerobic lagoons is now commonplace and can be used in conjunction with agricultural wastes.

The development of microalgal photobioreactors is perhaps the most exciting prospect for the next generation of wastewater treatment, combining resource recovery and carbon dioxide sequestration, with pollutant and pathogen reduction. Algae offer huge potential not only for the detoxification of wastewaters but also for their breakdown or recovery. Algal and macrophyte biomass can be used for a wide range of valuable bioproducts, from the building blocks of pharmaceuticals to biopolymers for bioplastics. Natural treatment systems allow for far greater circularity within wastewater treatment, making them potentially net zero or better treatment systems.

18.3 The Role of Natural Treatment in a Sustainable Water Industry

Global wastewater production grows as the population increases and was estimated to be 450×10^9 m^3 in 2010 (Florke *et al.*, 2013), of which 70% was domestic. This has now grown to now approximately 500×10^9 m^3 in 2021. This not only represents a huge potential threat to public and environmental health but also a major source of resources, in particular, water and its sustainability (Muthukrishna and Victor, 2016). Examples of wastewater production per country are given in Table 18.5, with the level of treatment varying from

Table 18.5. Municipal wastewater production and the level of treatment in various countries expressed in km^3 of wastewater. Those countries that have less than 50% treatment are highlighted in bold, all of which are countries that could benefit from the utilization of natural treatment systems. Adapted from (Mateo-Sagasta *et al.*, 2015). Reproduced with permission from Springer-Verlag.

Country	Wastewater produced (km^3)	Wastewater treated (km^3)	Percent treated (%)	Year data collected
United States	60.40	40.89	67.7	2008
China	37.98	26.61	70.1	2010/2009
Japan	16.93	11.56	68.3	2011
India	15.44	4.42	**28.6**	2011
Indonesia	14.28	unknown	—	2012
Russian Federation	12.32	unknown	—	2011
Brazil	9.73	2.51	**25.8**	2009
Korea, Rep.	7.84	6.58	83.9	2011
Mexico	7.46	3.08	**41.3**	2011
Egypt, Arab Rep.	7.08	3.71	52.4	2012
Canada	6.61	3.55	53.7	2009
Germany	5.30	5.18	97.7	2007
Thailand	5.11	1.17	**22.9**	2012
Malaysia	4.22	2.60	61.6	2009
United Kingdom	4.09	4.05	99.0	2011
Italy	3.93	3.9	99.2	2007
France	3.79	3.77	99.5	2008
Turkey	3.58	2.72	76.0	2010
Iran, Islamic Rep.	3.55	0.89	**25.1**	2012
South Africa	3.54	1.92	54.2	2009
Spain	3.18	3.16	99.4	2004
Pakistan	3.06	0.55	**18.0**	2011
Venezuela, RB	2.90	unknown	—	1996
Argentina	2.46	0.29	**11.8**	2010/2000
Colombia	2.40	0.60	**25.0**	2010
Poland	2.27	1.36	60.0	2011

(*Continued*)

Table 18.5. (*Continued*)

Country	Wastewater produced (km^3)	Wastewater treated (km^3)	Percent treated (%)	Year data collected
Vietnam	1.97	0.20	**10.2**	2012
Netherlands	1.93	1.88	97.4	2010
Australia	2.02	2.00	99.0	2013
Saudi Arabia	1.55	1.06	68.4	2010
Philippines	1.26	unknown	—	2011
Peru	1.00	0.28	**28.0**	2011/2012
Algeria	0.82	0.32	**39.0**	2012
Bangladesh	0.73	unknown	—	2000
Iraq	0.58	0.10	**17.2**	2012

country to country. In Europe, treatment levels are >90%, reaching 99% in France, Italy, Spain, and the United Kingdom. In contrast, many low-income countries have comparatively low levels of treatment but these are often the very countries that could benefit from using natural treatment systems, although the relatively high levels of urbanization can make this difficult (Mateo-Sagasta *et al.*, 2015).

The provision of adequate wastewater treatment is a global issue recognized by its inclusion in the UN Sustainable Development Goals, which are to be met by 2030. Sustainable wastewater treatment provision is central to ending poverty, disease and hunger. It also has a major role to play in protecting the environment, sustaining water supplies, and reducing greenhouse emissions. In the developed world, surface waters are increasingly sensitive to eutrophication, contaminants and climate change issues such as drought, flooding and water temperatures. All this has resulted not only in a significant reduction in both surface and groundwater quality but also massive losses in biodiversity. The situation in developing countries is worse, with 90% of its wastewater receiving no treatment, with raw wastewaters directly released to surface inland waters or directly into oceans (UNEP, 2010). Global-warming-induced climate change and an increasing global population are rapidly changing water availability, with water scarcity predicted to affect 60% of the world's population by 2025. Increasing urbanization is also leading to problems in sustaining sufficient supplies of water, and requires improved sustainable management of water resources, including the reuse of wastewater effluents (Sato *et al.*, 2013).

Natural treatment systems operate best where there is sufficient sunshine and adequate water temperatures, and where land and labor are relatively cheap (and so are ideal for use in low-income countries). Many conventional designs that are built in low-income countries and often funded by international aid rapidly fail due to a lack of maintenance, restricted access to spares, lack of funding, or more commonly due to overloading. Natural treatment systems, especially stabilization ponds, offer a flexible, robust, reliable, and easily extended treatment option. Unlike the developed world, there is often a high community engagement with wastewater treatment, especially at the smaller scale, with maturation ponds used for fish farming, duckweed production, fertilizer reuse and most importantly, as water scarcity increases, the option to supplement existing water supplies used for agriculture with treated effluent. Natural treatment systems can also be used in conjunction with climate action projects, such as the prevention of desertification, sustaining populations to prevent migration, and providing alternate income generation.

Low-energy, low-embedded carbon and no chemicals make natural treatment systems an important option to ensure good environmental and public health, with a high degree of resilience to cope with all the potential pressures that climate change may pose to the water industry.

While businesses cannot achieve circularity in isolation, wastewater treatment likewise needs the support of those who should see themselves as stakeholders rather than simply customers. This includes a better separation of products and materials that are flushed with sanitary products such as wipes, dental floss, tampons, plastic applicators, etc., which makes preliminary treatment difficult, resulting in frequent mechanical failures and contamination of biosolids. Wastewater contaminants, such as heavy metals and pharmaceuticals, all need to be reduced or eliminated from wastewaters at the point of use, with the sewer system not used for the disposal of wastes other than that from food preparation, personal washing and hygiene, and the normal use of the toilet. Better infrastructure is required for the proper disposal and recycling of those household items that currently are being flushed. Also, sanitary items and personal care products should be readily degradable and plastic-free.

Water use in high-income countries is falling due to the widespread adoption of water conservation measures. However, water usage is also changing, with toilets no longer the main use of water in the home, having been replaced by personal washing. This, in turn, reduces the amount of wastewater that needs to be pumped and treated. For the water industry to achieve net zero, new and innovative technologies and approaches are required. New treatment plants must embrace the use of low-carbon materials and construction techniques that can be most effective when used with natural treatment systems. Water companies are looking beyond net zero and looking at ways to reduce emissions further, which is where circularity will play an important part in turning traditional treatment plants into resource recovery and energy-generating plants.

References

Cossio, C., Norrman, J., McConville, J., Mercado, A. and Rauch, S. (2020) Indicators for sustainability assessment of small-scale wastewater treatment plants in low and lower-middle class income countries. *Environmental and Sustainability Indicators*, **6**, 100028.

Flores-Alsina, X., Gallego, A., Feijoo, G. and Rodriguez-Roda, I. (2010) Multiple-objective evaluation of wastewater treatment plant control alternatives. *Journal of Environmental Management*, **91**(5), 1193–1201.

Florke, M., Kynast E., Barlund I., Eisner S., Wimmer, F. and Alcamo J. (2013) Domestic and industrial water uses of the past 60 years as a mirror of socio-economic development: A global simulation study. *Global Environmental Change*, **23**, 144–156.

Gray, N.F. (1996) The use of an objective index for the assessment of the contamination of surface water and groundwater by acid mine drainage. *Water and Environment Journal*, **10**, 332–340.

Gray, N.F. (2021) *Fixed-Film Reactors in Wastewater Treatment.* World Scientific Publishing, Singapore.

Malik, O.A. Hsu, A., Johnson, L.A. and de Sherbinin, A. (2015) A global indicator of wastewater treatment to inform the Sustainable Development Goals (SDGs). *Environmental Science and Policy,* **48**, 172–185.

Mateo-Sagasta, J., Raschid-Sally, L. and Thebo, A. (2015) Global wastewater and sludge production treatment and use. In *Wastewater: Economic Asset in an Urbanizing World* (eds. P. Drechsel, M. Qadir and D. Wichelns). Springer, Dordrecht, pp. 15–38.

Molinos-Senante, M., Gómez, T., Garrido-Baserba, M., Caballero, R. and Sala-Garrido, R. (2014) Assessing the sustainability of small wastewater treatment systems: A composite indicator approach. *Science of the Total Environment,* **497–498**, 607–614.

Muthukrishna V.K. and Victor N.D. (2016) Study on recycling urban wastewater for non-potable uses for water conservation. *Polish Journal of Environmental Studies,* **25**(1), 167–171.

Sato T., Qadir M., Yamamoto S., Endo T. and Zahoor A. (2013) Global, regional, and country level need for data on wastewater generation, treatment, and use. *Agricultural Water Management,* **130**, 1–13.

Shah, A.I., Din Dar, M.J., Bhant, R.A., Singh, J.P., Singh, K. and Bhat, S.A. (2020) Prospectives and challenges of wastewater treatment technologies to combat contaminants of emerging concerns. *Ecological Engineering,* **152**, 105882. https://doi.org/10.1016/j.ecoleng.2020.105882

UNEP. (2010) *Sick Water? The Central Role of Wastewater Management in Sustainable Development.* United Nations Environment Programme. https://www.grida.no/publications/218

USEPA. (2013) Emerging technologies for wastewater treatment and in-plant wet weather management. EPA 832-R-12-11. US Environmental Protection Agency, Washington, DC.

van der Hoek, J.P., Duijff, R. and Reinstra, O. (2018) Nitrogen recovery from wastewater: Possibilities, competition with other resources, and adaptation pathways. *Sustainability,* **10**, 4605. https://www.mdpi.com/2071-1050/10/12/4605

Voulvoulis, N. (2018) Water reuse from a circular economy perspective and potential risks from an unregulated approach. *Current Opinion in Environmental Science and Health,* **2**, 32–45.

Index

Aeration lagoons, 465–473
Algal lipids, 271–275
Algal ponds, 68, 204, 251–282
Algal protein, 265–267
Anaerobic digestion, 487–488
Anaerobic lagoons, 69, 369–398
  Ammonium formation, 388–391
  Application, 378
  Biogas collection, 396–398
  Biogas production, 397–398
  C:N ratio, 379
  Case studies, 354–362
  Community structure, 371–373
  Covered lagoons, 396–398
  Denitrification, 394–395
  Design, 363–364, 369–371, 380–383, 397
  Gas emissions, 383
  Hydrogen sulphide, 391–394
  Layout, 353–362
  Methane production, 374–377, 383–388
  Microbiology, 371–373, 385
  N:P ratio, 379
  Odors, 391–394
  Pathogen inactivation, 379
  Process, 369–371, 397

Best available technologies, 35
Biocycle, 120
Biodegradable polymers, 46–47
Biodiesel production, 271–275
Biofilm, 206, 332
Biofuel, 68, 205, 222–223, 270–275, 484
Biogas, 49, 67, 204, 236–237, 396–398, 484
Biogas, digestibility of plants, 204, 236
Biomat ecology, 126
Biomat formation, 126–130
Biosolids, 49
Boron, 89–90

C:N ratio, 79, 379
Carbon capture, 58
Chlorophytes (green algae), 258–261
Circular economy, 33–50, 478
  Barriers, 36
  Definition, 33–39

Ellen MacArthur Foundation, 33–39
Energy pathway, 47–50
List of principles, 36–38
Materials pathway, 44–47
Natural capital, 37–38
Water industry, 39–50
Water pathway, 40–44
Water specific, 37–38
Climate effects, 488
Completely mixed aerated ponds, 467–473
Conductivity, 142–143
Constructed wetlands,
Classification of, 310
Design, 310–312
Free water surface (FWS), 68, 200–201, 226, 314–320
Layout, 310–312
Natural wetland, 68, 201, 289–303
Plants 199–202
Salinity, 313
Selection criteria, 199–200
Subsurface flow (SSF), 68, 199–202, 320–338
Surface flow (SF), 68, 199–202, 320–338
Tertiary treatment, 313, 330–332
Types, 299–202, 307
Types of wastewater, 313
Vegetation types, 199–201
Contaminants, 89, 91–95, 190–193, 210–213, 296, 454–456
Contaminants, removal mechanisms, 210–213, 296
Covered anaerobic lagoons, 396–398
Crop evaporation rate (ET_c), 81
Crop irrigation
Conductivity, 142–143
Crop irrigation efficiency (CIE), 148–149
Crop tolerance to salinity, 144–147, 150–155
Evapotranspiration – crop yields, 142
Fish ponds, 141
Leaching requirement (LR), 148–149
Net irrigation rate (NIR), 148–149
Performance, 140
Plant water requirement (ET_c), 148–149
Rice paddy fields, 140–141
Salinity intolerance, 142–155
Crop irrigation efficiency (CIE), 148–149
Crop tolerance to salinity, 144–147, 150–155
Cyanobacteria, 205, 261–263

Denitrification, see Nitrification–Denitrification
Digestibility, 204, 236
Digestion, 44, 46
Direct potable reuse (DPR), 44
Duckweed
Biological processes, 224–225
Biology, 225
Chemical composition of, 231
Design of ponds, 226, 229–233
Ecology, 225
Fish, 232–233
FWS, 226

Heavy metal accumulation, 231
Nutrient removal, 227
Sequestration of CO_2, 228
Species, 224

Economic dimension, 5, 21, 478–484
Economy–ecology–social nexus, 5, 39
Ellen MacArthur Foundation, 33–39
Emerging technologies, 484–492
Energy
Biodiesel, 271–275
Biogas, 49, 67, 204, 236–237, 396–398, 484
Biosolids, 49
Micro-turbines, 49
Solar, 49
Treatment requirements, 66, 482
Water efficiency, 49
Wind, 49
Energy pathway, 47–50
Environmental dimension, 5, 21, 478–484
Eutrophication, 203, 301, 488
Evapotranspiration, 80–84, 330
Crop yields, 142
Extracellular polymeric substances (EPS), 263

Facultative ponds, 69, 202, 350–362, 405–430
Algae 408–411
Application, 350–351
Bacteria, 407
Baffles, 418, 420–421
Basis of operation, 405–407
Case studies, 354–362
Cladocera, 414–415
Cold climate operation, 421–423
Community structure, 408–411
Definition, 351–352
Depth, 420, 423
Design, 425–427
Digestion of sludge, 407
Ecology, 407–417
Fish, 417
Hydrodynamics, 417–418
Invertebrates, 417
Layout, 353–362
Nitrogen removal, 413–414
Odors, 421
Operation, 427–430
Pathogen removal, 418, 445–454
Protozoa, 414
Purple photosynthetic sulfur bacteria, 414
Temperature, 418, 422–423
Fish ponds, 141, 207–208, 232–233, 270, 417, 435–439
Floating macrophytes, 210–238
Flood irrigation, 136, 138, 141
Flooding tolerance of trees, 161
Forestry
Fate of N and P, 158–159
Flooding tolerance of trees, 160
Pathogen risk, 162
Salt tolerance of trees, 160
Short rotation willow coppice (SRWC), 155–159
Vegetative filter strips, 159–160

Free water surface (FWS) systems, 68, 200–201, 226, 314–320
- Comparison of species, 317–318
- Design, 315
- Ecology, 320
- Nitrification–denitrification, 315
- Pathogen removal, 317, 338–340
- Performance, 316
- Phosphorus uptake, 315
- Plants used, 314
- Process, 314–315, 320
- Root zone, 308, 317

Global warming, 52–61, 477
Global warming potential, 53–54
Global water crisis, 13–15, 38, 488–492
Grass filtration, see Overland flow
Greenhouse gas protocol, 55
Greenhouse gases
- Carbon sequestering, 68, 205, 228
- GWP, 53–54
- Methane, see Methane
- Methodology, 55–56
- Nitrous oxide, 57, 60, 141
- Offsetting, 61
- Process emissions, 59
- Scoping, 55–56
- Standard emissions for treatment, 54

Groundwater, 9, 12, 130
Groundwater contamination, 178–179
Groundwater tracing, 130, 179–180
Groundwater viral indicators, 179–180
Guidelines, microbial, 100–105
GWP, see Global warming potential

Harvesting algae, 276–278
Health protection measures, 102–106
Heavy metal accumulation, 204, 223, 231, 296
Heavy metals, 91–93, 137, 186, 192, 204, 210–213, 223, 231, 259, 296, 326, 328, 442
High-rate algal ponds
- Algal lipids, 271–275
- Algal protein, 265–267
- Biochemical pathway, 257
- Biodiesel production, 271–275
- Biofuel, 270–275
- Chemical composition of algae, 260, 264–266
- Chlorohytes (green algae), 258–261
- Comparison between open and ponds, 255–256, 280–282
- Cyanobacteria (blue–green algae), 261–263
- Designs, 251–257, 261–264, 267
- Extracellular polymeric substances (EPS), 263
- Fish, 270
- Harvesting algae, 276–278
- Immobilized algal systems, 279–282
- Maturation ponds, 269–270
- Methane, algae, 269

Microalgal aggregates, 263–265
N:P ratio, 258
Removal of heavy metals, 259, 262
Removal of organic contaminants, 259, 261
Role in wastewater treatment, 257–258
Role of microalgae, 257–275
Sequestering CO_2, 256
Triacylglycerols (TAG), 272–275
High-rate irrigation, 167–168
HRAPs, see High-rate algal ponds
Horizontal reed bed systems, 323–331
Household GHG emissions, 56
Hybrid reed bed systems, 335–338
Hydrogen sulphide, 391–394
Hydrological cycle, 9–13, 38, 477

Immobilized algal systems, 279–282
Indicators of sustainability, 21–28, 70–71
Indirect potable reuse (IPR), 43
Infiltration–percolation, see Rapid infiltration basins (RIBs)
Institutional dimension, 482–484
IWA energy pathway model, 48
IWA material pathway model, 45
IWA water pathway model, 42

Kyoto protocol, 53

Land treatment
Basis of process, 78–79
Boron, 89–90
C:N ratio, 79
Cadmium, 93
Concerns, 77–78
Contaminants, 89, 91–95
Crop evaporation rate (ET_c), 81
DOM, 93–96
Evapotranspiration, 80–84
Heavy metals, 91–93, 186, 192
History, 75–78
Microbial guidelines, 100–105
Organic contaminants, 92, 94–96, 190–193
Pathogen removal, 96–100, 105
pH, 93–96
Problems, 77–79
Process of filtration, 78–80
Reference crop evaporation rate (ET_0), 81
Ridge and Furrow, 76, 136, 138
Scale of use, 76–77
Sewage farms, 75
Sodicity, 85–89
Sodium adsorption ratio (SAR), 85–88
Soil salinity, 84–89
Specific crop coefficient (K_c), 81–84
Subsurface infiltration, see Subsurface wastewater infiltration systems
Types, 77
Wildlife, 76
Leaching requirement (LR), 148

Macrophytes
- Biofilm, 206
- Biofuel, 222–223
- Byproducts, 222–223
- Chemical composition of, 222
- Contaminant removal mechanisms, 210–213
- Design of ponds, 211
- Duckweed, 210, 224– 233
- Fish, 207–208
- Floating macrophytes, 210–238
- Heavy metal accumulation, 223
- Nutrient cycling, 208–210, 221–222
- Phytoremediation, 211
- Pond design, 206–207, 217–221
- Role in FWS treatment, 206
- Submerged macrophytes, 205
- Tertiary treatment, 211
- Water ferns, 210, 233–238
- Water hyacinth, 210, 214–224

Materials pathway, 44–47

Maturation ponds, 69, 269–270, 354–362, 430–435
- Algae, 432
- Baffles, 430
- Biology, 432–433
- Case studies, 354–362
- Classification, 351–352
- Design, 364
- Function, 430
- Nitrogen removal, 433–434
- Pathogen reduction, 430–432, 439, 445–454

Melbourne WWTP, 84–85, 137, 356–359

Methane
- Algae, 269
- Greenhouse gas emissions, 57, 60, 296
- Production, 374–377, 383–388

Microalgae
- Agal ponds, 68, 204
- Biofuel, 205
- Biogas, 204
- Biology, 202–203
- Biomass production rates, 205, 260, 264–266
- Byproducts, 204–205
- Chemical composition of, 207
- Cyanobacteria, 205
- Digestibility of, 204
- Eutrophication, 203
- Facultative ponds, 202, 408–411
- Heavy metals, 204
- Maturation ponds, 432
- Phytoremediation, 205
- Protein content, 203, 207
- Role in wastewater treatment, 203, 257–258
- Sequestration of CO_2, 205
- Types, 202

Microalgal aggregates, 263–264

N:P, 200, 238, 258, 379

Natural capital, 37

Natural treatment systems
- Advantages and disadvantages, 64–66
- Comparison with conventional treatment, 64–65
- Concept, 63
- Energy usage, 64, 66

Net zero, 64–65, 484
Role, 488–492
Selection of processes, 63, 66–71
Types, 64–66
Natural wetlands, 68, 201, 289–303
Biology, 290–292
C:P ratio, 294
Contaminant removal, 296
Heavy metal removal, 296
Methane, 296
Nitrogen cycle, 294–295, 300–301
Organic contaminants, 297–298
Phosphate removal, 293, 300–302
Purification processes, 292–296
Ramsar Convention, 290
Root zone, 293–298
Sequestration of CO_2, 293–298
Types, 289–290
Wastewater treatment, 298–302
Net-zero, 47, 58–59, 64–65, 484
Nitrification–Denitrification
Ammonia forms, 440
Anammox, 395
Balance at WWTP, 486
Comammox, 176
Denitrification, 60, 176, 394–395
Fish ponds, 438–439
Hydroxylamine oxidation, 60
Incomplete nitrification, 60
Land treatment, 60
Natural wetlands, 294–295, 300–301
Nitrification, 60, 187
Nitrifier denitrification, 60, 176
Removal in RIBs, 173–176
Removal, process selection, 487
Waste stabilization ponds, 413–414, 433–434
Wetlands, 294–295, 300–301, 315
Nitrogen balance at WWTP, 486
Nitrogen removal, process selection, 487
Nitrous oxide
Formation, 141
Greenhouse gas emissions, 57, 60
Net irrigation rate (NIR), 148–149

Odors, 391–394, 421
Organic contaminants, 92, 94–96, 190–193, 210–213, 297–298, 454–456
Overland flow
Design 183–185, 194
Grass species used, 185–186
Heavy metals, 186, 192
Operation, 187–194
Organic contaminants, 190–193
Treatment process, 185–186

Partially mixed aerated ponds, 466–467
Pathogens
Algal anti–bacterial toxins, 445
Anaerobic lagoons, 379

Comparison of guidelines, 107–108
Factors affecting survival, 98
Forestry, 162
FWS, 317, 338–340
Health based targets, 103–104
Health protection measures, 102–106
Land application guidelines, 99–102, 106–108
Land treatment, 96–105
Multiple barrier technique, 98–99
Plant pathogens, 106–109
Removal, 96–100, 105, 139, 162, 179–180, 317, 338–340
Risks, 101, 162
Sludge, 453–454
SSF, 325, 338–340
Survival, 97–100
UV deactivation, 449–452
Viral indicators, 179–180
Waste stabilization ponds, 445–454
Wetlands, 317, 325, 338–340
Percolation areas, see Subsurface wastewater infiltration systems
Phosphate, 44, 293, 300–302, 315, 328
Photobioreactors, see High-rate algal ponds
Photosynthesis, 200, 257–258, 412–413, 435
Phytodegradation, 210–213
Phytoextraction, 210–213
Phytoremediation, 205, 211
Phytostabilization, 210–213
Phytovolatilization, 210–213
Plant pathogens, 106–109
Plant water requirement (ET_c), 148–149
Plants used for treatment, 68, 199–200, 291–292, 307, 313, 318–319, 321–322, 408–411
Polyhydroxyalkanoates, 46–47
Precipitation, 11–12

Radionuclides, 212
Ramsar Convention, 290
Rapid infiltration basins (RIBs)
Aquifer recharge, 168
Biofilm, 173–175
Biosorption activated media, 173–175
Breakdown of species, 176
Ground water contamination, 178–179
High-rate irrigation, 167–168
Nitrogen removal in RIBs, 173–176
Pathogen removal, 179–180
Problems, 178
Rapid infiltration basins (RIBs), 168–172
Soil aquifer treatment (SAT), 167, 172–178
Viral indicators, 179–180
Recovery of energy, 484
Reed bed treatment, see Subsurface flow systems
Reference crop evaporation rate (ET_0), 81
Renewable groundwater stress, 15
Reuse of nutrients, 484
Reuse of water, 484
Rhizofiltration, 210–213
RIBs, see Rapid infiltration basins

Ridge and furrow, 76, 136, 138
River purification lakes, 439–444
Root zone, 293–298, 308, 317, 320

Salinity intolerance, 142–155, 313
Salt tolerance of trees, 160
SAT, see Soil aquifer treatment
Sequestration of CO_2, 68, 205, 228, 256, 293–298
Sewage farms, 75
Short rotation willow coppice (SRWC), 67, 156–159
Slow-rate land treatment (SRLT)
  Conductivity, 142–143
  Crop irrigation, 140–154
  Crop irrigation efficiency (CIE), 148–149
  Crop tolerance to salinity, 144–147, 150–155
  Evapotranspiration crop yields, 142
  Fate of N and P in SRWC, 158–159
  Fish ponds, 141
  Flood irrigation, 136, 138, 141
  Flooding tolerance of trees, 161
  Forestry, 155–162
  Heavy metals, 139
  Irrigation rates, 137
  Leaching requirement (LR), 148–149
  Nitrous oxide formation, 141
  Net irrigation rate (NIR), 148–149
  Pathogen removal, 139
  Pathogen risk, 162
  Performance, 140
  Plant water requirement (ET_c), 148–149
  Process design, 137
  Rice paddy fields, 140–141
  Ridge and furrow irrigation, 136, 138
  Salinity intolerance, 142–155
  Salt tolerance of trees, 160
  Short rotation willow coppice (SRWC), 156–159
  Suitability of site, 135
  Surface irrigation, 136, 138
  Vegetative filter strips, 159–160
Social dimension, 5, 22, 478–484
Sodicity, 85–89
Sodium adsorption ratio (SAR), 85–88
Soil aquifer treatment (SAT), 67, 167, 172–178
Soil salinity, 84–89
Specific crop coefficient (K_c), 81–84
SRWC, see Short rotation willow coppice
Struvite, 44
Submerged macrophytes, 205
Subsurface drip irrigation, 119–120
Subsurface flow systems (SSF), 68, 199–202, 320–338
  Aerenchyma tissue, 324, 327
  Biology, 308–309
  Byproducts, 331
  Design, 323–330, 332–334
  Ecology, 321–323
  Horizontal systems, 323–331

Hybrid reed bed systems, 335–338
Metal removal, 326, 328
Pathogen removal, 325, 338–340
Performance, 327–330, 333–335, 337
Phosphorus, 328
Plants, 321–322
Process, 308–309, 320–323
Removal mechanisms, 326
Role of medium, 329
Tertiary treatment, 330, 332
Vertical systems, 331–335
Subsurface wastewater infiltration systems (SWISs)
Biological oxidation, 117–118
Biomat ecology, 126
Biomat formation, 126–130
Design, 120–126
Groundwater pollution, 130
Indicator plants for drainage, 123–124
Long term acceptance rate (LTAR), 118, 130
Percolation rates, 121
Percolation test, 121–122
Percolation trench, 117–118, 123
Raised percolation area, 126
Sand filters, 125
Subsurface drip irrigation, 119–120
Surface flowsystems (SF), 68, 199–202, 320–338
Sustainability
Circular economy, 478
Comparison of systems, 478–483
Definition, 3
Economic dimension, 5, 21, 478–484
Environmental dimension, 5, 21, 478–484
Global warming effects, 477
Hydrological cycle, 477
Indicators, 21–28, 70–71, 478–484
Institutional dimension, 482–484
Nexus, 5–7, 39
Options, 477
Social dimension, 5, 22, 478–484
Sustainability indices, 478–484
Technical dimension, 482–484
Water demand management, 478
Water Framework Directive, 477
Water industry, 9–19
Wellbeing, 5, 8, 477
Sustainability indicators, 21–28, 70–71, 478–484
Sustainable development, see Sustainability
Sustainable Development Goals (SDGs), 7–8, 14, 482, 488
Sustainable urban drainage systems (SuDSs), 41, 77

Technical dimension, 482–484
Tertiary treatment, 211, 313, 330, 332, 430–435
Triacylglycerols (TAG), 272–275

UN Sustainability goals, 7–8, 14, 482, 488

Vegetative filter strips, 67, 159–160
Vertical reed bed systems, 331–335
Virtual water, 16–20

Wastewater treatment
- Circular economy, 33–50
- Economic indicators, 21
- Energy pathway, 47–50
- Environmental indicators, 21
- GHG standard values, 54, 59
- Greenhouse gas emissions, 55–61
- Materials pathway, 44–47
- Net zero, 47, 58–59
- Process emissions, 59
- Social indicators, 22
- Standard GHG emissions for treatment, 54
- Sustainability, 20–28, 70–71
- Sustainability indicators, 21–28, 70–71, 478–484
- UK Net Zero Route Map, 58–59
- Water pathway, 40–44

Waste stabilization ponds
- Anaerobic lagoons, 369–398
- Case studies, 354–362
- Classification, 351–352
- Design, 353, 362–365
- Development, 349–351
- Facultative ponds, 69, 350–362
- Layout, 353
- Selection, 68–70

Water balance at WWTP, 485
Water demand management, 39, 478
Water energy efficiency, 47–50
Water ferns, 210, 233–238
- *Azolla*, 233–238
- Biogas, 236–237
- Biology, 234, 238
- Digestibility, 236
- N:P ratio, 238
- Species, 233–234

Water footprint, 15–20
Water Framework Directive, 477
Water hyacinth, 210, 214–224
Water industry
- Circular economy, 33–50
- Energy pathway, 47–50
- Greenhouse gas emissions, 53–61
- Household GHG emissions, 56
- Materials pathway, 44–47
- Net zero, 47, 58–59
- Process emissions, 59
- Standard GHG emissions for treatment, 54
- Sustainability, 9–19, 70–71
- UK Net Zero Route Map, 58–59
- Water pathway, 40–44

Water pathway, 40–44
Water reuse, 41–43
Water scarcity, 14
Water services cycle, 13, 40
Water stress, 14–15
Water use, 39–41
Water used in production, 16–18
Wellbeing, 5, 8, 477

Made in the USA
Las Vegas, NV
15 September 2026

57f5aa5f-61de-4cda-b489-7fe1a73fe633R01